Collins
INTERNATIONAL
PRIMARY
MATHS

Teacher's Guide 5

William Collins' dream of knowledge for all began with the publication of his first book in 1819. A self-educated mill worker, he not only enriched millions of lives, but also founded a flourishing publishing house. Today, staying true to this spirit, Collins books are packed with inspiration, innovation and practical expertise. They place you at the centre of a world of possibility and give you exactly what you need to explore it.

Collins. Freedom to teach.

An imprint of HarperCollins*Publishers*
The News Building
1 London Bridge Street
London
SE1 9GF

Browse the complete Collins catalogue at
www.collins.co.uk

Publishing manager Fiona McGlade
Series editor Peter Clarke
Managing editor Caroline Green
Editor Kate Ellis
Project managed by Emily Hooton
Developed by Joan Miller, Tracy Thomas and Karen Williams
Edited by Tanya Solomons
Proofread by Tracy Thomas
Cover design by Amparo Barrera
Cover artwork by sylv1rob1/Shutterstock
Internal design by Ken Vail Graphic Design
Typesetting by QBS
Illustrations by Ken Vail Graphic Design, Advocate Art and Beehive Illustration
Production by Robin Forrester

Printed and bound by CPI Group (UK) Ltd

Contents

Introduction

Key principles of Collins International Primary Maths

Collins International Primary Maths is a mathematics course that ensures complete coverage of the Cambridge Primary Mathematics Curriculum Framework.

The course offers:

- a rigorous and cohesive scope and sequence of the Cambridge Primary Mathematics Curriculum Framework, while at the same time allowing for schools' own curriculum design
- a problem-solving and discovery approach to the teaching and learning of mathematics
- lesson plans following a highly effective and proven lesson structure
- a bank of practical hands-on learning activities
- controlled, manageable differentiation with activities and suggestions for at least three different ability groups in every lesson

- extensive teacher support through materials which:
  - promote the most effective pedagogical methods in the teaching of mathematics
  - are sufficiently detailed to aid confidence
  - are rich enough to be varied and developed
  - take into account issues of pace and classroom management
  - give careful consideration to the key skill of appropriate and effective questioning
  - provide a careful balance of teacher intervention and learner participation
  - encourage communication of methods and foster mathematical rigor
  - are aimed at raising levels of attainment for every learner
- manageable strategies for effective monitoring and record-keeping, to inform planning and teaching.

How Collins International Primary Maths supports Cambridge Primary and the Cambridge Primary Mathematics Curriculum Framework

Cambridge Primary is typically for learners aged 5 to 11 years. It develops learner skills and understanding through the primary years in English, Mathematics and Science. It provides a flexible framework that can be used to tailor the curriculum to the needs of individual schools.

Cambridge Primary Mathematics Curriculum Framework:

- provides a comprehensive set of learning objectives in Mathematics for each year of primary education
- focuses on developing knowledge and skills which form an excellent foundation for future study
- focuses on learners' development in each year
- provides a natural progression throughout the years of primary education
- is compatible with other curricula, internationally relevant and sensitive to different needs and cultures
- is suitable for learners whose first language is not English
- provides schools with international benchmarks.

The Cambridge approach supports schools to develop learners who are:

- **confident** in working with information and ideas – their own and those of others
- **responsible** for themselves, responsive to and respectful of others
- **reflective** as learners, developing their ability to learn
- **innovative** and equipped for new and future challenges
- **engaged** intellectually and socially, and ready to make a difference in the world.

The Cambridge Primary Mathematics Curriculum Framework is organised into six stages. Each stage reflects the teaching targets for a year group. Broadly speaking, Stage 1 covers the first year of primary teaching, when learners are approximately five years old. Stage 6 covers the final year of primary teaching when learners are approximately 11 years old.

The Mathematics framework is presented in five content areas. The first four content areas are all underpinned by Problem solving. Mental strategies are also a key part of the Number content.

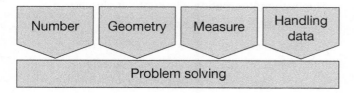

Collins International Primary Maths is designed to fulfill the requirements of the Cambridge Primary Mathematics Curriculum. All of the components emphasise, and provide guidance on, the importance of the cyclical nature of teaching in order to best promote learning and raise attainment.

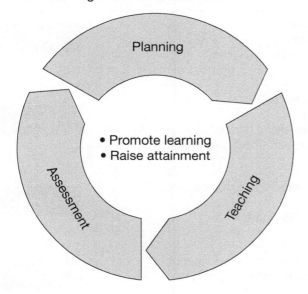

Planning

Collins International Primary Maths supports teachers in planning a successful mathematics programme for their unique teaching context and ensures:

- a clear understanding of learners' pre-requisite skills before undertaking particular tasks and learning new concepts
- considered progression from one lesson to another
- regular revisiting and extension of previous learning
- a judicious balance of objectives, and the time dedicated to each one
- the use of a consistent format and structure.

The elements of Collins International Primary Maths that form the basis for planning can be summarised as follows:

Long-term plans

The Cambridge Primary Mathematics Curriculum Framework constitutes the long-term plan for schools to follow. By closely reflecting the Curriculum Framework and the Cambridge Primary Schemes of Work, the Collins International Primary Maths course embodies this long-term plan.

Medium-term plans

The Collins International Primary Maths units and recommended teaching sequence (see pages xvii–xxxvi) show termly outlines of units of work with Cambridge Primary Mathematics Curriculum Framework references (including the Curriculum Framework codes). Using the Collins International Primary Maths online Planning tool via Collins Connect, these plans can be easily adapted to meet the specific needs of individual schools and teachers.

Short-term plans

Individual lesson plans and accompanying Additional practice activities represent the majority of each Teacher's Guide. The lessons provide short-term plans that can easily be followed closely, or used as a 'springboard' and varied to suit specific needs of particular classes. An editable 'Weekly Planning Grid' is also provided on Collins Connect, which individual teachers can fully adapt.

Teaching

The most important role of teaching is to promote learning and to raise learners' attainment. To best achieve these goals, Collins International Primary Maths believes in the importance of teachers:

- having high expectations for *all* learners
- systematically and effectively checking learners' understanding throughout lessons, anticipating where they may need to intervene, and doing so with notable impact on the quality of learning
- generating high levels of engagement and commitment to learning
- consistently providing high quality marking and constructive feedback to ensure that learners make rapid gains
- offering sharply focused and timely support and intervention that matches learners' individual needs.

Introduction

To help teachers achieve these goals, Collins International Primary Maths provides:

- highly focused and clearly defined learning objectives
- examples of targeted questioning, using appropriate mathematical vocabulary, that is aimed at both encouraging and checking learner progress
- a proven lesson structure that provides clear and accurate directions, instructions and explanations

- meaningful and well-matched activities for learners at all levels of understanding to practise and consolidate their learning
- highly effective models and images to clearly illustrate mathematical concepts, including interactive digital resources.

Each lesson plan in Collins International Primary Maths has a specific learning objective derived from the Cambridge Primary Mathematics Curriculum Framework, and follows the same teaching and learning:

Whole class	**Refresh**	• Consolidate fluency of previously learned facts and/or methods and further develop oral skills • Refresh the prerequisites of the lesson	5 min
	Discover	• Introduce the 'big idea' of the lesson • Acts as a springboard for discussion and exploration • Aimed at getting the learner engaged in the lesson	5 min
	Teach & Learn	• Direct teaching of the learning objective(s) • Promote mathematical vocabulary • Interactive whole-class teaching involving learners through planned questioning, learner demonstration and sharing of methods	10 min
Groups / Pairs / Individuals	**Practise**	• Practical and written activities to consolidate the learning objective(s) • Content covering three levels of differentation	15 min
	Apply	• Investigation/problem/puzzle/cross-curricular application where the learner uses and applies knowledge, skills and understanding in a problem-solving context	10 min
Whole class	**Review**	• Provide opportunities for learners to feedback to the rest of the class • Review questions to establish that learning is secure • Bring lesson to a close and evaluate its success	5 min

NOTE: Timings are approximate recommendations only.

Assessment

Assessment, record-keeping and reporting continue the teaching and learning cycle and are used to form the basis for adjustments to the teaching programme. Collins International Primary Maths offers manageable and meaningful assessment on two of the following three levels:

Short-term 'on-going' assessment

Short-term assessments are an informal part of every lesson. A combination of carefully crafted recall, observation and thought questions are provided in each lesson of Collins International Primary Maths and are linked to specific learning objective(s). They are designed to provide immediate feedback to learners and to gauge learners' progress in order to adapt teaching.

Shared Success Criteria are also provided in each lesson to assist learners in identifying the steps required to achieve the learning objective.

Each unit in Collins International Primary Maths also begins with a Unit overview. One of the features of the Unit overview is 'Common difficulties and remediation'. This feature can be used to help identify why learners do not understand, or have difficulty with, a topic or concepts and to use this information to take appropriate action to correct mistakes or misconceptions.

Medium-term 'formative' assessment

Medium-term assessments are used to review and record the progress learners are making over time in relation to the learning objectives of the Cambridge Primary Mathematics Curriculum Framework. They are used to establish whether learners have met the learning objectives or are on track to do so.

The class record-keeping charts on pages 299–306 are intended to be a working document that teachers start to complete at the beginning of the academic year. They can be continually updated and amended throughout the course of the year.

Teachers use their own professional judgment of each learner's level of mastery throughout the term in each of the Strands and Sub-strands, taking into account the following:

- mastery of the learning objectives associated with each particular Strand and Sub-strand
- performance in whole-class discussions
- participation in group work
- work presented in exercise books
- any other written evidence.

The degree of mastery achieved by a learner in each Strand and Sub-strand is shown by writing the learner's name (or initials) in the appropriate column:

A: Exceeding expectations in this Strand/Sub-strand

B: Meeting expectations in this Strand/Sub-strand

C: Below expectations in this Strand/Sub-strand

'Assessment *for* learning' is the term generally used to describe the conceptual approach to the two levels of assessment described above. Assessment *for* learning involves both learners and teachers finding out about the specific strengths and weaknesses of individual learners, and the class as a whole, and using this to inform future teaching and learning.

Assessment *for* learning:

- is part of the planning process
- is informed by learning objectives
- engages learners in the assessment process
- recognises the achievements of *all* learners
- takes account of how learners learn
- motivates learners.

Long-term 'summative' assessment

Long-term assessment is the third level of assessment. It is used at the end of the school year in order to track progress and attainment against school and external targets, and to report to other establishments and to parents on the actual attainments of learners. By ensuring complete and thorough coverage of the Cambridge Primary Mathematics Curriculum Framework, Collins International Primary Maths provides an excellent foundation for the Cambridge Primary end of stage tests (Progression Tests) as well as the end of primary Cambridge Primary Checkpoint.

The components of Collins International Primary Maths

Teacher's Guide

The Teacher's Guide comprises the following elements:

- a bank of **Refresh activities** for the first 'warm up' stage of the lesson
- **Teaching and Learning units** which consist of Unit overviews, Lesson plans and Additional practice activities
- **Resource sheets** for use with particular lessons and activities
- **Answers** to the learner practice and consolidation components.
- **Progression charts** that track back and forward through the Cambridge Primary Mathematics Curriculum Framework.

A key aim of Collins International Primary Maths is to support teachers in planning a successful mathematics programme of work in line with the Cambridge Primary Mathematics Curriculum Framework.

To ensure complete curriculum coverage and adequate revision of the learning objectives, for each stage the learning objectives from the Cambridge Primary Mathematics Curriculum Framework have been grouped into topic areas or 'units'. For a more detailed explanation of the 23 units in Collins International Primary Maths Stage 5, including a recommended teaching sequence (Scheme of work), see pages xviii and xx.

The Medium-term plans on pages xxi–xli show how each unit of Collins International Primary Maths Stage 5 matches the Cambridge Primary Mathematics Curriculum Framework.

Icons used in Collins International Primary Maths

- 👤 work individually
- 👥 work in pairs
- 👪 work in groups
- 👨‍👩‍👧‍👦 work as a whole class
- **SB** refer to the Student's Book
- **WB** refer to the Workbook
- interactive digital resource
- 🖥 slide
- **Challenge 1** suitable for less-able learners who require additional support with prerequisites knowledge
- **Challenge 2** suitable for the majority of learners to practise and consolidate the lesson objective(s)
- **Challenge 3** suitable for more-able learners who require further enrichment and/or extension

Refresh

A bank of 5-minute 'warm-up' or 'starter' activities is provided for teachers to use at this first stage of the mathematics lesson. Reference is given in each lesson plan to appropriate Refresh activities.

All the activities are for whole-class work. However, where individual learners are involved in demonstrating something to the rest of the class, or when an activity or game is undertaken in pairs or groups, this is indicated with a relevant icon.

1 Title

Each activity is given a title. This is designed to help both teacher and learners identify a particularly favourite activity.

2 Classroom organisation

Icons are used to indicate whether the activity is designed to be used by the whole class working together, or for learners working in groups, pairs or individually.

3 Strand and Sub-strand

The relevant Cambridge Primary Mathematics Curriculum Framework Strand and Sub-strand covered is in the sidebar.

Unit **1** Whole numbers 1

Refresh activities

Sequences

Learning objectives	
Code	Learning objective
5Nn1	Count on and back in steps of constant size, extending beyond zero.

Resources

paper (per learner) or mini whiteboard and pen if available

What to do

- Learners sit in a circle. Distribute paper and pencils. Suggest a number sequence beginning from zero that counts on in steps of 2, 3, 4, 5, 9 or 10.
- Say: **The number sequence will begin on my left and move around the circle clockwise. As the turn passes, say the next number in the sequence.**
- Give learners 15 seconds before the game begins to predict and write their number in the sequence.
- The game begins, the turn passing around the circle. Learners say the number that they have written on their paper, holding it up. If it is correct, they score a point. The next round begins with a different sequence and a different starting position, shifted one place to the right of the previous position.

Variations

- Start counting on from a number other than zero.
- Introduce backward sequences that extend beyond zero.

Reverse

Learning objective

er – Numbers and the number system

4 Learning objective(s)

Each activity has clearly defined learning objective(s) to assist teachers in choosing the most appropriate activity for the concept they want the learners to practise and consolidate.

5 Resources

To aid preparation, any resources required are listed along with whether they are for the whole class, per group, per pair or per learner. Most of these resources are readily available in classrooms.

6 What to do

The activity is broken down into clear steps to support teachers in achieving the objective(s) of the activity, and facilitate interactive whole-class teaching.

7 Variations

Where appropriate, variations are included. Variations may be designed to make the activity easier or more difficult, or change the focus of the activity completely.

Collins International Primary Maths Stage 5 units

There are 23 units in Collins International Primary Maths Stage 5, each consisting of a Unit overview, Lesson plans and Additional practice activities.

Unit overview
A one-page introduction to each unit in the Teacher's Guide is designed to help teachers plan that teaching unit.

1 Collins International Primary Maths unit number and title

2 Cambridge Primary Mathematics Curriculum Framework Strand and Sub-strand

3 Learning objectives
The Cambridge Primary Mathematics Curriculum Framework learning objectives covered in the unit, including those objectives from Strand 5: Problem solving.

4 Unit overview
General description of the knowledge, skills and understanding taught in the unit.

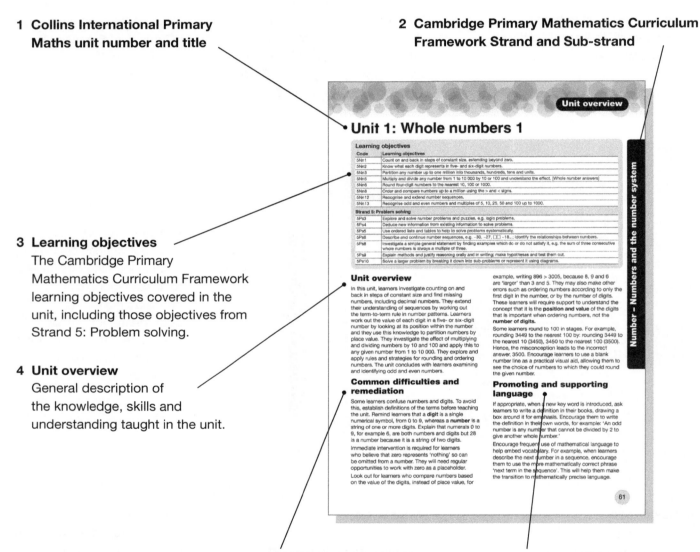

5 Common difficulties and remediation
Common errors and misconceptions, along with useful remediation hints are offered where appropriate.

6 Promoting and supporting language
Key strategy or idea to help learners access the mathematics of the unit and overcome any barriers that the language of mathematics may present.

Lesson plans

There are two different types of teaching and learning guidance provided for each of the 23 units in Collins International Primary Maths Stage 5.

For a unit where the recommended teaching time is one week there are:

• four lesson plans

• two Additional practice activities.

For a unit where the recommended teaching time is two weeks there are:

• eight lesson plans

• four Additional practice activities.

The lesson plans in each week provide a clear, structured, step-by-step approach to teaching mathematics according to the learning objective(s) being covered throughout the unit. Each of these lessons has been written in a comprehensive way in order to give teachers maximum support. It is intended, however, that these lessons will act as a model to be varied to the particular needs of each class.

1 **Collins International Primary Maths unit number and title**

2 **Cambridge Primary Mathematics Curriculum Framework Strand and Sub-strand**

3 **Reference to accompanying Student's Book page and Workbook page**

4 **Lesson number**

5 **Lesson title**

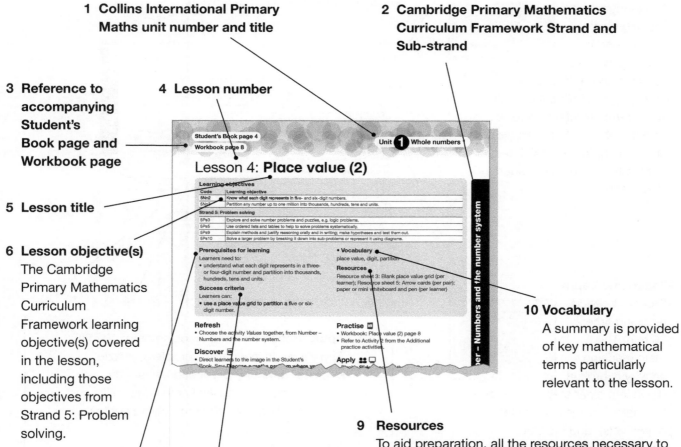

6 **Lesson objective(s)**
The Cambridge Primary Mathematics Curriculum Framework learning objective(s) covered in the lesson, including those objectives from Strand 5: Problem solving.

10 **Vocabulary**
A summary is provided of key mathematical terms particularly relevant to the lesson.

9 **Resources**
To aid preparation, all the resources necessary to teach the lesson are listed. Each resource clearly states whether this is for the whole class, per group, per pair or per learner. Icons are displayed within the lesson plan to indicate any Collins Connect or other digital resource used in the lesson.
NOTE: Although some lessons make use of dice, these are not always readily available in some countries. Where dice are unavailable or the use of dice is not appropriate, we suggest using a spinner, and have provided spinners on several Resource sheets.

7 **Prerequisites for learning**
List of knowledge, skills and understanding that are prerequisites for learning in the lesson. This list is particularly useful for diagnostic assessment.

8 **Success criteria**
Success criteria are provided to help both teachers and learners identify what learners are required to know, understand and do in order to achieve the learning objective(s).

11 Refresh (Recommended teaching time: 5 min) A bank of Refresh activities can be found on pages 1–60. In some lessons, the Refresh activity has been written specifically for that lesson. In these lessons, a brief description of the activity is provided.

12 Discover (Recommended teaching time: 5 min) Introduces the 'big idea' of the lesson, drawing learners' attention to the 'Discover' feature in the Student's Book. This feature is intended to act as a springboard for discussion and exploration and is aimed at getting the learners engaged in the lesson.

13 Teach and Learn (Recommended teaching time: 10 min) The main teaching activity is broken down into clear steps to support teachers in achieving the lesson objective(s), and facilitate interaction with the whole class. Suggested statements and questions to ask are provided to support the teacher. Where relevant, model answers are given (in brackets). Teachers should draw learners' attention to the 'Learn' feature in the Student's Book.

14 Practise (Recommended teaching time: 15 min) 'Teach & Learn' is followed by pupil practice and consolidation, which provides an opportunity for all learners to focus on their newly acquired knowledge. Pupil practice and consolidation consists of both written exercises and practical hands-on activities, with reference to the relevant Workbook page and bank of Additional practice activities. All of the tasks are differentiated into three different ability levels:

Challenge 1 suitable for less-able learners who require additional support with prerequisites knowledge

Challenge 2 suitable for the majority of learners to practise and consolidate the lesson objective(s)

Challenge 3 suitable for more-able learners who require further enrichment and / or extension.

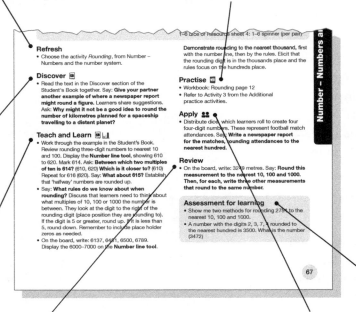

15 Apply (Recommended teaching time: 10 min) Apply is an investigation, problem, puzzle or cross-curricular application where the learner uses and applies knowledge, skills and understanding in a problem-solving context.

16 Review (Recommended teaching time: 5 min) The all-important conclusion to the lesson offers an opportunity for learners to make reflective comments about their learning, as well as to discuss misconceptions and common errors, and summarise what they have learned.

17 Assessment for learning Specific questions designed to assist teachers in checking learners' understanding of the lesson objective(s). These questions can be used at any time throughout the lesson.

NOTE: Timings are approximate recommendations only.

Additional practice activities

As well as four (one week unit) or eight (two week unit) lesson plans, each unit in Collins International Primary Maths provides two (one week unit) or four (two week unit) Additional practice activities.

These activities are intended to be used at any time throughout the week, not just as the 'fifth' lesson of the week. They are designed to provide teachers with a bank of practical activities that they can offer to learners to further practise and consolidate their understanding of the lesson objectives being taught throughout the unit.

1 Strand and Sub-strand
The relevant Cambridge Primary Mathematics Curriculum Framework Strand and Sub-strand covered is stated in the sidebar.

3 Classroom organisation
Icons are used to indicate whether the activity is suitable for learners working as a whole class, in groups, pairs or individually.

4 Challenge level
The challenge level for each activity is given:

Challenge 1 suitable for less-able learners who require additional support with prerequisites knowledge

Challenge 2 suitable for the majority of learners to practise and consolidate the lesson objective(s)

Challenge 3 suitable for more-able learners who require further enrichment and/or extension.

5 Learning objective(s)
Each activity has clearly defined objective(s) to assist teachers in choosing the most appropriate activity for the concept they want the learners to practise and consolidate.

2 Collins International Primary Maths unit number and title

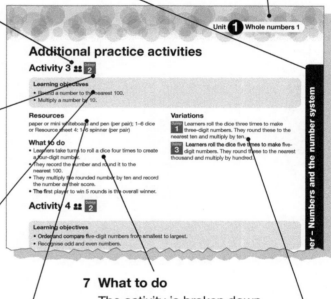

6 Resources
To aid preparation, the resources needed for each activity are listed. Each resource clearly states whether this is per group, per pair or per learner.
NOTE: Although some activities make use of dice, these are not always readily available in some countries. Where dice are unavailable or the use of dice is not appropriate, we suggest using a spinner, and have provided spinners on various Resource sheets.

7 What to do
The activity is broken down into clear steps to support teachers in explaining the activity to the learners.

8 Variations
Where appropriate, variations are included. Variations may be designed to make the activity easier or more difficult, or change the focus of the activity completely. Where the variation affects the challenge level of the activity, the new challenge level is given.

Introduction

Resource sheets

Where specific paper-based resources are needed for individual lesson plans or Additional practice activities, these are provided as Resource sheets in the Teacher's Guide. Use of Resource sheets is indicated in the resources list of the relevant lesson plan or Additional practice activity.

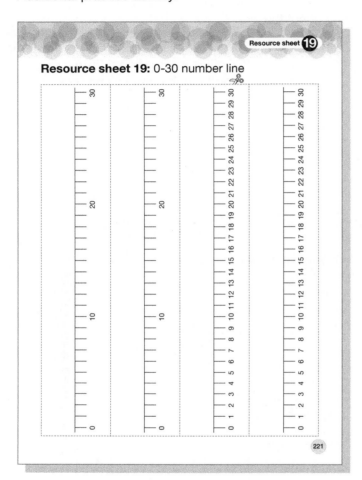

Answers

Answers are provided for all the Workbook pages.

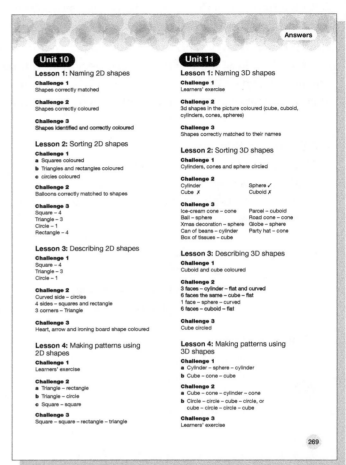

Tracking back and forward through the Cambridge Primary Mathematics Curriculum Framework

The progression charts on pages 288–306 track back and forward through the Cambridge Primary Mathematics Curriculum Framework.

The charts are structured using the Strands and Sub-strands in the Stage 5 Cambridge Primary Mathematics Curriculum Framework:

Strand	Sub-strand
Number	Numbers and the number system
	Calculation – Addition and subtraction, including Mental strategies (MS)
	Calculation – Multiplication and division, including Mental strategies (MS)
Geometry	Shapes and geometric reasoning
	Position and movement
Measure	Length, mass and capacity
	Time
	Area and perimeter
Handling data	Organising, categorising and representing data
	Probability
Problem solving	Using techniques and skills in solving mathematics problems
	Using understanding and strategies in solving problems

The learning objectives (and Curriculum Framework codes) are, in most instances, arranged in the order in which they appear in the Stage 5 Cambridge Primary Mathematics Curriculum Framework, e.g. 5Nn1, 5Nn2, 5Nn3, ..., with the equivalent learning objective(s) for Stages 1, 2, 3, 4 and 6 in adjacent columns. Learning objectives for Stages 1, 2, 3, 4 and 6 therefore are not in the order in which they appear in the Curriculum Framework, and in some cases may appear more than once. This occurs where a learning objective has relevance to more than one Stage 5 learning objective.

These progression charts serve several purposes:

1 To show the continuity and progression both *within* and *between* stages. Teachers are able to choose any learning objective and trace its pathway through the six stages of the Cambridge Primary Mathematics Curriculum Framework.

2 To assist teachers and learners in making connections in mathematics in order to deepen the learner's knowledge of concepts and procedures and to ensure that what is learned is sustained over time. However, the connections made are not intended to be exhaustive and teachers should seek to support learners in making other connections.

3 To provide teachers with the 'big picture' of the knowledge, skills and understanding that learners are developing in Stage 5, and also understand and appreciate their 'piece of the puzzle' and the associated learning in previous and future stages.

4 To identify the pre-requisites for learning that learners need to know, apply and understand before being introduced to a new learning objective.

5 To support teachers in planning, providing and adapting learning experiences for:

– those learners who are working *below* Stage 5 expectations

– those learners who are working *above* Stage 5 expectations.

6 For those learners who are working either *below* or *above* Stage 5 expectations, to assist teachers in determining the Stage that the learner is currently working.

Student's Book

There is one Student's Book for each stage in Collins International Primary Maths with one page provided for each lesson plan.

The content provided in the Student's Book is designed to be used during the following two stages of a typical Collins International Primary Maths lesson:

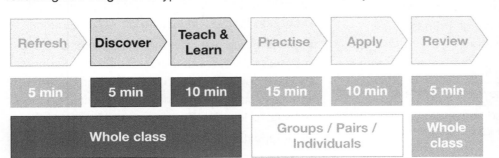

Workbook

All Workbook page exercises reinforce and build upon the main teaching points and learning objective(s) of a particular lesson in the Teacher's Guide. The work is intended to allow *all* learners in the class to practise and consolidate their newly acquired knowledge, skills and understanding.

The content provided in the Workbook is designed to be used during the following stage of a typical Collins International Primary Maths lesson:

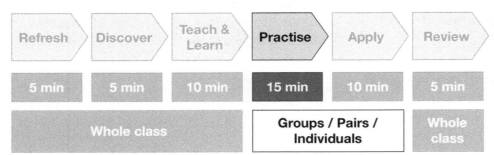

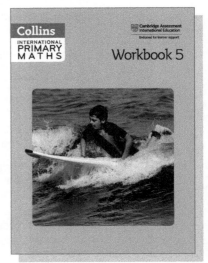

In Stage 5, one Workbook page is provided for each lesson plan. There is no Workbook page for the Additional practice activities.

Each Workbook page has three levels of challenge designed to cater, not only for the different abilities that occur in a mixed-ability or mixed-aged class, but also to assist those schools who 'set' or 'stream' their learners into ability groups. The three different levels of challenge are identified as follows:

Challenge 1 suitable for less-able learners who require additional support with prerequisites knowledge

Challenge 2 suitable for the majority of learners to practise and consolidate the lesson objective(s)

Challenge 3 suitable for more-able learners who require further enrichment and/ or extension.

NOTES:

- Timings are approximate recommendations only.
- Although some pages in the Student's Book and/or Workbook require learners to use dice, these are not always readily available in some countries. Where dice are unavailable or the use of dice is not appropriate, we suggest using a spinner, and have provided spinners on various Resource sheets.

Collins Connect

Collins Connect is an innovative online learning platform designed to support teachers and learners by providing interactive content and tools – ideal as a front-of-class teaching and learning tool.

As well as being the home for all the digital teaching resources provided by Collins International Primary Maths, Collins Connect also includes the following features.

Teach

The Teach section contains all of the teaching content, organised into units. This includes Unit overview, Weekly planning grids, Lesson plans, Additional practice activities, Resource sheets, Slideshows*, links to Tools (interactive digital resources). All components are editable to provide maximum flexibility for teachers.

Within Teach, the planning tool allows schools and individual teachers to customise the sequence of units in Collins International Primary Maths within and across all stages. This allows schools and individual teachers to develop their own unique Scheme of work.

Record

The Record section enables teachers to record results, teacher judgements and comments easily, and produce reports.

Interact

The Interact section contains the 32 flexible interactive whiteboard teaching tools and 16 interactive mathematical games. The audio glossary of terms for all stages is also located here.

Support

The Support section contains useful documents for the teacher, such as the medium-term plan, record-keeping formats, the introduction to the Teacher's Guide.

DVD

A DVD has been included as a component in Collins International Primary Maths for those schools who have no, or limited, access to the Internet.

The DVD contains:

- 32 interactive whiteboard mathematical tools
- comprehensive audio glossary
- all components of the Teacher's Guide in editable word documents and PDFs
- accompanying resources for lessons including slideshows*.

* The slideshows are provided as visual aids to be shown at the front of class at various stages, as directed, throughout the lesson.

Collins International Primary Maths Stage 5 units and recommended teaching sequence

The Stage 5 learning objectives from the Cambridge Primary Mathematics Curriculum Framework have been grouped into the following topic areas or 'units':

Cambridge Primary Mathematics Curriculum		Collins International Primary Maths	
Strand	Sub-strand	Unit number / title	
Number	Numbers and the number system	1	Whole numbers 1
		2	Whole numbers 2
		3	Decimals 1
		4	Decimals 2
		5	Fractions
		6	Percentages
		7	Ratio and proportion
	Calculation: – Mental strategies – Addition and subtraction	8	Addition and subtraction 1
		9	Addition and subtraction 2
		10	Addition and subtraction 3
	Calculation – Mental strategies – Multiplication and division	11	Multiplication and division 1
		12	Multiplication and division 2
		13	Multiplication and division 3
Geometry	Shapes and geometric reasoning	14	2D shape, including symmetry
		15	3D shape
		16	Angles
	Position and movement	17	Position and movement
Measure	Length, mass and capacity	18	Length
		19	Mass
		20	Capacity
	Time	21	Time
	Area and perimeter	22	Area and perimeter
Handling data	Organising, categorising and representing data	23	Handling data
	Probability		

Strand 5: Problem solving, is incorporated throughout each unit.

The Cambridge Primary Scheme of Work – Mathematics Stage 5 offers a suggested sequence for covering the content of the Stage 5 curriculum. An overview of this sequence can be seen in the table below.

Term 1	Term 2	Term 3
Number and Problem solving	Number and Problem solving	Number and Problem solving
Geometry and Problem solving	Handling data and Problem solving	Geometry and Problem solving
Measure and Problem solving	Measure and Problem solving	Measure and Problem solving

The 23 units in Collins International Primary Maths Stage 5 have been broadly organised into the sequence recommended above. However, as with the Cambridge Primary Scheme of Work, schools and individual teachers are free to teach the learning objectives in any order to best meet the needs of individual schools, teachers and learners.

The table on page xx shows a recommended teaching sequence for the 23 Collins International Primary Maths Stage 5 units.

Again, as with the Cambridge Primary Scheme of Work, Collins International Primary Maths has assumed an academic year of three terms, each of 10 weeks' duration. This is the minimum length of a school year and thereby allows flexibility for schools to add in more teaching time as necessary to meet the needs of the learners, and also to comfortably cover the content of the curriculum into an individual school's specific term times.

Term 1

Strand: Number

Sub-strand: Numbers and the number system
CIPM Unit: 1 – Whole numbers 1
Recommended time allocation: 2 weeks

Sub-strand: Calculation:
– Mental strategies
– Addition and subtraction
CIPM Unit: 8 – Addition and subtraction 1
Recommended time allocation: 1 week

Sub-strand: Calculation:
– Mental strategies
– Multiplication and division
CIPM Unit: 11 – Multiplication and division 1
Recommended time allocation: 2 weeks

Strand: Geometry

Sub-strand: Shapes and geometric reasoning
CIPM Unit: 14 – 2D shape, including symmetry
Recommended Time Allocation: 1 week

Sub-strand: Shapes and geometric reasoning
CIPM Unit: 15 – 3D shape
Recommended time allocation: 1 week

Sub-strand: Position and movement
CIPM Unit: 17 – Position and movement
Recommended time allocation: 2 weeks

Strand: Measure

Sub-strand: Time
CIPM Unit: 21 – Time
Recommended time allocation: 1 week

Term 2

Strand: Number

Sub-strand: Numbers and the number system
CIPM Unit: 2 – Whole numbers 2
Recommended time allocation: 1 week

Sub-strand: Numbers and the number system
CIPM Unit: 3 – Decimals 1
Recommended time allocation: 1 week

Sub-strand: Calculation:
– Mental strategies
– Addition and subtraction
CIPM Unit: 9 – Addition and subtraction 2
Recommended time allocation: 2 weeks

Sub-strand: Calculation:
– Mental strategies
– Multiplication and division
CIPM Unit: 12 – Multiplication and division 2
Recommended time allocation: 2 weeks

Strand: Handling data

Sub-strands:
– Organising, categorising and representing data
– Probability
CIPM Unit: 23 – Handling data
Recommended time allocation: 2 weeks

Strand: Measure

Sub-strand: Length, mass and capacity
CIPM Unit: 18 – Length
Recommended time allocation: 1 week

Sub-strand: Length, mass and capacity
CIPM Unit: 19 – Mass
Recommended time allocation: 1 week

Term 3

Strand: Number

Sub-strand: Numbers and the number system
CIPM Unit: 4 – Decimals 2
Recommended time allocation: 1 week

Sub-strand: Numbers and the number system
CIPM Unit: 5 – Fractions
Recommended time allocation: 1 week

Sub-strand: Numbers and the number system
CIPM Unit: 6 – Percentages
Recommended time allocation: 1 week

Sub-strand: Numbers and the number system
CIPM Unit: 7 – Ratio and proportion
Recommended time allocation: 1 week

Sub-strand: Calculation:
– Mental strategies
– Addition and subtraction
CIPM Unit: 10 – Addition and subtraction 3
Recommended time allocation: 1 week

Sub-strand: Calculation:
– Mental strategies
– Multiplication and division
CIPM Unit: 13 – Multiplication and division 3
Recommended time allocation: 2 weeks

Strand: Geometry

Sub-strand: Shapes and geometric reasoning
CIPM Unit: 16 – Angles
Recommended time allocation: 1 week

Strand: Measure

Sub-strand: Length, mass and capacity
CIPM Unit: 20 – Capacity
Recommended time allocation: 1 week

Sub-strand: Area and perimeter
CIPM Unit: 22 – Area and perimeter
Recommended time allocation: 1 week

Strand 5: Problem solving, is incorporated throughout each unit.
Recommended time allocation based on a total of 30 weeks – 10 weeks per term.

Collins International Primary Maths Stage 5 units link to Cambridge Primary Mathematics Curriculum Framework

Unit 1 – Whole numbers 1			Recommended time allocation: 2 weeks

Cambridge Primary Mathematics Curriculum Framework

Strand	Sub-strand	Code	Learning objective
Number	Numbers and the number system	5Nn1	Count on and back in steps of constant size[, extending beyond zero].
		5Nn2	Know what each digit represents in five- and six-digit numbers.
		5Nn3	Partition any number up to one million into thousands, hundreds, tens and units.
		5Nn5	Multiply and divide any number from 1 to 10 000 by 10 or 100 and understand the effect.
		5Nn6	Round four-digit numbers to the nearest 10, 100 or 1000.
		5Nn8	Order and compare numbers up to a million using the > and < signs.
		5Nn12	Recognise and extend number sequences.
		5Nn13	Recognise odd and even numbers and multiples of 5, 10, 25, 50 and 100 up to 1000.
Problem solving	Using understanding and strategies in solving problems	5Ps3	Explore and solve number problems and puzzles, e.g. logic problems.
		5Ps4	Deduce new information from existing information to solve problems.
		5Ps5	Use ordered lists and tables to help to solve problems systematically.
		5Ps6	Describe and continue number sequences[, e.g. −30, −27, ☐, ☐, −18…]; identify the relationships between numbers.
		5Ps8	Investigate a simple general statement by finding examples which do or do not satisfy it, e.g. the sum of three consecutive whole numbers is always a multiple of three.
		5Ps9	Explain methods and justify reasoning orally and in writing; make hypotheses and test them out.
		5Ps10	Solve a larger problem by breaking it down into sub-problems or represent it using diagrams.

Unit 2 – Whole numbers 2			Recommended time allocation: 1 week

Cambridge Primary Mathematics Curriculum Framework

Strand	Sub-strand	Code	Learning objective
Number	Numbers and the number system	5Nn1*	Count on and back in steps of constant size, extending beyond zero.*
		5Nn3*	Partition any number up to one million into thousands, hundreds, tens and units.*
		5Nn9	Order and compare negative and positive numbers on a number line and temperature scale.
		5Nn10	Calculate a rise or fall in temperature.
		5Nn12*	Recognise and extend number sequences.*
		5Nn14	Make general statements about sums, differences and multiples of odd and even numbers.

*Learning objective revised and consolidated

Problem solving	Using techniques and skills in solving mathematical problems	5Pt1	Understand everyday systems of measurement in [length, weight, capacity,] temperature [and time] and use these to perform simple calculations.
		5Pt7	Consider whether an answer is reasonable in the context of a problem.
	Using understanding and strategies in solving problems	5Ps1	Understand everyday systems of measurement in [length, weight, capacity,] temperature [and time] and use these to perform simple calculations.
		5Ps2	Choose an appropriate strategy for a calculation and explain how they worked out the answer.
		5Ps3	Explore and solve number problems and puzzles, e.g. logic problems.
		5Ps4	Deduce new information from existing information to solve problems.
		5Ps5	Use ordered lists and tables to help to solve problems systematically.
		5Ps6	Describe and continue number sequences, e.g. −30, −27, ☐, ☐, −18...; identify the relationships between numbers.
		5Ps8	Investigate a simple general statement by finding examples which do or do not satisfy it, e.g. the sum of three consecutive whole numbers is always a multiple of three.
		5Ps9	Explain methods and justify reasoning orally and in writing; make hypotheses and test them out.
		5Ps10	Solve a larger problem by breaking it down into sub-problems or represent it using diagrams.

*Learning objective revised and consolidated

Unit 3 – Decimals 1			Recommended time allocation: 1 week
Cambridge Primary Mathematics Curriculum Framework			
Strand	**Sub-strand**	**Code**	**Learning objective**
Number	Numbers and the number system	5Nn4	Use decimal notation for tenths and hundredths and understand what each digit represents.
		5Nn5	Multiply and divide any number from 1 to 10 000 by 10 or 100 and understand the effect.
Problem solving	Using techniques and skills in solving mathematical problems	5Pt7	Consider whether an answer is reasonable in the context of a problem.
	Using understanding and strategies in solving problems	5Ps3	Explore and solve number problems and puzzles, e.g. logic problems.
		5Ps4	Deduce new information from existing information to solve problems.
		5Ps8	Investigate a simple general statement by finding examples which do or do not satisfy it[, e.g. the sum of three consecutive whole numbers is always a multiple of three].
		5Ps9	Explain methods and justify reasoning orally and in writing; make hypotheses and test them out.
		5Ps10	Solve a larger problem by breaking it down into sub-problems or represent it using diagrams.

Unit 4 – Decimals 2			**Recommended time allocation:** 1 week
Cambridge Primary Mathematics Curriculum Framework			
Strand	**Sub-strand**	**Code**	**Learning objective**
Number	Numbers and the number system	5Nn4*	Use decimal notation for tenths and hundredths and understand what each digit represents.*
		5Nn7	Round a number with one or two decimal places to the nearest whole number.
		5Nn11	Order numbers with one or two decimal places and compare using the > and < signs.
Problem solving	Using techniques and skills in solving mathematical problems	5Pt7	Consider whether an answer is reasonable in the context of a problem.
	Using understanding and strategies in solving problems	5Ps3	Explore and solve number problems and puzzles, e.g. logic problems.
		5Ps4	Deduce new information from existing information to solve problems.
		5Ps8	Investigate a simple general statement by finding examples which do or do not satisfy it[, e.g. the sum of three consecutive whole numbers is always a multiple of three].
		5Ps9	Explain methods and justify reasoning orally and in writing; make hypotheses and test them out.
		5Ps10	Solve a larger problem by breaking it down into sub-problems or represent it using diagrams.

*Learning objective revised and consolidated

Unit 5 – Fractions			**Recommended time allocation:** 1 week
Cambridge Primary Mathematics Curriculum Framework			
Strand	**Sub-strand**	**Code**	**Learning objective**
Number	Numbers and the number system	5Nn15	Recognise equivalence between: $\frac{1}{2}$, $\frac{1}{4}$ and $\frac{1}{8}$; $\frac{1}{3}$ and $\frac{1}{6}$; $\frac{1}{5}$ and $\frac{1}{10}$.
		5Nn16	Recognise equivalence between the decimal and fraction forms of halves, tenths and hundredths and use this to help order fractions, e.g. 0.6 is more than 50% and less than $\frac{7}{10}$.
		5Nn17	Change an improper fraction to a mixed number, e.g. $\frac{7}{4}$ to $\frac{13}{4}$; order mixed numbers and place between whole numbers on a number line.
		5Nn18	Relate finding fractions to division and use to find simple fractions of quantities.

Problem solving	Using techniques and skills in solving mathematical problems	5Pt2	Solve single [and multi]-step word problems (all four operations); represent them, e.g. with diagrams or a number line.
		5Pt7	Consider whether an answer is reasonable in the context of a problem.
	Using understanding and strategies in solving problems	5Ps2	Choose an appropriate strategy for a calculation and explain how they worked out the answer.
		5Ps3	Explore and solve number problems and puzzles, e.g. logic problems.
		5Ps4	Deduce new information from existing information to solve problems.
		5Ps8	Investigate a simple general statement by finding examples which do or do not satisfy it[, e.g. the sum of three consecutive whole numbers is always a multiple of three].
		5Ps9	Explain methods and justify reasoning orally and in writing; make hypotheses and test them out.
		5Ps10	Solve a larger problem by breaking it down into sub-problems or represent it using diagrams.

Unit 6 – Percentages			Recommended time allocation: 1 week
Cambridge Primary Mathematics Curriculum Framework			
Strand	**Sub-strand**	**Code**	**Learning objective**
Number	Numbers and the number system	5Nn19	Understand percentage as the number of parts in every 100 and find simple percentages of quantities.
		5Nn20	Express halves, tenths and hundredths as percentages.
Problem solving	Using techniques and skills in solving mathematical problems	5Pt2	Solve single [and multi]-step word problems (all four operations); represent them, e.g. with diagrams or a number line.
		5Pt7	Consider whether an answer is reasonable in the context of a problem.
	Using understanding and strategies in solving problems	5Ps2	Choose an appropriate strategy for a calculation and explain how they worked out the answer.
		5Ps3	Explore and solve number problems and puzzles, e.g. logic problems.
		5Ps4	Deduce new information from existing information to solve problems.
		5Ps8	Investigate a simple general statement by finding examples which do or do not satisfy it [e.g. the sum of three consecutive whole numbers is always a multiple of three].
		5Ps9	Explain methods and justify reasoning orally and in writing; make hypotheses and test them out.
		5Ps10	Solve a larger problem by breaking it down into sub-problems or represent it using diagrams.

Unit 7 – Ratio and proportion			Recommended time allocation: 1 week
Cambridge Primary Mathematics Curriculum Framework			
Strand	**Sub-strand**	**Code**	**Learning objective**
Number	Numbers and the number system	5Nn21	Use fractions to describe and estimate a simple proportion, e.g. $\frac{1}{5}$ of the beads are yellow.
		5Nn22	Use ratio to solve problems, e.g. to adapt a recipe for 6 people to one for 3 or 12 people.

Problem solving	Using techniques and skills in solving mathematical problems	5Pt2	Solve single [and multi]-step word problems (all four operations); represent them, e.g. with diagrams or a number line.
		5Pt7	Consider whether an answer is reasonable in the context of a problem.
	Using understanding and strategies in solving problems	5Ps2	Choose an appropriate strategy for a calculation and explain how they worked out the answer.
		5Ps3	Explore and solve number problems and puzzles, e.g. logic problems.
		5Ps4	Deduce new information from existing information to solve problems.
		5Ps8	Investigate a simple general statement by finding examples which do or do not satisfy it [e.g. the sum of three consecutive whole numbers is always a multiple of three].
		5Ps9	Explain methods and justify reasoning orally and in writing; make hypotheses and test them out.
		5Ps10	Solve a larger problem by breaking it down into sub-problems or represent it using diagrams.

Unit 8 – Addition and subtraction 1			**Recommended time allocation:** 1 week
Cambridge Primary Mathematics Curriculum Framework			
Strand	**Sub-strand**	**Code**	**Learning objective**
Number	Calculation – Mental strategies	5Nc8	Count on or back in thousands, hundreds, tens and ones to add or subtract.
		5Nc10	Use appropriate strategies to add or subtract pairs of two- and three-digit numbers [and numbers with one decimal place], using jottings where necessary.
	Calculation – Addition and subtraction	5Nc18	Find the total of more than three two- or three-digit numbers using a written method.
Problem solving	Using techniques and skills in solving mathematical problems	5Pt2	Solve single and multi-step word problems (all four operations); represent them, e.g. with diagrams or a number line.
		5Pt3	Check with a different order when adding several numbers or by using the inverse when adding or subtracting a pair of numbers.
		5Pt6	Estimate and approximate when calculating, e.g. using rounding, and check working.
		5Pt7	Consider whether an answer is reasonable in the context of a problem.
	Using understanding and strategies in solving problems	5Ps2	Choose an appropriate strategy for a calculation and explain how they worked out the answer.
		5Ps3	Explore and solve number problems and puzzles, e.g. logic problems.
		5Ps4	Deduce new information from existing information to solve problems.
		5Ps5	Use ordered lists and tables to help to solve problems systematically.
		5Ps9	Explain methods and justify reasoning orally and in writing; make hypotheses and test them out.
		5Ps10	Solve a larger problem by breaking it down into sub-problems or represent it using diagrams.

Unit 9 – Addition and subtraction 2			Recommended time allocation: 2 weeks
Cambridge Primary Mathematics Curriculum Framework			
Strand	Sub-strand	Code	Learning objective
Number	Calculation – Mental strategies	5Nc1	Know by heart pairs of one-place decimals with a total of 1, e.g. 0.8 + 0.2.
		5Nc2	Derive quickly pairs of decimals with a total of 10, and with a total of 1.
		5Nc8*	Count on or back in thousands, hundreds, tens and ones to add or subtract.*
		5Nc9	Add or subtract near multiples of 10 or 100, e.g. 4387 – 299.
		5Nc10*	Use appropriate strategies to add or subtract pairs of two- and three-digit numbers* [and numbers with one decimal place, using jottings where necessary].
		5Nc11	Calculate differences between near multiples of 1000, e.g. 5026 – 4998, or near multiples of 1, e.g. 3.2 – 2.6.
	Calculation – Addition and subtraction	5Nc18*	Find the total of more than three two- or three-digit numbers using a written method.*
		5Nc19	Add or subtract any pair of three- and/or four-digit numbers, with the same number of decimal places, including amounts of money.
Problem solving	Using techniques and skills in solving mathematical problems	5Pt2	Solve single and multi-step word problems (all four operations); represent them, e.g. with diagrams or a number line.
		5Pt3	Check with a different order when adding several numbers or by using the inverse when adding or subtracting a pair of numbers.
		5Pt6	Estimate and approximate when calculating, e.g. using rounding, and check working.
		5Pt7	Consider whether an answer is reasonable in the context of a problem.
	Using understanding and strategies in solving problems	5Ps2	Choose an appropriate strategy for a calculation and explain how they worked out the answer.
		5Ps3	Explore and solve number problems and puzzles, e.g. logic problems.
		5Ps4	Deduce new information from existing information to solve problems.
		5Ps5	Use ordered lists and tables to help to solve problems systematically.
		5Ps8	Investigate a simple general statement by finding examples which do or do not satisfy it, e.g. the sum of three consecutive whole numbers is always a multiple of three.
		5Ps9	Explain methods and justify reasoning orally and in writing; make hypotheses and test them out.
		5Ps10	Solve a larger problem by breaking it down into sub-problems or represent it using diagrams.

*Learning objective revised and consolidated

Unit 10 – Addition and subtraction 3			Recommended time allocation: 1 week
Cambridge Primary Mathematics Curriculum Framework			
Strand	**Sub-strand**	**Code**	**Learning objective**
Number	Calculation – Mental strategies	5Nc9*	Add or subtract near multiples of 10 or 100, e.g. 4387 – 299.*
		5Nc10	Use appropriate strategies to add or subtract pairs of [two- and three-digit numbers and] numbers with one decimal place, using jottings where necessary.
		5Nc11*	Calculate differences between near multiples of 1000, e.g. 5026 – 4998, or near multiples of 1, e.g. 3.2 – 2.6.*
	Calculation – Addition and subtraction	5Nc18*	Find the total of more than three two- or three-digit numbers using a written method.*
		5Nc19*	Add or subtract any pair of three- and/or four-digit numbers, with the same number of decimal places, including amounts of money.*
Problem solving	Using techniques and skills in solving mathematical problems	5Pt2	Solve single and multi-step word problems (all four operations); represent them, e.g. with diagrams or a number line.
		5Pt3	Check with a different order when adding several numbers or by using the inverse when adding or subtracting a pair of numbers.
		5Pt6	Estimate and approximate when calculating, e.g. using rounding, and check working.
		5Pt7	Consider whether an answer is reasonable in the context of a problem.
	Using understanding and strategies in solving problems	5Ps2	Choose an appropriate strategy for a calculation and explain how they worked out the answer.
		5Ps3	Explore and solve number problems and puzzles, e.g. logic problems.
		5Ps4	Deduce new information from existing information to solve problems.
		5Ps5	Use ordered lists and tables to help to solve problems systematically.
		5Ps8	Investigate a simple general statement by finding examples which do or do not satisfy it, e.g. the sum of three consecutive whole numbers is always a multiple of three.
		5Ps9	Explain methods and justify reasoning orally and in writing; make hypotheses and test them out.
		5Ps10	Solve a larger problem by breaking it down into sub-problems or represent it using diagrams.

*Learning objective revised and consolidated

Unit 11 – Multiplication and division 1			Recommended time allocation: 2 weeks
Cambridge Primary Mathematics Curriculum Framework			
Strand	**Sub-strand**	**Code**	**Learning objective**
Number	Calculation – Mental strategies	5Nc3	Know multiplication and division facts for the 2x to 10x tables.
		5Nc4	Know and apply tests of divisibility by 2, 5, 10 and 100.
		5Nc5	Recognise multiples of 6, 7, 8 and 9 up to the 10th multiple.
		5Nc6	Know squares of all numbers to 10 × 10.
		5Nc7	Find factors of two-digit numbers.

	Calculation – Multiplication and division	5Nc20	Multiply or divide three-digit numbers by single-digit numbers.
		5Nc21	Multiply two-digit numbers by two-digit numbers.
		5Nc23	Divide three-digit numbers by single-digit numbers, including those with a remainder (answers no greater than 30).
		5Nc25	Decide whether to group (using multiplication facts and multiples of the divisor) or to share (halving and quartering) to solve divisions.
Problem solving	Using techniques and skills in solving mathematical problems	5Pt2	Solve single and multi-step word problems (all four operations); represent them, e.g. with diagrams or a number line.
		5Pt4	Use multiplication to check the result of a division[, e.g. multiply 3.7 × 8 to check 29.6 ÷ 8].
		5Pt6	Estimate and approximate when calculating, e.g. using rounding, and check working.
		5Pt7	Consider whether an answer is reasonable in the context of a problem.
	Using understanding and strategies in solving problems	5Ps2	Choose an appropriate strategy for a calculation and explain how they worked out the answer.
		5Ps3	Explore and solve number problems and puzzles, e.g. logic problems.
		5Ps4	Deduce new information from existing information to solve problems.
		5Ps5	Use ordered lists and tables to help to solve problems systematically.
		5Ps9	Explain methods and justify reasoning orally and in writing; make hypotheses and test them out.
		5Ps10	Solve a larger problem by breaking it down into sub-problems or represent it using diagrams.

Unit 12 – Multiplication and division 2			Recommended time allocation: 2 weeks
Cambridge Primary Mathematics Curriculum Framework			
Strand	**Sub-strand**	**Code**	**Learning objective**
Number	Calculation – Mental strategies	5Nc3*	Know multiplication and division facts for the 2x to 10x tables.*
		5Nc4*	Know and apply tests of divisibility by 2, 5, 10 and 100.*
		5Nc5*	Recognise multiples of 6, 7, 8 and 9 up to the 10th multiple.*
		5Nc7*	Find factors of two-digit numbers.*
		5Nc12	Multiply multiples of 10 to 90, and multiples of 100 to 900, by a single-digit number.
		5Nc13	Multiply by 19 or 21 by multiplying by 20 and adjusting.
		5Nc14	Multiply by 25 by multiplying by 100 and dividing by 4.
		5Nc15	Use factors to multiply, e.g. multiply by 3, then double to multiply by 6.
		5Nc16	Double any number up to 100 and halve even numbers to 200 [and use this to double and halve numbers with one or two decimal places, e.g. double 3.4 and half of 8.6].
		5Nc17	Double multiples of 10 to 1000 and multiples of 100 to 10 000, e.g. double 360 or double 3600, and derive the corresponding halves.
	Calculation – Multiplication and division	5Nc20*	Multiply or divide three-digit numbers by single-digit numbers.*
		5Nc21*	Multiply two-digit numbers by two-digit numbers.*
		5Nc23*	Divide three-digit numbers by single-digit numbers, including those with a remainder (answers no greater than 30).*
		5Nc26	Decide whether to round an answer up or down after division, depending on the context.

Problem solving	Using techniques and skills in solving mathematical problems	5Pt2	Solve single and multi-step word problems (all four operations); represent them, e.g. with diagrams or a number line.
		5Pt4	Use multiplication to check the result of a division[, e.g. multiply 3.7 × 8 to check 29.6 ÷ 8].
		5Pt6	Estimate and approximate when calculating, e.g. using rounding, and check working.
		5Pt7	Consider whether an answer is reasonable in the context of a problem.
	Using understanding and strategies in solving problems	5Ps2	Choose an appropriate strategy for a calculation and explain how they worked out the answer.
		5Ps3	Explore and solve number problems and puzzles, e.g. logic problems.
		5Ps4	Deduce new information from existing information to solve problems.
		5Ps5	Use ordered lists and tables to help to solve problems systematically.
		5Ps8	Investigate a simple general statement by finding examples which do or do not satisfy it[, e.g. the sum of three consecutive whole numbers is always a multiple of three].
		5Ps9	Explain methods and justify reasoning orally and in writing; make hypotheses and test them out.
		5Ps10	Solve a larger problem by breaking it down into sub-problems or represent it using diagrams.

*Learning objective revised and consolidated

Unit 13 – Multiplication and division 3			Recommended time allocation: 2 weeks
Cambridge Primary Mathematics Curriculum Framework			
Strand	**Sub-strand**	**Code**	**Learning objective**
Number	Calculation – Mental strategies	5Nc12*	Multiply multiples of 10 to 90, and multiples of 100 to 900, by a single-digit number.*
		5Nc13*	Multiply by 19 or 21 by multiplying by 20 and adjusting.*
		5Nc14*	Multiply by 25 by multiplying by 100 and dividing by 4.*
		5Nc15*	Use factors to multiply, e.g. multiply by 3, then double to multiply by 6.*
		5Nc16*	Double any number up to 100 and halve even numbers to 200* and use this to double and halve numbers with one or two decimal places, e.g. double 3.4 and half of 8.6.
		5Nc17*	Double multiples of 10 to 1000 and multiples of 100 to 10 000, e.g. double 360 or double 3600, and derive the corresponding halves.*
	Calculation – Multiplication and division	5Nc22	Multiply two-digit numbers with one decimal place by single-digit numbers, e.g. 3.6 × 7.
		5Nc24	Start expressing remainders as a fraction of the divisor when dividing two-digit numbers by single-digit numbers.
		5Nc26*	Decide whether to round an answer up or down after division, depending on the context.*
		5Nc27	Begin to use brackets to order operations and understand the relationship between the four operations and how the laws of arithmetic apply to multiplication.

Problem solving	Using techniques and skills in solving mathematical problems	5Pt2	Solve single and multi-step word problems (all four operations); represent them, e.g. with diagrams or a number line.
		5Pt4	Use multiplication to check the result of a division, e.g. multiply 3.7×8 to check $29.6 \div 8$.
		5Pt6	Estimate and approximate when calculating, e.g. using rounding, and check working.
		5Pt7	Consider whether an answer is reasonable in the context of a problem.
	Using understanding and strategies in solving problems	5Ps2	Choose an appropriate strategy for a calculation and explain how they worked out the answer.
		5Ps3	Explore and solve number problems and puzzles, e.g. logic problems.
		5Ps4	Deduce new information from existing information to solve problems.
		5Ps5	Use ordered lists and tables to help to solve problems systematically.
		5Ps8	Investigate a simple general statement by finding examples which do or do not satisfy it[, e.g. the sum of three consecutive whole numbers is always a multiple of three].
		5Ps9	Explain methods and justify reasoning orally and in writing; make hypotheses and test them out.
		5Ps10	Solve a larger problem by breaking it down into sub-problems or represent it using diagrams.

*Learning objective revised and consolidated

Unit 14 – 2D shape, including symmetry			Recommended time allocation: 1 week
Cambridge Primary Mathematics Curriculum Framework			
Strand	**Sub-strand**	**Code**	**Learning objective**
Geometry	Shapes and geometric reasoning	5Gs1	Identify and describe properties of triangles and classify as isosceles, equilateral or scalene.
		5Gs2	Recognise reflective and rotational symmetry in regular polygons.
		5Gs3	Create patterns with two lines of symmetry, e.g. on a pegboard or squared paper.
		5Gs5	Recognise perpendicular and parallel lines in 2D shapes, drawings and the environment.
Problem solving	Using techniques and skills in solving mathematical problems	5Pt5	Recognise the relationships between different 2D [and 3]D shapes, e.g. a face of a cube is a square.
	Using understanding and strategies in solving problems	5Ps4	Deduce new information from existing information to solve problems.
		5Ps7	Identify simple relationships between shapes, e.g. these triangles are all isosceles because ...
		5Ps9	Explain methods and justify reasoning orally and in writing; make hypotheses and test them out.

Unit 15 – 3D shape			**Recommended time allocation:** 1 week
Cambridge Primary Mathematics Curriculum Framework			
Strand	**Sub-strand**	**Code**	**Learning objective**
Geometry	Shapes and geometric reasoning	5Gs4	Visualise 3D shapes from 2D drawings and nets, e.g. different nets of an open or closed cube.
Problem solving	Using techniques and skills in solving mathematical problems	5Pt5	Recognise the relationships between different [2D and] 3D shapes, e.g. a face of a cube is a square.
	Using understanding and strategies in solving problems	5Ps4	Deduce new information from existing information to solve problems.
		5Ps7	Identify simple relationships between shapes[, e.g. these triangles are all isosceles because …].
		5Ps9	Explain methods and justify reasoning orally and in writing; make hypotheses and test them out.

Unit 16 – Angles			**Recommended time allocation:** 1 week
Cambridge Primary Mathematics Curriculum Framework			
Strand	**Sub-strand**	**Code**	**Learning objective**
Geometry	Shapes and geometric reasoning	5Gs6	Understand and use angle measure in degrees; measure angles to the nearest 5°; identify, describe and estimate the size of angles and classify them as acute, right or obtuse.
		5Gs7	Calculate angles in a straight line.
Problem solving	Using techniques and skills in solving mathematical problems	5Pt7	Consider whether an answer is reasonable in the context of a problem.
	Using understanding and strategies in solving problems	5Ps4	Deduce new information from existing information to solve problems.
		5Ps7	Identify simple relationships between shapes [e.g. these triangles are all isosceles because …]
		5Ps8	Investigate a simple general statement by finding examples which do or do not satisfy it [e.g. the sum of three consecutive whole numbers is always a multiple of three].
		5Ps9	Explain methods and justify reasoning orally and in writing; make hypotheses and test them out.

Unit 17 – Position and movement			**Recommended time allocation:** 2 weeks
Cambridge Primary Mathematics Curriculum Framework			
Strand	**Sub-strand**	**Code**	**Learning objective**
Geometry	Position and movement	5Gp1	Read and plot co-ordinates in the first quadrant.
		5Gp2	Predict where a polygon will be after reflection where the mirror line is parallel to one of the sides, including where the line is oblique.
		5Gp3	Understand translation as movement along a straight line, identify where polygons will be after a translation and give instructions for translating shapes.

Problem solving	Using understanding and strategies in solving problems	5Ps4	Deduce new information from existing information to solve problems.
		5Ps7	Identify simple relationships between shapes [e.g. these triangles are all isosceles because ...].
		5Ps8	Investigate a simple general statement by finding examples which do or do not satisfy it [e.g. the sum of three consecutive whole numbers is always a multiple of three].
		5Ps9	Explain methods and justify reasoning orally and in writing; make hypotheses and test them out.

Unit 18 – Length			Recommended time allocation: 1 week
Cambridge Primary Mathematics Curriculum Framework			
Strand	**Sub-strand**	**Code**	**Learning objective**
Measure	Length, mass and capacity	5Ml1	Read, choose, use and record standard units to estimate and measure length[, mass and] capacity to a suitable degree of accuracy.
		5Ml2	Convert larger to smaller metric units (decimals to one place) [e.g. change 2.6 kg to 2600 g].
		5Ml3	Order measurements in mixed units.
		5Ml4	Round measurements to the nearest whole unit.
		5Ml7	Draw and measure lines to the nearest centimetre and millimetre.
Problem solving	Using techniques and skills in solving mathematical problems	5Pt1	Understand everyday systems of measurement in length [weight, capacity, temperature and time] and use these to perform simple calculations.
		5Pt2	Solve single and multi-step word problems (all four operations); represent them, e.g. with diagrams or a number line.
		5Pt7	Consider whether an answer is reasonable in the context of a problem.
	Using understanding and strategies in solving problems	5Ps1	Understand everyday systems of measurement in length [weight, capacity, temperature and time] and use these to perform simple calculations.
		5Ps2	Choose an appropriate strategy for a calculation and explain how they worked out the answer.
		5Ps4	Deduce new information from existing information to solve problems.
		5Ps5	Use ordered lists and tables to help to solve problems systematically.
		5Ps9	Explain methods and justify reasoning orally and in writing; make hypotheses and test them out.
		5Ps10	Solve a larger problem by breaking it down into sub-problems or represent it using diagrams.

Unit 19 – Mass			**Recommended time allocation:** 1 week
Cambridge Primary Mathematics Curriculum Framework			
Strand	**Sub-strand**	**Code**	**Learning objective**
Measure	Length, mass and capacity	5Ml1	Read, choose, use and record standard units to estimate and measure [length,] mass [and capacity] to a suitable degree of accuracy.
		5Ml2	Convert larger to smaller metric units (decimals to one place), e.g. change 2.6 kg to 2600 g.
		5Ml3	Order measurements in mixed units.
		5Ml4	Round measurements to the nearest whole unit.
		5Ml5	Interpret a reading that lies between two unnumbered divisions on a scale.
		5Ml6	Compare readings on different scales.
Problem solving	Using techniques and skills in solving mathematical problems	5Pt1	Understand everyday systems of measurement in [length,] weight, [capacity, temperature and time] and use these to perform simple calculations.
		5Pt2	Solve single and multi-step word problems (all four operations); represent them, e.g. with diagrams or a number line.
		5Pt7	Consider whether an answer is reasonable in the context of a problem.
	Using understanding and strategies in solving problems	5Ps1	Understand everyday systems of measurement in [length,] weight, [capacity, temperature and time] and use these to perform simple calculations.
		5Ps2	Choose an appropriate strategy for a calculation and explain how they worked out the answer.
		5Ps4	Deduce new information from existing information to solve problems.
		5Ps5	Use ordered lists and tables to help to solve problems systematically.
		5Ps9	Explain methods and justify reasoning orally and in writing; make hypotheses and test them out.
		5Ps10	Solve a larger problem by breaking it down into sub-problems or represent it using diagrams.

Unit 20 – Capacity			**Recommended time allocation:** 1 week
Cambridge Primary Mathematics Curriculum Framework			
Strand	**Sub-strand**	**Code**	**Learning objective**
Measure	Length, mass and capacity	5Ml1	Read, choose, use and record standard units to estimate and measure [length, mass and] capacity to a suitable degree of accuracy.
		5Ml2	Convert larger to smaller metric units (decimals to one place) [e.g. change 2.6 kg to 2600 g].
		5Ml3	Order measurements in mixed units.
		5Ml4	Round measurements to the nearest whole unit.
		5Ml5	Interpret a reading that lies between two unnumbered divisions on a scale.
		5Ml6	Compare readings on different scales.

Problem solving	Using techniques and skills in solving mathematical problems	5Pt1	Understand everyday systems of measurement in [length, weight,] capacity, [temperature and time] and use these to perform simple calculations.
		5Pt2	Solve single and multi-step word problems (all four operations); represent them, e.g. with diagrams or a number line.
		5Pt7	Consider whether an answer is reasonable in the context of a problem.
	Using understanding and strategies in solving problems	5Ps1	Understand everyday systems of measurement in [length, weight,] capacity, [temperature and time] and use these to perform simple calculations.
		5Ps2	Choose an appropriate strategy for a calculation and explain how they worked out the answer.
		5Ps4	Deduce new information from existing information to solve problems.
		5Ps5	Use ordered lists and tables to help to solve problems systematically.
		5Ps9	Explain methods and justify reasoning orally and in writing; make hypotheses and test them out.
		5Ps10	Solve a larger problem by breaking it down into sub-problems or represent it using diagrams.

Unit 21 – Time			Recommended time allocation: 1 week
Cambridge Primary Mathematics Curriculum Framework			
Strand	**Sub-strand**	**Code**	**Learning objective**
Measure	Time	5Mt1	Recognise and use the units for time (seconds, minutes, hours, days, months and years).
		5Mt2	Tell and compare the time using digital and analogue clocks using the 24-hour clock.
		5Mt3	Read timetables using the 24-hour clock.
		5Mt4	Calculate time intervals in seconds, minutes and hours using digital or analogue formats.
		5Mt5	Use a calendar to calculate time intervals in days and weeks (using knowledge of days in calendar months).
		5Mt6	Calculate time intervals in months or years.
Problem solving	Using techniques and skills in solving mathematical problems	5Pt1	Understand everyday systems of measurement in [length, weight, capacity, temperature and] time and use these to perform simple calculations.
		5Pt2	Solve single and multi-step word problems (all four operations); represent them, e.g. with diagrams or a number line.
		5Pt7	Consider whether an answer is reasonable in the context of a problem.
	Using understanding and strategies in solving problems	5Ps1	Understand everyday systems of measurement in [length, weight, capacity, temperature and] time and use these to perform simple calculations.
		5Ps2	Choose an appropriate strategy for a calculation and explain how they worked out the answer.
		5Ps4	Deduce new information from existing information to solve problems.
		5Ps5	Use ordered lists and tables to help to solve problems systematically.
		5Ps9	Explain methods and justify reasoning orally and in writing; make hypotheses and test them out.
		5Ps10	Solve a larger problem by breaking it down into sub-problems or represent it using diagrams.

Unit 22 – Area and perimeter			Recommended time allocation: 1 week
Cambridge Primary Mathematics Curriculum Framework			
Strand	**Sub-strand**	**Code**	**Learning objective**
Measure	Area and perimeter	5Ma1	Measure and calculate the perimeter of regular and irregular polygons.
		5Ma2	Understand area measured in square centimetres (cm^2).
		5Ma3	Use the formula for the area of a rectangle to calculate the rectangle's area.
Problem solving	Using techniques and skills in solving mathematical problems	5Pt1	Understand everyday systems of measurement in length[, weight, capacity, temperature and time] and use these to perform simple calculations.
		5Pt2	Solve single and multi-step word problems (all four operations); represent them, e.g. with diagrams or a number line.
		5Pt7	Consider whether an answer is reasonable in the context of a problem.
	Using understanding and strategies in solving problems	5Ps1	Understand everyday systems of measurement in length [weight, capacity, temperature and time] and use these to perform simple calculations.
		5Ps2	Choose an appropriate strategy for a calculation and explain how they worked out the answer.
		5Ps4	Deduce new information from existing information to solve problems.
		5Ps5	Use ordered lists and tables to help to solve problems systematically.
		5Ps9	Explain methods and justify reasoning orally and in writing; make hypotheses and test them out.
		5Ps10	Solve a larger problem by breaking it down into sub-problems or represent it using diagrams.

Unit 23 – Handling data			Recommended time allocation: 2 weeks
Cambridge Primary Mathematics Curriculum Framework			
Strand	**Sub-strand**	**Code**	**Learning objective**
Handling data	Organising, categorising and representing data	5Dh1	Answer a set of related questions by collecting, selecting and organising relevant data; draw conclusions from their own and others' data and identify further questions to ask.
		5Dh2	Draw and interpret frequency tables, pictograms and bar line charts, with the vertical axis labelled for example in twos, fives, tens, twenties or hundreds. Consider the effect of changing the scale on the vertical axis.
		5Dh3	Construct simple line graphs, e.g. to show changes in temperature over time.
		5Dh4	Understand where intermediate points have and do not have meaning, e.g. comparing a line graph of temperature against time with a graph of class attendance for each day of the week.
		5Dh5	Find and interpret the mode of a set of data.
	Probability	5Db1	Describe the occurrence of familiar events using the language of chance or likelihood.

Problem solving	Using techniques and skills in solving mathematical problems	5Pt2	Solve single and multi-step word problems (all four operations); represent them, e.g. with diagrams or a number line.
		5Pt7	Consider whether an answer is reasonable in the context of a problem.
	Using understanding and strategies in solving problems	5Ps4	Deduce new information from existing information to solve problems.
		5Ps5	Use ordered lists and tables to help to solve problems systematically.
		5Ps8	Investigate a simple general statement by finding examples which do or do not satisfy it [e.g. the sum of three consecutive whole numbers is always a multiple of three].
		5Ps9	Explain methods and justify reasoning orally and in writing; make hypotheses and test them out.

Cambridge Primary Mathematics Curriculum Framework Stage 5 link to Collins International Primary Maths units

Cambridge Primary Mathematics Curriculum Framework				Collins International Primary Maths Unit(s)
Strand	Sub-strand	Code	Learning objective	
Number	Numbers and the number system	5Nn1	Count on and back in steps of constant size, extending beyond zero.	1, 2
		5Nn2	Know what each digit represents in five- and six-digit numbers.	1
		5Nn3	Partition any number up to one million into thousands, hundreds, tens and units.	1, 2
		5Nn4	Use decimal notation for tenths and hundredths and understand what each digit represents.	3, 4
		5Nn5	Multiply and divide any number from 1 to 10 000 by 10 or 100 and understand the effect.	1, 3
		5Nn6	Round four-digit numbers to the nearest 10, 100 or 1000.	1
		5Nn7	Round a number with one or two decimal places to the nearest whole number.	4
		5Nn8	Order and compare numbers up to a million using the > and < signs.	1
		5Nn9	Order and compare negative and positive numbers on a number line and temperature scale.	2
		5Nn10	Calculate a rise or fall in temperature.	2
		5Nn11	Order numbers with one or two decimal places and compare using the > and < signs.	4
		5Nn12	Recognise and extend number sequences.	1, 2
		5Nn13	Recognise odd and even numbers and multiples of 5, 10, 25, 50 and 100 up to 1000.	1
		5Nn14	Make general statements about sums, differences and multiples of odd and even numbers.	2
		5Nn15	Recognise equivalence between: $\frac{1}{2}$, $\frac{1}{4}$ and $\frac{1}{8}$; $\frac{1}{3}$ and $\frac{1}{6}$; $\frac{1}{5}$ and $\frac{1}{10}$.	5
		5Nn16	Recognise equivalence between the decimal and fraction forms of halves, tenths and hundredths and use this to help order fractions, e.g. 0.6 is more than 50% and less than $\frac{7}{10}$.	5
		5Nn17	Change an improper fraction to a mixed number, e.g. $\frac{7}{4}$ to $\frac{13}{4}$; order mixed numbers and place between whole numbers on a number line.	5
		5Nn18	Relate finding fractions to division and use to find simple fractions of quantities.	5
		5Nn19	Understand percentage as the number of parts in every 100 and find simple percentages of quantities.	6
		5Nn20	Express halves, tenths and hundredths as percentages.	6

| | | 5Nn21 | Use fractions to describe and estimate a simple proportion, e.g. $\frac{1}{5}$ of the beads are yellow. | 7 |
| | | 5Nn22 | Use ratio to solve problems, e.g. to adapt a recipe for 6 people to one for 3 or 12 people. | 7 |

Cambridge Primary Mathematics Curriculum Framework				Collins International Primary Maths Unit(s)
Strand	**Sub-strand**	**Code**	**Learning objective**	
Number	Calculation - Mental strategies	5Nc1	Know by heart pairs of one-place decimals with a total of 1, e.g. 0.8 + 0.2.	9
		5Nc2	Derive quickly pairs of decimals with a total of 10, and with a total of 1.	9
		5Nc3	Know multiplication and division facts for the 2x to 10x tables.	11, 12
		5Nc4	Know and apply tests of divisibility by 2, 5, 10 and 100.	11, 12
		5Nc5	Recognise multiples of 6, 7, 8 and 9 up to the 10th multiple.	11, 12
		5Nc6	Know squares of all numbers to 10 × 10.	11
		5Nc7	Find factors of two-digit numbers.	11, 12
		5Nc8	Count on or back in thousands, hundreds, tens and ones to add or subtract.	8, 9
		5Nc9	Add or subtract near multiples of 10 or 100, e.g. 4387–299.	9, 10
		5Nc10	Use appropriate strategies to add or subtract pairs of two- and three-digit numbers and numbers with one decimal place, using jottings where necessary.	8, 9, 10
		5Nc11	Calculate differences between near multiples of 1000, e.g. 5026–4998, or near multiples of 1, e.g. 3.2–2.6.	9, 10
		5Nc12	Multiply multiples of 10 to 90, and multiples of 100 to 900, by a single-digit number.	12, 13
		5Nc13	Multiply by 19 or 21 by multiplying by 20 and adjusting.	12, 13
		5Nc14	Multiply by 25 by multiplying by 100 and dividing by 4.	12, 13
		5Nc15	Use factors to multiply, e.g. multiply by 3, then double to multiply by 6.	12, 13
		5Nc16	Double any number up to 100 and halve even numbers to 200 and use this to double and halve numbers with one or two decimal places, e.g. double 3.4 and half of 8.6	12, 13
		5Nc17	Double multiples of 10 to 1000 and multiples of 100 to 10 000, e.g. double 360 or double 3600, and derive the corresponding halves.	12, 13

Cambridge Primary Mathematics Curriculum Framework				Collins International Primary Maths Unit(s)
Strand	**Sub-strand**	**Code**	**Learning objective**	
Number	Calculation – Addition and subtraction	5Nc18	Find the total of more than three two- or three-digit numbers using a written method.	8–10
		5Nc19	Add or subtract any pair of three- and/or four-digit numbers, with the same number of decimal places, including amounts of money.	9, 10
	Calculation – Multiplication and division	5Nc20	Multiply or divide three-digit numbers by single-digit numbers.	11, 12
		5Nc21	Multiply two-digit numbers by two-digit numbers.	11, 12
		5Nc22	Multiply two-digit numbers with one decimal place by single-digit numbers, e.g. 3.6×7.	13
		5Nc23	Divide three-digit numbers by single-digit numbers, including those with a remainder (answers no greater than 30).	11, 12
		5Nc24	Start expressing remainders as a fraction of the divisor when dividing two-digit numbers by single-digit numbers.	13
		5Nc25	Decide whether to group (using multiplication facts and multiples of the divisor) or to share (halving and quartering) to solve divisions.	11
		5Nc26	Decide whether to round an answer up or down after division, depending on the context.	12, 13
		5Nc27	Begin to use brackets to order operations and understand the relationship between the four operations and how the laws of arithmetic apply to multiplication.	13
Geometry	Shapes and geometric reasoning	5Gs1	Identify and describe properties of triangles and classify as isosceles, equilateral or scalene.	14
		5Gs2	Recognise reflective and rotational symmetry in regular polygons.	14
		5Gs3	Create patterns with two lines of symmetry, e.g. on a pegboard or squared paper.	14
		5Gs4	Visualise 3D shapes from 2D drawings and nets, e.g. different nets of an open or closed cube.	15
		5Gs5	Recognise perpendicular and parallel lines in 2D shapes, drawings and the environment.	14
		5Gs6	Understand and use angle measure in degrees; measure angles to the nearest 5°; identify, describe and estimate the size of angles and classify them as acute, right or obtuse.	16
		5Gs7	Calculate angles in a straight line.	16
	Position and movement	5Gp1	Read and plot co-ordinates in the first quadrant.	17
		5Gp2	Predict where a polygon will be after reflection where the mirror line is parallel to one of the sides, including where the line is oblique.	17
		5Gp3	Understand translation as movement along a straight line, identify where polygons will be after a translation and give instructions for translating shapes.	17

Cambridge Primary Mathematics Curriculum Framework				Collins International Primary Maths Unit(s)
Strand	**Sub-strand**	**Code**	**Learning objective**	
Measure	Length, mass and capacity	5Ml1	Read, choose, use and record standard units to estimate and measure length, mass and capacity to a suitable degree of accuracy.	18–20
		5Ml2	Convert larger to smaller metric units (decimals to one place), e.g. change 2.6 kg to 2600 g.	18–20
		5Ml3	Order measurements in mixed units.	18–20
		5Ml4	Round measurements to the nearest whole unit.	18–20
		5Ml5	Interpret a reading that lies between two unnumbered divisions on a scale.	19, 20
		5Ml6	Compare readings on different scales.	19, 20
		5Ml7	Draw and measure lines to the nearest centimetre and millimetre.	18
	Time	5Mt1	Recognise and use the units for time (seconds, minutes, hours, days, months and years).	21
		5Mt2	Tell and compare the time using digital and analogue clocks using the 24-hour clock.	21
		5Mt3	Read timetables using the 24-hour clock.	21
		5Mt4	Calculate time intervals in seconds, minutes and hours using digital or analogue formats.	21
		5Mt5	Use a calendar to calculate time intervals in days and weeks (using knowledge of days in calendar months).	21
		5Mt6	Calculate time intervals in months or years.	21
	Area and perimeter	5Ma1	Measure and calculate the perimeter of regular and irregular polygons.	22
		5Ma2	Understand area measured in square centimetres (cm^2).	22
		5Ma3	Use the formula for the area of a rectangle to calculate the rectangle's area.	22
Handling data	Organising, categorising and representing data	5Dh1	Answer a set of related questions by collecting, selecting and organising relevant data; draw conclusions from their own and others' data and identify further questions to ask.	23
		5Dh2	Draw and interpret frequency tables, pictograms and bar line charts, with the vertical axis labelled for example in twos, fives, tens, twenties or hundreds. Consider the effect of changing the scale on the vertical axis.	23
		5Dh3	Construct simple line graphs, e.g. to show changes in temperature over time.	23
		5Dh4	Understand where intermediate points have and do not have meaning, e.g. comparing a line graph of temperature against time with a graph of class attendance for each day of the week.	23
		5Dh5	Find and interpret the mode of a set of data.	23
	Probability	5Db1	Describe the occurrence of familiar events using the language of chance or likelihood.	23

Cambridge Primary Mathematics Curriculum Framework				Collins International Primary Maths Unit(s)
Strand	**Sub-strand**	**Code**	**Learning objective**	
Problem solving	Using techniques and skills in solving mathematical problems	5Pt1	Understand everyday systems of measurement in length, weight, capacity, temperature and time and use these to perform simple calculations.	2, 18–22
		5Pt2	Solve single and multi-step word problems (all four operations); represent them, e.g. with diagrams or a number line.	5, 6–13, 18–23
		5Pt3	Check with a different order when adding several numbers or by using the inverse when adding or subtracting a pair of numbers.	8–10
		5Pt4	Use multiplication to check the result of a division, e.g. multiply 3.7×8 to check $29.6 \div 8$.	11–13
		5Pt5	Recognise the relationships between different 2D and 3D shapes, e.g. a face of a cube is a square.	14, 15
		5Pt6	Estimate and approximate when calculating, e.g. using rounding, and check working.	8–13
		5Pt7	Consider whether an answer is reasonable in the context of a problem.	2 –13, 16, 18–23
	Using understanding and strategies in solving problems	5Ps1	Understand everyday systems of measurement in length, weight, capacity, temperature and time and use these to perform simple calculations.	2, 18–22
		5Ps2	Choose an appropriate strategy for a calculation and explain how they worked out the answer.	2, 5–13, 18–22
		5Ps3	Explore and solve number problems and puzzles, e.g. logic problems.	1–13
		5Ps4	Deduce new information from existing information to solve problems.	1–23
		5Ps5	Use ordered lists and tables to help to solve problems systematically.	1, 2, 8–13, 18–23
		5Ps6	Describe and continue number sequences, e.g. –30, –27, □, □, –18…; identify the relationships between numbers.	1, 2
		5Ps7	Identify simple relationships between shapes, e.g. these triangles are all isosceles because …	14–17
		5Ps8	Investigate a simple general statement by finding examples which do or do not satisfy it, e.g. the sum of three consecutive whole numbers is always a multiple of three.	1–7, 9, 10, 12, 13, 16, 17, 23
		5Ps9	Explain methods and justify reasoning orally and in writing; make hypotheses and test them out.	1–23
		5Ps10	Solve a larger problem by breaking it down into sub-problems or represent it using diagrams.	1–13, 18–22

Refresh activities

Sequences 👥

Learning objectives

Code	Learning objective
5Nn1	Count on and back in steps of constant size [, extending beyond zero].

Resources

paper (per learner) or mini whiteboard and pen if available

What to do

- Learners sit in a circle. Distribute paper and pencils. Suggest a number sequence beginning from zero that counts on in steps of 2, 3, 4, 5, 9 or 10.
- Say: **The number sequence will begin on my left and move around the circle clockwise. As the turn passes, say the next number in the sequence.**
- Give learners 15 seconds before the game begins to predict and write their number in the sequence.
- The game begins, the turn passing around the circle. Learners say the number that they have written on their paper, holding it up. If it is correct, they score a point. The next round begins with a different sequence and a different starting position, shifted one place to the right of the previous position.

Variations

- Start counting on from a number other than zero.
- Introduce backward sequences that extend beyond zero.

Reverse 👥

Learning objective

Code	Learning objective
5Nn12	Recognise and extend number sequences.

What to do

- Learners sit in a circle.
- The teacher announces a sequence, for example: adding 4, subtracting 5 or multiples of 3.
- The teacher also announces a 'reverse' number, for example: 'any number ending in 4' or 'any two-digit number in which both digits are identical'.
- When it is the turn of a learner to say a 'reverse' number, they must say 'reverse' instead. The turn immediately reverses, passing the opposite way around the circle.
- The teacher begins the game and the turn passes clockwise. If a learner makes a mistake, or fails to use the 'reverse' number at the correct point, they sit out of the game. The winner is the last learner in the game.

Variation

- The game can be played without the 'reverse' action, focusing on the sequence announced.

Dice digits :

Learning objectives

Code	Learning objective
5Nn2	Know what each digit represents in five- and six-digit numbers.
5Nn3	Partition any number up to one million into thousands, hundreds, tens and units.

Resources

1–6 dice or Resource sheet 4: 1–6 spinner (per learner); mini whiteboard and pen or paper (per learner)

What to do

- Learners roll a dice five times and write the numbers on their whiteboard.
- Ask learners to create the largest and smallest numbers they can from their set of five digits.
- As a class, work out who has the largest and smallest number. Ask: **Are you sure these are the largest and smallest numbers?** Learners check to make sure.
- Choose individual learners to display their numbers and ask the class: **How would we partition this number?** Take feedback. Point to a digit and ask: **What is the place value of this digit?** Continue until you are sure all learners can recall how to partition numbers and are confident with place value.
- Repeat with six digit numbers.

Variation

- Learners tell their partner the value of each digit in the number.

Values together :

Learning objectives

Code	Learning objective
5Nn2	Know what each digit represents in five- and six-digit numbers.
5Nn3	Partition any number up to one million into thousands, hundreds, tens and units.

Resources

1–6 dice or Resource sheet 4: 1–6 spinner (per learner); mini whiteboard and pen or paper (per learner); Resource sheet 5: Arrow cards (per learner)

What to do

- Learners roll a dice four times, each number representing the number of thousands, hundreds, tens or ones digit.
- The learners then write their numbers on their whiteboards in expanded notation, for example: 2000, 300, 10, 9.
- Ask individual learners to demonstrate what they have done and say the place value of each digit in their number.
- Next, ask five learners to each roll one number. Say: **We are going to partition this five-digit number using the Place value tool.** Ask individual learners to say which arrow cards you need to drag on to the screen. Repeat with different five- and six-digit numbers.
- Learners roll their dice five (or six) times and use the place value arrow cards to display their numbers in expanded notation form. If time allows, let individual learners use the **Place value tool** to partition their five- or six-digit number.

Variation

- Less able learners can work with three- or four-digits only.

Number – Numbers and the number system

Multiply or divide 👥

Learning objective

Code	Learning objective
5Nn5	Multiply and divide any number from 1 to 10000 by 10 or 100 and understand the effect.

Resources

1–6 dice or Resource sheet 4: 1–6 spinner (per pair); mini whiteboard and pen or paper (per pair); a counter with a multiplication symbol drawn or taped on one side and a division symbol on the other (per pair); Resource sheet 3: Blank place value grid (per learner)

What to do

- Remind learners they already know that when they multiply a number by 10 the digits move one place value to the left and when multiplying by 100 the digits move two place values to the left. In division, when they divide a number by 10 the digits move one place value to the right and when dividing by 100 the digits move two place values to the right. If necessary, demonstrate this by drawing a place value grid on the board and modelling some simple two-digit by three-digit calculations.
- In pairs, learners roll a dice up to five times. They write the number in their copy book or on a mini whiteboard.
- Without telling their partner, each learner should decide whether to multiply or divide the number by 10 or 100. Or, to make this a more random choice, learners can flip a counter with a multiplication symbol drawn or taped on one side and a division symbol on the other. Each learner then does their calculation individually.
- Learners use Resource sheet 3: Blank place value grid, to help calculate their answer.
- Learners reveal their multiplied or divided figure to their partner who must say whether they have multiplied or divided by 10 or 100.

Variation

- Less able learners use a five-digit number, with zeros in the tens and ones places.

Rounding 👥

Learning objective

Code	Learning objective
5Nn6	Round four-digit numbers to the nearest 10, 100 or 1000.

Resources

1–6 dice or Resource sheet 4: 1–6 spinner (per class); paper (per learner)

What to do

- Roll a dice four times. Write the digits on the board to form a three-digit number.
- Ask learners to round it to the nearest 10 or 100.
- Learners compare their answer with a partner.
- Repeat.

Variation

- For less able learners, roll the dice three times to create a three-digit number.

Comparing numbers 👥

Learning objective

Code	Learning objective
5Nn8	Order and compare numbers up to a million using the > and < signs.

Resources

1–6 dice or Resource sheet 4: 1–6 spinner (per class); paper (per learner)

What to do

- Roll the dice to form two three-digit numbers.
- Write the numbers on the board.
- Learners copy the numbers and write the correct symbol of comparison, < or >, between them.
- They suggest a number that comes between the two numbers.
- Repeat.

Variations

- Roll the dice to create five three-digit numbers, then place these numbers in order of size.
- Create four- or five-digit numbers to compare and order.

Odd or even 👥

Learning objective

Code	Learning objective
5Nn13	Recognise odd and even numbers and multiples of 5, 10, 25, 50 and 100 up to 1000.

Resources

1–6 dice or Resource sheet 4: 1–6 spinner (per class); paper (per learner)

What to do

- Explain that on the command you will ask the learners to arrange themselves in four random groups of any size.
- Give the command: Ready, set, go!
- After learners have settled in their groups, ask them to state the total number in their group and say whether it is an odd or an even number.

Variation

- Ask two groups to join together and say whether the sum of the groups is odd or even. Repeat for groups that are: odd + odd; even + even; odd + even. Ask: **Do you get the same odd/even result every time? Learners investigate and discuss the results.**

Guess the steps 👥

Learning objective

Code	Learning objective
5Nn1	Count on and back in steps of constant size, extending beyond zero.

What to do

- Learners sit in a circle.
- A volunteer is chosen to be the 'guesser'. They leave the circle and stand at a distance, with their back to the other learners.
- The teacher whispers a step sequence to the learners, for example 'forwards in steps of four' or 'backwards in steps of six'.
- The guesser is invited to return to the circle.
- The teacher indicates for the sequence to begin.
- The guesser has to work out the rule of the step sequence.

Variation

- Introduce forwards or backwards sequences that extend beyond zero.

Up above, down below 👥

Learning objective

Code	Learning objective
5Nn9	Order and compare negative and positive numbers on a number line and temperature scale.

What to do

- Divide the class into two teams.
- Say: **Think of an island. The trees on the island are above sea level. The fish swimming around the island are below sea level. I will tell you the height of the top of a tree, a positive number of metres, and the depth of a fish, a negative number of metres. You have to work out the distance between the top of the tree and the fish.**
- Begin the game with: 'The tree is five metres tall; the fish is at minus four metres.'
- Invite the first learners to put up their hands to respond. If correct, they score a point for their team.
- Repeat with different heights above and below sea level. The team with higher score wins.

Variation

- Provide a number line to assist learners.

Temperature change 🏃👥

Learning objective

Code	Learning objective
5Nn9	Order and compare negative and positive numbers on a number line and temperature scale.

What to do 📊

- Display the **Thermometer tool**, with two thermometers both set with the scale from −20 to 30.
- Divide the class into two teams.
- Place the markers of the two thermometers at different positive readings.
- Ask: **What is the difference between the two readings?** Repeat with different questions.
- Teams take turns to answer, scoring a point each time they are correct.
- The team with the higher score is the winner.

Variation

- Use both negative and positive scale readings.

Odd and even guess 🏃👥

Learning objective

Code	Learning objective
5Nn14	Make general statements about sums, differences and multiples of odd and even numbers.

What to do 📊

- Display the **Spinner tool**, with the sectors labelled 1 to 8.
- Learners work with a partner.
- Click 'spin' for a start number.
- Learners tell their partner their prediction for whether the sum of the start number and the next number spun will be odd or even. They score a point if correct.
- The game continues with the teacher spinning new numbers and learners predicting the cumulative sum. The learner with the higher score wins the game.

Variation

- Reduce the range of sector numbers to 1 to 4.

Fraction finder 👥👥

Learning objective

Code	Learning objective
5Nn4	Use decimal notation for tenths and hundredths and understand what each digit represents.

What to do 📊

- Display the **Fractions tool**, set to 10 or 100 parts (no fraction label).
- Shade one or more parts.
- Ask: **What fraction of the shape has been shaded?**
- Learners take turns to tell their partner the fraction shown.
- Click 'show fraction' to confirm the answer.
- Repeat with different fractions.

Variation

- Display fraction bars divided into tenths or hundredths. Shade to show mixed numbers, for example: $1\frac{3}{10}$.

Between the numbers 👥👥

Learning objective

Code	Learning objective
5Nn4	Use decimal notation for tenths and hundredths and understand what each digit represents.

What to do 📊

- Display the **Number line tool**, set from 0 to 10 in increments of 0·1; sub-divide: 0·1 or 0·01. (Tick 'show only end numbers' if 0·1 is chosen.)
- Position pointers above two divisions.
- Ask: **What two decimals are highlighted? Name a decimal that comes between them.**
- Repeat with different decimal positions.

Variation

- Learners name the decimal position of each pointer, not numbers between.

Number – Numbers and the number system

Counting in tenths and hundredths

Learning objective

Code	Learning objective
5Nn4	Use decimal notation for tenths and hundredths and understand what each digit represents.

What to do

- Display the **Place value tool**, from 27·1.
- Learners count forwards from 27·1 as the teacher displays the corresponding decimals.
- Stop at 27·9 and ask: **What comes next?** (28)
- Continue counting to 30, then count back to 27·1.
- Repeat for different decimal number steps.

Variation

- Learners begin at 27·9 and count in hundredths to 27·99, stopping to discuss counting through 28 and continuing the sequence to 28·3.

Place value

Learning objective

Code	Learning objective
5Nn4	Use decimal notation for tenths and hundredths and understand what each digit represents.
5Nn5	Multiply and divide any number from 1 to 10 000 by 10 or 100 and understand the effect. [Decimals answers to one and two decimal places]

What to do

- Divide the class into two teams. Display the **Number line tool**, set 0 to 2 with divisions of 0·1.
- Say: **I am thinking of a 'secret number', a decimal tenth between zero and two** (for example, 1·7).
- Each team is allowed to ask one question and make one guess at the 'secret number' per turn. Allow up to ten questions.
- All questions must be of the closed type, demanding only a 'yes' or 'no' response, for example: Is the number between 0·5 and 0·9? Is it less than 0·8?
- The first team to guess the 'secret' number scores points equivalent to 11 *minus* the number of questions asked.

Variation

- Choose a decimal tenth as the 'secret number'. Use an interval of 0·2, for example 0·45 to 0·65 or 1·21 to 1·41.

Decimal sequences

Learning objectives

Code	Learning objective
5Nn4	Use decimal notation for tenths and hundredths and understand what each digit represents.
5Nn5	Multiply and divide any number from 1 to 10 000 by 10 or 100 and understand the effect.

What to do

- Provide short decimal sequences and ask learners to find the next two terms.
- Begin with forwards counting sequences of tenths, for example: 4·3 … 4·4 … 4·5 …
- Move on to numbers with two decimal places, for example: 7·11 … 7·12 … 7·13 …
- Learners discuss sequences with a partner and put up their hands when they have the next two terms.

Variation

- Include backwards counting sequences, for example: 2·49 … 2·48 … 2·47 …

Comparing tenths

Learning objective

Code	Learning objective
5Nn11	Order numbers with one or two decimal places and compare using the > and < signs.

What to do

- Divide the class into two teams, A and B.
- Say: **I am going to give each team a decimal number. Compare the two decimals and indicate whether your number is greater or smaller than the other by holding your hand up high or down low.**
- Start with a pair of decimal tenths; for example: Team A: 4·3; Team B: 4·1.
- Repeat with different pairs of decimal tenths. Look out for learners who are unable to compare numbers correctly.

Variation

- Provide a number line to help learners make the comparison between pairs of numbers.

Number – Numbers and the number system

Comparing hundredths 👥👥

Learning objective

Code	Learning objective
5Nn11	Order numbers with one or two decimal places and compare using the > and < signs.

What to do

- Divide the class into two teams, A and B.
- Say: **I am going to give each team a number with two decimal places. Compare the two decimals and indicate whether your number is greater or smaller than the other by holding your hand up high or down low.**
- Start with a pair of numbers in which the decimal parts use the same digits but in a different order, such as Team A: 2·17; Team B: 2·71.
- Repeat with different pairs of decimals, again using the same digits in a different order. Look out for learners who are unable to compare numbers correctly.

Variation

- Ask learners to compare a number with one decimal place to a number with two decimal places.

50 or 60 👥👥

Learning objective

Code	Learning objective
5Nn7	Round a number with one or two decimal places to the nearest whole number.

Resources

1–9 number cards from Resource sheet 2: 0–100 number cards (per class)

What to do 📊

- Display the **Number line tool** set from 50 to 60. Shuffle the number cards and place them face down in a pile.
- Divide the class into two teams, A and B.
- On the board, write: 5__. Say: **I am going to choose a random number card for each team. The digit on the card will be placed with the five tens to make a two-digit number that, when rounded to the nearest ten, will become 50 or 60. Decide whether the number is rounded up or down.**
- Begin by choosing a number card and writing a number of ones alongside five tens on the board. Team A responds with '50' or '60'. If they are correct they score that number of points. Repeat for Team B.
- The winner is the team with the higher score.

Variation

- Construct three-digit numbers and round to the nearest 100.

Fraction dice 👥

Learning objective

Code	Learning objective
5Nn15	Recognise equivalence between: $\frac{1}{2}$, $\frac{1}{4}$ and $\frac{1}{8}$; $\frac{1}{3}$ and $\frac{1}{6}$; $\frac{1}{5}$ and $\frac{1}{10}$.

Resources

1–6 dice, alternately use Resource sheet 4: 1–6 spinners (per pair)

What to do

- Players take turns to throw a dice.
- The number thrown is the denominator of a unit fraction. For example, if a 5 is thrown, the fraction is $\frac{1}{5}$.
- The player who makes the greater fraction wins.
- Remind learners of the order: $\frac{1}{1} > \frac{1}{2} > \frac{1}{3} > \frac{1}{4} > \frac{1}{5} > \frac{1}{6}$.

Variation

- Players throw two dice and use the sum as the denominator of a unit fraction.

Secret fraction 👥👥

Learning objective

Code	Learning objective
5Nn15	Recognise equivalence between: $\frac{1}{2}$, $\frac{1}{4}$ and $\frac{1}{8}$; $\frac{1}{3}$ and $\frac{1}{6}$; $\frac{1}{5}$ and $\frac{1}{10}$.

What to do

- Write a fraction on the board, for example: $\frac{1}{4}$, $\frac{1}{5}$, $\frac{3}{4}$, $\frac{2}{3}$.
- Say: **I have a fraction in my head. The fraction I am thinking of is greater/less than the fraction on the board.**
- Learners try to guess the 'secret' fraction. They raise their hands to respond.
- Repeat with a different 'secret' fraction.

Variation

- Learners are given two fractions and have to find a 'secret' fraction that lies between the two fractions on the number line.

Number – Numbers and the number system

In-betweens

Learning objective

Code	Learning objective
5Nn17	Change an improper fraction to a mixed number, e.g. $\frac{7}{4}$ to $1\frac{3}{4}$; order mixed numbers and place between whole numbers on a number line.

What to do

- Display the **Fraction wall tool** to show: $\frac{1}{2}, \frac{1}{3}, \frac{1}{4}, \frac{1}{5}, \frac{1}{10}$.
- Say: **I am thinking of a fraction that is in between** $\frac{2}{10}$ **and** $\frac{8}{10}$**. What is my fraction?**
- Learners use the wall to identify a fraction in between the two fractions given, for example: $\frac{3}{10}, \frac{1}{4}, \frac{1}{3}, \frac{2}{3}$.
- Learners score a point if they guess correctly.
- Repeat with two more fractions.

Variation

- Reduce rows in the fraction wall to halves, quarters and fifths. Ask questions, for example: **I am thinking of a fraction that is between** $\frac{1}{5}$ **and** $\frac{3}{5}$**.** $\left(\frac{2}{5}, \frac{1}{4}, \frac{1}{2}\right)$

Addition wall

Learning objective

Code	Learning objective
5Nn17	Change an improper fraction to a mixed number, e.g. $\frac{7}{4}$ to $1\frac{3}{4}$; order mixed numbers and place between whole numbers on a number line.

What to do

- Display the **Fraction wall tool** to show: $\frac{1}{2}, \frac{1}{4}, \frac{1}{6}, \frac{1}{8}, \frac{1}{10}$.
- Say: **Give me three fractions that add together to make one half.** $\left(\frac{1}{6} + \frac{1}{6} + \frac{1}{6}\right)$
- Say: **Give me four fractions that make one half.** $\left(\frac{1}{8} + \frac{1}{8} + \frac{1}{8} + \frac{1}{8}\right)$
- Say: **Give me two fractions that make** $\frac{1}{4}$**.** $\left(\frac{1}{8} + \frac{1}{8}\right)$
- Repeat for other fractions, for example fifths and tenths.

Variation

- Reduce the number of fraction bars, for example: $\frac{1}{2}, \frac{1}{4}$ and $\frac{1}{8}$.

How many hundredths?

Learning objective

Code	Learning objective
5Nn19	Understand percentage as the number of parts in every 100 and find simple percentages of quantities.
5Nn20	Express halves, tenths and hundredths as percentages.

What to do

- Say: **A hundred children are split between two sides of a hall. On one side, ten children wear red T-shirts; on the other side, 13 children wear red T-shirts. All the other children wear blue T-shirts. Tell me how many children wear red T-shirts as a fraction of the 100 children?**
- Learners, in pairs, solve the word problem and raise their hands when they have the answer. $\left(\frac{23}{100}\right)$
- Select a pair to share their solution.
- Repeat with a different '100 children split between two sides of a hall' question, for which the answer is expressed as a fraction of 100.

Variation

- Present the problem as children wearing two colours of T-shirt. The teacher provides the number wearing one colour and learners determine the fraction wearing the other colour.

Fraction stories

Learning objective

Code	Learning objective
5Nn20	Express halves, tenths and hundredths as percentages.

What to do

- Describe a fraction in a story, for example: **I have four plastic cups. One is green; the others are blue. What fraction of the cups are green?** $\left(\frac{1}{4}\right)$
- Learners work in pairs, taking turns to name the fraction described.
- Repeat with a different 'fraction' story.

Variation

- Describe fractions that can be simplified, for example: **A farmer has 20 sheep. Two of them have been sheared. What fraction have been sheared?** $\left(\frac{2}{20} = \frac{1}{10}\right)$

Number – Numbers and the number system

Fractions and percentages

Learning objective

Code	Learning objective
5Nn19	Understand percentage as the number of parts in every 100 and find simple percentages of quantities.
5Nn20	Express halves, tenths and hundredths as percentages.

Resources

mini whiteboard and pen (per learner)

What to do

- Say: **I am going to read out a list of fractions and percentages. If I read a percentage I want you to write it down as a fraction; if I read a fraction I want you to write it down as a percentage.**
- Say the list: **10%, $\frac{1}{100}$, $\frac{1}{2}$, 50%, 100%, 1%.** Learners write their responses on their mini whiteboards.
- Ask learners, in turn, to read out their lists and agree the order: $\frac{1}{10}$, 1%, 50%, $\frac{1}{2}$, $\frac{100}{100}$, $\frac{1}{100}$.
- Repeat with a different list of simple percentages.

Variation

- Include 25%, $\frac{1}{4}$, 30%, $\frac{3}{10}$, 70%, $\frac{7}{10}$, 75%, $\frac{3}{4}$.

50%, 10%, 1%

Learning objective

Code	Learning objective
5Nn20	Express halves, tenths and hundredths as percentages.

What to do

- Divide the class into three groups, A, B and C.
- Say: **Group A, you are the '50% group'; you will find 50%, or half, of any number. Group B, you are the '10%' group; you will find 10%, or a tenth, of any number. Group C, you are the '1%' group; you will find 1%, or a hundredth, of any number.**
- Call out a four-digit number that is a multiple of 100, then a group name, for example: **2300, Group B.** Group B has to find their percentage, that is 10%, of the four-digit number. The group calls out their answer and scores a point if successful.
- Repeat with other numbers and groups.

Variation

- Allocate a broader set of percentages to groups: 1%, 10%, 25%, 50%, 75%.

Boys and girls 👥

Learning objective

Code	Learning objective
5Nn21	Use fractions to describe and estimate a simple proportion, e.g. $\frac{1}{5}$ of the beads are yellow.

What to do

- Say: **When I say 'Go!', I want you to get into groups of any size you choose. Raise your hand when your group is complete.**
- Ask learners to say the numbers of girls and boys as fractions of the whole group.
- Repeat, asking learners to change their group.

Variation

- Say: **If the fractions of boys and of girls in your group are always the same as they are now, tell me how many boys/girls there will be in your group if the size increases to [x] children.**

Snap! 👥

Learning objective

Code	Learning objective
5Nn21	Use fractions to describe and estimate a simple proportion, e.g. $\frac{1}{5}$ of the beads are yellow.

What to do

- On the board, write: $\frac{2}{3}$.
- Call out a list of fractions.
- Learners call out 'Snap!' when they hear an equivalent fraction. For example: 'One-third … three-quarters … four-fifths … four-sixths.' 'Snap!'
- Repeat with a different fraction.

Variation

- Write fractions on the board, one at a time, rather than calling them out.

Simplify the fraction 👥

Learning objective

Code	Learning objective
5Nn22	Use ratio to solve problems, e.g. to adapt a recipe for 6 people to one for 3 or 12 people.

What to do

- On the board write $\frac{4}{8}$.
- Call out a list of fractions.
- Learners call out 'Snap' when they hear a simplified fraction.
- For example: 'Two-thirds ... three-fifths ... two-eighths ... one-quarter.' 'Snap!'
- Repeat with a different fraction.

Variation

- Write each fraction one at a time on the board rather than calling them out.

Squash 👥

Learning objective

Code	Learning objective
5Nn22	Use ratio to solve problems, e.g. to adapt a recipe for 6 people to one for 3 or 12 people.

Resources

- three plastic cups labelled A, B and C

What to do

- Show learners three plastic cups.
- Say: **Imagine that cup A has one part orange to three parts water, cup B has one part orange to two parts water, cup C has one part orange to four parts water.**
- Ask: **In which cup is the squash strongest, most concentrated?** (B) **Why?**
- Repeat for different concentrations of squash.

Variation

- Introduce more complicated ratios of parts squash to parts water, for example 2 : 3.

Swapping places 🏃🏃

Learning objective

Code	Learning objective
5Nc8	Count on or back in thousands, hundreds, tens and ones to add or subtract.

Resources

0–9 number cards from Resource sheet 2: 0–100 number cards (one per learner)

What to do

- Shuffle the cards and ask each learner to take one.
- On the command 'Go!' learners arrange themselves into groups of five and create a five-digit number by standing in a line holding out their cards.
- Go around the groups and ask them to say their number name.
- Explain to the learners that you are going to call out a place position, for example 'thousands'. On hearing the command, all learners holding a digit in the thousands position swap places with the learner in the same place value position in another group. After completing the swap, each group must call out its new number.
- Choose any place position, from ten thousands to ones, to begin the game.

Variation

- Work with three-digit numbers only.

Ones down the line 🏃🏃

Learning objective

Code	Learning objective
5Nc10	Use appropriate strategies to add or subtract pairs of two- and three-digit numbers [and numbers with one decimal place], using jottings where necessary.

What to do

- Organise learners in small groups of around five or six.
- Ask them to stand in a line.
- The first learner comes to the board and writes two three-digit numbers at the top, one below the other. The two numbers should have a difference of no more than ten.
- The second learner determines the difference between the two numbers, then calls it out. For example, if the two numbers are 347 and 351, the difference is four.
- The third learner comes to the board and writes a new three-digit number, no smaller than ten less than the previous number. The fourth learner calls out the difference and the procedure continues, each learner taking a turn. Any learner who is not successful joins the back of the line for another turn.

Variation

- Work with four- or five-digit numbers.

Function machines

Learning objective

Code	Learning objective
5Nc10	Use appropriate strategies to add or subtract pairs of two- and three-digit numbers [and numbers with one decimal place], using jottings where necessary.

Resources

mini whiteboard and pen or paper (per pair)

What to do

- Draw three function machines on the board: −10, −30 and −60. These should be boxes with input and output arrows, with the function written inside the box.
- On the board, write two three-digit numbers, one as the 'start' number and the other as the 'target' number, for example a start number of 310 and a target number of 130. Each number should be a multiple of ten.
- Learners must try to reach the target number from the start number by passing it through the function machines as few times as possible.
- Learners share their solutions. Praise learners who find the shortest 'route'.
- Repeat for other numbers.

Variation

- Work with four-digit 'start' and 'target' numbers and three-digit functions.

Guess the answer

Learning objective

Code	Learning objective
5Nc18	Find the total of more than three two- or three-digit numbers using a written method.

Resources

1–6 dice, alternately use Resource sheet 4: 1–6 spinners (per class); mini whiteboard and pen or paper (per pair)

What to do

- Write on the board the layout for the addition of two three-digit numbers with the appropriate headings of H, T and U.
- Roll the dice three times to create a three-digit number. Write the number in the top row of the layout.
- Say: **Before I roll to create a second three-digit number that will be added to the first number, I want you to estimate the answer to the addition.**
- Learners write their estimates on their mini whiteboards or paper.
- Roll the dice three more times, for a second three-digit number, and write it in the layout.
- Ask learners to work out the answer to the addition, then the difference between the answer and their estimate. The winner is the learner who has the smallest difference.

Variation

- Use two-digit instead of three-digit numbers.

Crossing boundaries 👥

Learning objectives

Code	Learning objective
5Nc8	Count on or back in thousands, hundreds, tens and ones to add or subtract.
5Nc11	Calculate differences between near multiples of 1000, e.g. 5026 – 4998, or near multiples of 1, e.g. 3.2 – 2.6.

What to do 📊

- Display the **Place value tool** showing the number 2678.
- Place the number 3 below the digit 8 in 2678.
- Say: **If three was added to the number 2678 what would happen to the thousands digit? The hundreds digit? The tens digit? The units digit?**
- Agree that only the tens and units digits will change: 8 becomes 1 and 7 becomes 8. Ask learners to explain why.
- Repeat the procedure twice more with numbers 50 and 700. Ask what digits would change and why.

Variation

- Begin with the number 87 678 and discuss the addition of the numbers 9, 89, 789 and 6789.

Changing digits 👥

Learning objectives

Code	Learning objective
5Nc8	Count on or back in thousands, hundreds, tens and ones to add or subtract.
5Nc9	Add or subtract near multiples of 10 or 100, e.g. 4387 – 299.

What to do 📊

- Display the **Place value tool** showing the number 3465 twice.
- Say: **I am going to change one or more of the digits in this number and you have to say how the value of the digit has changed and why.**
- Demonstrate changing 6 to 9 and showing that this is the result of adding 30 to the number.
- Ask learners to turn their backs while you change the digit 4 to 7.
- Expect learners to say that value of the digit has changed from 400 to 700 as the result of adding 300 to the number.
- Repeat, changing 5 to 7 and 6 to 8. Expect learners to say that this is the result of adding 22.
- Repeat for other digit changes and ask learners to say what has been added to the number.

Variation

- Restrict changes to single digits with crossing tens/hundreds/thousands boundaries.

Bonds to 100 👥

Learning objectives

Code	Learning objective
5Nc1	Know by heart pairs of one-place decimals with a total of 1, e.g. $0.8 + 0.2$.
5Nc2	Derive quickly pairs of decimals with a total of 10, and with a total of 1.
5Nc10	Use appropriate strategies to add or subtract pairs of two- and three-digit numbers and numbers with one decimal place, using jottings where necessary.

Resources

mini whiteboard and pen or paper (per learner)

What to do

- Before the lesson, write on the board the numbers 47, 16, 8, 84, 63, 29, 92, 3, 78, 35, 54. Hide the numbers under a piece of paper.
- Say: **I have written 11 numbers, each less than 100, on the board and hidden them. I will reveal the numbers. When I do, for each number, write the number that you can add to it to make 100.**
- Explain that it is a race, so learners should work as fast as possible.
- Reveal the numbers and ask learners to begin.
- Praise the fastest in the class.

Variation

- Give learners three-digit numbers and ask them to find complements of 1000.

Card pairs 👥

Learning objective

Code	Learning objective
5Nc2	Derive quickly pairs of decimals with a total of 10, and with a total of 1.
5Nc11	Calculate differences between near multiples of 1000, e.g. $5026 - 4998$, or near multiples of 1, e.g. $3.2 - 2.6$.

Resources

0·1–0·9 from Resource sheet 11: One-place decimal digit cards (per learner)

What to do

- Shuffle the cards and ask each learner to take one.
- Say: **You each have a card that displays a decimal tenth, a one-place decimal. When I say 'Go' I want you to hold your card against your chest with the number displayed. Then move around the classroom and stand next to a learner with whom you can make the number 1.**
- On the board, write an example: $0.1 + 0.9 = 1$.
- Say 'Go' and check that learners know the decimal pairs that make 1.

Variation

- Ask learners to come together in groups with more than two learners, for example: $0.1 + 0.4 + 0.5$.

Secret values 👥

Learning objective

Code	Learning objective
5Nc10	Use appropriate strategies to add or subtract pairs of two- and three-digit numbers [and numbers with one decimal place], using jottings where necessary.

Resources

1–6 dice, alternately use Resource sheet 4: 1–6 spinners (per pair); mini whiteboard and pen or paper (per pair)

What to do

- Player 1 rolls a dice three times to create a three-digit number and records it on paper.
- They roll the dice once more to create a number with a mystery purpose.
- Player 1 secretly decides whether the place value of the new number will be ones, tens or hundreds. They mentally add or subtract the new number to/from the three-digit number, record the new number and show it to Player 2.
- Player 2 has to say what place value the digit represents.
- Players swap roles.

Variation

- Play the game with two-digit numbers.

Find the sum 👥👥

Learning objective

Code	Learning objective
5Nc18	Find the total of more than three two- or three-digit numbers using a written method.

Resources

mini whiteboard and pen or paper (per pair)

What to do 📊

- Display the **Spinner tool** with ten sectors labelled 0 to 9. Spin twice to create a two-digit number.
- Draw a column addition layout on the board.
- Write the number from the spinner in the top row of the layout.
- Spin two more times to create a second two-digit number.
- Write the new number in the second row of the layout.
- Ask learners to use column addition to calculate the sum of the numbers.
- Repeat with different numbers.

Variation

- Create three-digit numbers.

Missing digits (1)

Learning objective

Code	Learning objective
5Nc19	Add or subtract any pair of three- and/or four-digit numbers, with the same number of decimal places, including amounts of money.

Resources

mini whiteboard and pen or paper (per pair)

What to do

- On the board, draw the layout of a column addition for two two-digit numbers, for example, 27 + 39 = 66, but leave out some of the digits, for example, the 6 in the ones column of the answer and the 2 representing tens in the top number and replace the missing digits with a square.
- Ask learners to say what the missing digits are. They may use their mini whiteboards or paper to record the layout and determine the missing digits.
- Repeat with a different calculation.

Variation

- Use column addition layouts for the sum of two three-digit numbers. Remove one digit from each of the three place value columns.

Missing digits (2)

Learning objective

Code	Learning objective
5Nc19	Add or subtract any pair of three- and/or four-digit numbers, with the same number of decimal places, including amounts of money.

Resources

mini whiteboard and pen or paper (per pair)

What to do

- On the board, draw the column layout for a subtraction of one two-digit number from another, for example, 64 – 27 = 37 but leave out some of the digits, for example, the 7 in the ones column of the answer and the 6 tens in the top number and replace the missing digits with a square.
- Ask learners to say what the missing digits are. They may use their mini whiteboards or paper to record the layout and determine the missing digits.
- Repeat with a different calculation.

Variation

- Use column subtraction layouts for the subtraction of three-digit numbers. Remove one digit from each of the three place value columns.

Spot the facts 👥

Learning objectives

Code	Learning objective
5Nc1	Know by heart pairs of one-place decimals with a total of 1, e.g. 0·8 + 0·2.
5Nc2	Derive quickly pairs of decimals with a total of 10, and with a total of 1.
5Nc11	Calculate differences between near multiples of 1000, e.g. 5026 – 4998, or near multiples of 1, e.g. 3·2 – 2·6.

What to do

- On the board, write: 2·3, 4·9, 6·4, 1·8, 5·1, 8·2, 7·7, 2·5, 3·6, 7·5.
- Remind learners of decimal number bonds to 10, for example 3·7 + 6·3, and their relationship to whole-number bonds to 100: 37 + 63.
- Say: **Using your knowledge of decimal number bonds, add all the numbers on the board as quickly as you can.**
- Answer is 40: 2·3 + 7·7 (10), 4·9 + 5·1 (10), 1·8 + 8·2 (10), 7·5 + 2·5 (10)

Variation

- Learners add decimals with two decimal places: 0·33, 0·67, 0·48, 0·52, 0·24, 0·76, 0·11, 0·89, 0·25, 0·75 arranged in random order.

All the 9s 👥

Learning objectives

Code	Learning objective
5Nc9	Add or subtract near multiples of 10 or 100, e.g. 4387 – 299.
5Nc11	Calculate differences between near multiples of 1000, e.g. 5026 – 4998, or near multiples of 1, e.g. 3·2 – 2·6.

What to do

- Say: **Beginning with 17, repeatedly add nine by adding ten then subtracting one. Stop when you make a number greater than 100. How quickly did you do it?**
- Say: **Beginning with the number 153, repeatedly subtract nine by subtracting ten then adding one. Stop when you reach a number less than ten. How quickly did you do it?**

Variations

- Learners begin with the number 117 and repeatedly add 99 by adding 100 and subtracting one, stopping when they make a number greater than 1000.
- For subtraction, they begin at 953 and repeatedly subtract 99 by subtracting 100 and adding one, stopping when they reach a number less than 100.

Numbers on the brain 👥

Learning objective

Code	Learning objective
5Nc18	Find the total of more than three two- or three-digit numbers using a written method.

What to do

- Say: **I am going to read out a list of one-digit numbers. Add the numbers and keep a mental total.**
- Explain that at some point you will stop to ask what the total is.
- Begin with a set of ten numbers and build up to over 15. For example: 6, 5, 8, 3, 9, 7, 6, 9, 5, 8 (66). Leave a two second gap before saying the next number.
- Ask learners to explain the strategies they used to add the numbers quickly.

Variation

- Include more numbers below six in the sequence.

Money goes up 👥

Learning objective

Code	Learning objective
5Nc19	Add or subtract any pair of three- and/or four-digit numbers, with the same number of decimal places, including amounts of money.

Resources

Mini whiteboard and pen or paper (per learner)

What to do 📊

- Display the **Money tool**, showing five dollars, a 50 cent coin and a 25 cent coin.
- Say: **How much money is there in total?** Elicit five dollars and 75 cents.
- Ask: **How do you write this?** Ask a learner to come to the board and write $5.75.
- Say: **I am going to add a note or a coin, one at a time, and I want you to write the new total on your whiteboards, using correct money notation.**
- Choose a bank note or coin to add to the amount shown. Ask learners to hold up the amounts they record each round to confirm they are correct.

Variation

- Add bank notes and ten cent coins only.

Mixed up multiples 👥

Learning objectives

Code	Learning objective
5Nc3	Know multiplication and division facts for the 2× to 10× tables.
5Nc4	Know and apply tests of divisibility by 2, 5, 10 and 100.
5Nc5	Recognise multiples of 6, 7, 8 and 9 up to the 10th multiple.
5Nc7	Find factors of two-digit numbers.

Resources

mini whiteboard and pen or paper (per learner)

What to do

- Write the multiples of six, seven, eight and nine, up to the tenth multiple, randomly on the board.
- Add four numbers that do not appear in any of the tables.
- Say: **The numbers on the board are all multiples of 6, 7, 8 or 9, except for four of the numbers. When I say 'Go!' I want you to write the numbers as a list of multiples for each table. You do not have to put them in order.**
- Give the command '**Go!**'. Ask learners to identify the four numbers that are not multiples of 6, 7, 8 or 9.

Variation

- Include 11th, 12th and 13th multiples.

Where's the finish? 👥

Learning objectives

Code	Learning objective
5Nc4	Know and apply tests of divisibility by 2, 5, 10 and 100.
5Nc5	Recognise multiples of 6, 7, 8 and 9 up to the 10th multiple.

What to do 📊

- Display the **Number line tool**, set from 0 to 100.
- Say: **I am going to place the pointer next to a start number on the number line. Then I am going to give you a skip counting number, the number of skips and the direction, forwards or backwards. You have to work out the finish number.**
- Start the game with start number 32, skip count number 8, three skips forward.
- Say: **Fold your arms as soon as you have the finish number.** (56)
- Invite a learner to state the finish number and confirm with the class that the answer is correct.
- Repeat for other skip count multiples. Demonstrate steps on the number line.

Variation

- Feature multiples that cross the 100 boundary and multiples 11 to 15.

Trios

Learning objective

Code	Learning objective
5Nc7	Find factors of two-digit numbers.

Resources

Resource sheet 2: 0–100 number cards from which trios of related number facts are selected, for example: 12, 2, 6 (12 = 2 × 6); 28, 4, 7 (28 = 4 × 7) (one card per learner)

What to do

- Write examples of number facts on the board for example, 2 × 6 = 12 and 4 × 7 = 28.
- Shuffle the trios together so that they are mixed up.
- Deal out one card to each learner.
- Say: **Hold up your card so that everyone can see it. On the command 'Go!' I want you to get into groups of three numbers that form a number fact.**
- Say: **Go!** See how quickly learners can get into number fact groups.
- Discuss the facts displayed.

Variation

- Introduce trios that represent number facts beyond the tenth multiple, for example: 39 = 3 × 13.

Multiplication ball

Learning objective

Code	Learning objective
5Nc3	Know multiplication and division facts for the 2× to 10× tables.
5Nc20	Multiply or divide three-digit numbers by single-digit numbers.

Resources

medium-sized ball (per class)

What to do

- Ask learners to sit in a circle.
- Begin the game by saying a multiplication question out loud, for example: **Six times four.**
- Roll the ball to any learner in the circle. They receive the ball and call out the answer to the multiplication.
- If successful, they call out a new multiplication question and roll the ball to another learner. If unsuccessful, they lose one of two 'lives'. Once both lives are gone they are out of the game.
- The game continues with the ball being passed from one learner to the next, answering the multiplication question each time.

Variation

- Display the **Multiplication square tool** to help learners answer the questions.

Multiplication bingo 👥

Learning objective

Code	Learning objective
5Nc3	Know multiplication and division facts for the 2× to 10× tables.
5Nc21	Multiply two-digit numbers by two-digit numbers.

Resources

mini whiteboard and pen or paper (per learner)

What to do

- Ask learners to draw a three by three grid and fill the squares with multiplication questions covering the times tables, 2 to 10, for example: 3 × 9, 5 × 8.
- Call out answers to multiplication questions for the times tables 2 to 10.
- Learners cross through any question on their grid that gives the answer called out.
- The first learner to cross through a line of three questions down, across or diagonally calls out 'Bingo!' and wins the game.

Variation 📊

- Display the **Multiplication square tool** to help learners answer the questions.

Doubles and halves 👥

Learning objective

Code	Learning objective
5Nc21	Multiply two-digit numbers by two-digit numbers.
5Nc23	Divide three-digit numbers by single-digit numbers, including those with a remainder (answers no greater than 30).
5Nc25	Decide whether to group (using multiplication facts and multiples of the divisor) or to share (halving and quartering) to solve divisions.

Resources

mini whiteboard and pen or paper (per learner)

What to do

- Distribute paper.
- Write this set of numbers in random places across the board: 2, 4, 3, 6, 4, 8, 5, 10, 6, 12, 7, 14, 8, 16, 9, 18, 10, 20, 24, 48, 25, 50, 28, 56, 33, 66, 35, 70, 39, 68, 43, 86, 45, 90, 51, 102, 56, 112.
- Learners race to find all the half/double pairs and record them, for example (2, 4), (3, 6).
- The first learner to raise their hand and read all the pairs is the winner.

Variation

- Include triples and thirds, for example (3, 9), (5, 15), (12, 36).

Multiplication and division dice 👥

Learning objective

Code	Learning objective
5Nc3	Know multiplication and division facts for the 2× to 10× tables.

Resources

1–6 dice, alternately use Resource sheet 4: 1–6 spinners (per pair)

What to do

- Learners take turns to roll a dice four times. They add two rolls together to make a number.
- They record their two numbers and use them to create two multiplication and two division facts.
- For 8, 6, for example, two multiplication facts would be 48 = 8 × 6 and 48 = 6 × 8; two division facts would be 48 ÷ 8 = 6 and 48 ÷ 6 = 8.
- Learners play four rounds, adding the solutions to each multiplication to make a total score.
- The player with the higher score is the winner.

Variation

- Learners take turns to roll two numbers with a dice, each the sum of four separate rolls.

Division facts 👥

Learning objective

Code	Learning objective
5Nc3	Know multiplication and division facts for the 2× to 10× tables.

Resources

mini whiteboard and pen or paper (per learner)

What to do

- Write the numbers 2 to 10 in a column down the board.
- Learners race to work down the list, copying the numbers from the board, one by one, and writing division facts for which that number is the quotient.
- For 6, for example, they may write: = 18 ÷ 3 or = 24 ÷ 4.
- The first learner to raise their hand and read out a full set of division facts – one for every number in the list – is the winner.

Variation

- Include the numbers 10 to 20 in the column on the board.

How many skips? 👥👥

Learning objectives

Code	Learning objective
5Nc3	Know multiplication and division facts for the 2× to 10× tables.
5Nc6	Know squares of all numbers to 10 × 10.

What to do 📊

- Display the **Number line tool**, set from 0 to 100.
- Say: **I am going to place the pointer next to a target number on the number line. Then I am going to give you a skip counting number. Starting from zero, I want you to skip count in your head from zero to the target number.**
- Start the game with the pointer pointing to 18 and the skip counting number 3.
- Say: **Fold your arms as soon as you have the number of skips.** (6)
- Invite a learner to state the number of skips and confirm with the class that the answer is correct emphasising the related multiplication and division facts, $3 \times 6 = 18$ / $18 \div 3 = 6$.
- Repeat for other multiples and steps. Demonstrate steps on the number line.

Variation

- Keep both the target numbers and steps small, for example target number 12 and steps of 3.

Multiple actions 👥👥

Learning objective

Code	Learning objective
5Nc12	Multiply multiples of 10 to 90, and multiples of 100 to 900, by a single-digit number.

Resources

small slip of paper (per learner); sticky tape (per class)

What to do

- Ask learners to write a multiple between the third and ninth multiples for the three times to nine times tables.
- They attach the number to their chest, using a piece of sticky tape, so that the number can be seen.
- Learners begin the game with two points each.
- Ask a set of questions to which the response is an action, for example:
  ◊ Put your hands on your head if your number is a multiple of seven.
  ◊ Sit down if your number is the answer to eight times six.
  ◊ Fold your arms if your number is an odd multiple of three.
- Learners who do not complete the action lose a point.
- Learners who lose both points are out of the game.
- Continue the game until there is just one learner left.

Variation

- Increase the complexity of the questions, for example: **Sit down if your number is a multiple of six and a multiple of nine.**

Multiply by 20

Learning objectives

Code	Learning objective
5Nc13	Multiply by 19 or 21 by multiplying by 20 and adjusting.
5Nc14	Multiply by 25 by multiplying by 100 and dividing by 4.

Resources

mini whiteboard and pen or paper (per learner)

What to do

- Learners draw a row of four boxes and write a one-digit number in each box.
- Say: **I am going to call out a multiple of 20. If you have the number that, when multiplied by 20, gives the number I just called out, then cross it through. For example, if I call out 120 you should cross through six, as six multiplied by 20 is 120.**
- Call out a number that is the answer to a one-digit number multiplied by 20, i.e. 20, 40, 60, …180.
- Explain that the first learner to cross out all four numbers should raise their hand. If correct, they win the game.

Variation

- Learners draw three boxes only and use digits 1 to 5 (20 to 100).

Factor pairs

Learning objective

Code	Learning objective
5Nc15	Use factors to multiply, e.g. multiply by 3, then double to multiply by 6.

Resources

Resource sheet 2: 0–100 number cards (using numbers 2–10 only, one card per learner)

What to do

- Randomly distribute sets of number cards from 2 to 10 so every learner has a card.
- Ask them to hold the card against their chest so that it can be seen.
- Say: **When I say a number, anyone holding a number that is a factor must get together with their factor partner as quickly as possible. For example, if I call out 12, numbers two and six and numbers three and four should stand together.**
- Call out any number from four to 100.
- Play the game with a different number each time.

Variation

- Use number cards 2 to 15 and include factors of numbers to 150.

Halves and doubles 👥

Learning objectives

Code	Learning objective
5Nc16	Double any number up to 100 and halve even numbers to 200 [and use this to double and halve numbers with one or two decimal places, e.g. double 3·4 and half of 8·6].
5Nc17	Double multiples of 10 to 1000 and multiples of 100 to 10 000, e.g. double 360 or double 3600, and derive the corresponding halves.

Resources

Resource sheet 2: 0–100 number cards using cards 2–10, 12, 14, 16, 18, 20 (one per learner)

What to do

- Randomly distribute number cards so that every learner has a card.
- Say: **The number you are holding is either a half or a double of the number someone else is holding. When I say 'Go!' I want you to get into halves and doubles partners as quickly as possible.**
- Repeat the game with a new set of cards. Make sure learners recognise their half or double partner each time.

Variation

- Use number cards that cover halves and doubles sets to 40.

Addition table 👥

Learning objective

Code	Learning objective
5Nc20	Multiply or divide three-digit numbers by single-digit numbers.

Resources

mini whiteboard and pen or paper (per learner)

What to do

- Ask learners to draw a table with three rows and six columns.
- Say: **Write six numbers that are multiples of ten across the top row. There should be three two-digit numbers and three three-digit numbers. Beside the rows below write '+ 33', '+ 67'.**
- Ask learners to write the sum of each addition in the table.
- Say: **See how quickly you can complete the table.**

Variation

- Use two-digit numbers only for the column headings.

Number – Calculation: Mental strategies, Multiplication and division

Magic number 👥

Learning objective

Code	Learning objective
5Nc20	Multiply or divide three-digit numbers by single-digit numbers.
5Nc21	Multiply two-digit numbers by two-digit numbers.

Resources

mini whiteboard and pen or paper (per learner)

What to do

- On the board, write: 116.
- Ask learners to think of a one-digit start number.
- Call out a one-digit number and ask learners to add it to their start number.
- Call out a one-digit number every five seconds and ask learners to keep a running total in their heads.
- Say: **Put your hand up if you get a total that is the magic number on the board.**
- If no one claims the magic number then write a bigger number and continue.

Variation

- Read the numbers at a faster rate and use a bigger magic number.

Factor bingo 👥

Learning objectives

Code	Learning objective
5Nc3	Know multiplication and division facts for the 2× to 10× tables.
5Nc20	Multiply or divide three-digit numbers by single-digit numbers.
5Nc23	Divide three-digit numbers by single-digit numbers, including those with a remainder (answers no greater than 30).
5Nc26	Decide whether to round an answer up or down after division, depending on the context.

Resources

mini whiteboard and pen or paper (per learner)

What to do

- Ask learners to draw a two by three grid.
- They fill the grid squares with any numbers from two to ten.
- Call out division questions for the three times to ten times tables, for example: **Forty divided by five is ...**
- Learners cross out a number if it is the answer to the question.
- The first player to cross out three numbers in a row calls out 'Bingo!' and is the winner.

Variation

- Include division questions for the 11 times to 15 times tables.

Target 👥

Learning objective

Code	Learning objective
5Nc12	Multiply multiples of 10 to 90, and multiples of 100 to 900, by a single-digit number.

Resources

mini whiteboard and pen or paper (per pair)

What to do

- On the board, draw a circle divided into eight sectors. Insert the numbers 2 to 9.
- Say: **Imagine I am playing a darts game. I throw two darts at the target and my score is equal to the product of the two numbers I score. I am going to tell you where my darts land and you have to tell me the score.**
- Begin by saying: **Three and seven.**
- Learners write the answer (21).
- Repeat, with learners showing their answers each time.
- Ask individual learners to give the answer.

Variation

- Play the game with three darts. Learners have to multiply three single-digit numbers each time.

Build the facts 👥

Learning objectives

Code	Learning objective
5Nc13	Multiply by 19 or 21 by multiplying by 20 and adjusting.
5Nc14	Multiply by 25 by multiplying by 100 and dividing by 4.

Resources

mini whiteboard and pen or paper (per pair)

What to do

- On the board, write a multiplication fact up to 10 × 10, for example: 6 × 7 = 42.
- Learners write at least three related facts, for example: 60 × 7 = 420, 6 × 70 = 420, 600 × 7 = 4200.
- After each round, invite one learner to share their list of facts.

Variation

- Use division facts, for example: 72 ÷ 8 = 9.

Find the factor

Learning objective

Code	Learning objective
5Nc15	Use factors to multiply, e.g. multiply by 3, then double to multiply by 6.

Resources

mini whiteboard and pen or paper (per pair)

What to do

- On the board, write the numbers: 18, 36, 45.
- Ask learners to list all the factors of each number.
- Ask: **Which number has the largest list of factors?** (36)

Variation

- Include numbers 54 and 90. Ask: **Which two factors do all the numbers have in common?** (6, 9)

Half or double

Learning objective

Code	Learning objective
5Nc16	Double any number up to 100 and halve even numbers to 200 and use this to double and halve numbers with one or two decimal places, e.g. double 3.4 and half of 8.6.
5Nc17	Double multiples of 10 to 1000 and multiples of 100 to 10 000, e.g. double 360 or double 3600, and derive the corresponding halves.

Resources

mini whiteboard and pen or paper (per pair)

What to do

- Write ten numbers on the board, between 10 and 30.
- Divide the class into two teams, A and B.
- Team A starts the game by calling out double or half one of the numbers on the board, for example, 36. Team B scores a point if they identify 36 as double 18, calling out 'double 18'.
- Give learners another example. Ask: **If I said 12, what would you say?** (half 24)
- Play several rounds of the game. The team with the higher score wins the game.

Variation

- Use ten numbers in the range 1 to 20.

Related halves and doubles 👥

Learning objective

Code	Learning objective
5Nc16	Double any number up to 100 and halve even numbers to 200 and use this to double and halve numbers with one or two decimal places, e.g. double 3.4 and half of 8.6.
5Nc17	Double multiples of 10 to 1000 and multiples of 100 to 10 000, e.g. double 360 or double 3600, and derive the corresponding halves.

Resources

mini whiteboard and pen or paper (per pair)

What to do

- On the board, write a half or double fact for numbers between 2 and 20, for example: Double 12 is 24.
- Learners write at least one related fact, for example: Double 120 is 240.
- Demonstrate an example of a related fact for a half, for example: Half 20 is 10, half 200 is 100.
- Each round, invite one learner to read out their list of facts.

Variation

- Expect learners to write at least one decimal fact, for example:
  Double 18 is 36, double 1·8 is 3·6.

Factor pairs 👥

Learning objective

Code	Learning objective
5Nc12	Multiply multiples of 10 to 90, and multiples of 100 to 900, by a single-digit number.
5Nc22	Multiply two-digit numbers with one decimal place by single-digit numbers, e.g. 3·6 × 7.

Resources

mini whiteboard and pen or paper (per pair)

What to do

- On the board, write these numbers in a row: 3, 4, 7, 9. Then write another set of numbers in a row below: 20, 50, 60, 80.
- Ask learners to copy the two rows of numbers.
- Say: **I am going to multiply a top number by a bottom number and say the answer. Join the two factors that multiply to make this product. For example, if I were to call out 240, you could join three and 80 or four and 60.**
- Call out a list of six products.
- Invite learners to read out their factor pairs.

Variation

- Use a second row of numbers that comprise single-digit numbers: 2, 5, 6, 8.

Number – Calculation: Mental strategies, Multiplication and division

Division pairs

Learning objectives

Code	Learning objective
5Nc24	Start expressing remainders as a fraction of the divisor when dividing two-digit numbers by single-digit numbers.
5Nc26	Decide whether to round an answer up or down after division, depending on the context.

Resources

mini whiteboard and pen or paper (per pair)

What to do

- On the board, write these numbers in a row: 72, 54, 36, 24. Write another set of numbers in a row below: 9, 8, 6, 4, 3.
- Ask learners to copy the two rows of numbers.
- Say: **I am going to divide a top number by a bottom number and say the answer. Join two numbers that I could be thinking of. For example, if I were to call out six, you could join 36 and six, as 36 divided by six equals six.**
- Call out a list of six quotients.
- Invite learners to share their division pairs.

Variation

- Use a first row of numbers comprising: 720, 540, 360, 240.

Spoke diagrams

Learning objective

Code	Learning objective
5Nc27	Begin to use brackets to order operations and understand the relationship between the four operations and how the laws of arithmetic apply to multiplication.

Resources

mini whiteboard and pen or paper (per pair)

What to do

- On the board, draw a circle with spokes radiating out, terminating in six larger circles.
- Ask learners to draw four similar diagrams.
- Say: **I am going to call out a number and I want you to write it in the centre circle of one of your 'spoke' diagrams.** Then write at least six number facts, a mixture of addition, subtraction, multiplication and division, where the answer is the number in the centre.
  Say: **The larger the numbers you use the better.**
- Begin the activity with 8.
- Invite learners to discuss their spoke diagrams.

Variation

- User larger numbers for the spoke centres.

Guess the triangle 👥

Learning objective

Code	Learning objective
5Gs1	Identify and describe properties of triangles and classify as isosceles, equilateral or scalene.

What to do 📊

- Display the **Geometry set tool**.
- Say: **I will describe a triangle by its properties. I want you to identify which triangle I describe.**
  - ◊ **The triangle has one right angle.** (pink right-angled isosceles triangle)
  - ◊ **The triangle is regular.** (orange equilateral triangle)
  - ◊ **The triangle has only two sides of equal length.** (yellow isosceles triangle)
  - ◊ **The triangle has three sides of different lengths.** (green scalene triangle)
  - ◊ **The triangle is irregular.** (yellow, green or pink triangle)
- Invite a learner to come to the board and identify the triangle that matches the description.

Variation

- Introduce the term 'angle' in the questions, for example: **Which triangle has angles all the same size?**

Guess the shape 👥

Learning objective

Code	Learning objective
5Gs2	Recognise reflective and rotational symmetry in regular polygons.

Resources

bag containing a selection of 2D shapes including triangles (equilateral and right-angled), square, rectangle, pentagon, hexagon, heptagon and octagon (per class)

What to do

- Choose a 2D shape from the bag and hide it from the class.
- Invite learners to ask questions in order to identify the shape. They can only ask questions that will have a 'yes' or 'no' answer (closed questions).
- Encourage learners to use correct shape vocabulary in their questions, for example 'sides', 'vertices'.
- The learner who correctly identifies the shape becomes the 'teacher' and secretly selects the next shape. They invite questions from the class.

Variation

- Limit the range of shapes to right-angled triangle, square, rectangle and hexagon.

Geometry – Shapes and geometric reasoning

Shape sort 👥

Learning objective

Code	Learning objective
5Gs3	Create patterns with two lines of symmetry, e.g. on a pegboard or squared paper.

What to do 📊

- Display the **Venn diagram tool**. Label the sets: '4 sides or more, 'Regular shapes'.
- Remind learners that regular shapes have sides of the same length.
- Point to a shape from one of the following tabs: 2D shapes, Triangles, Quadrilaterals.
- Invite learners to come to the board in turn to sort a shape into the correct set. Ask them to explain why they have placed the shape in the set.
- Confirm with the class that the shape is sorted correctly. Repeat for other shapes.

Variation

- Change set labels to '4 sides or fewer', '5 or more sides'

Do the lines cross? 👥

Learning objective

Code	Learning objective
5Gs5	Recognise perpendicular and parallel lines in 2D shapes, drawings and the environment.

Resources

paper (per learner); ruler (per pair)

What to do

- Ask learners to use a ruler to lightly draw two long straight lines that may or may not cross.
- They turn over the paper and use the ruler to trace a short section of each line so that the two sections are far apart (they should not trace over any intersection of the lines).
- They swap papers with a partner and determine whether the lines will cross or not.
- They state their decision. Then they use a ruler to extend the lines to see if they are correct.

Variation

- Learners use three lines. They determine which line will be crossed by the other two lines.

Secret shapes 👥

Learning objective

Code	Learning objective
5Gs4	Visualise 3D shapes from 2D drawings and nets, e.g. different nets of an open or closed cube.

Resources

set of 3D shapes: cube, cuboid, triangular based pyramid, square based pyramid, triangular prism, pentagonal prism, hexagonal prism, tetrahedron, octahedron (per class)

What to do

- Select a 3D shape and hold it behind your back so that the class is unable to see it.
- Say: **I am holding a shape behind my back. I will give you a set of clues to help you identify the shape. The clues will be difficult to start with, then get easier. The first clue is worth three points, the second two points and the third one point.**
- Begin the game by describing one property of the 'secret shape'. For example, if you are holding a square-based pyramid, you might say: **This shape has one non-triangular face**. Then include some easier clues.
- Repeat with different shapes.

Variation

- Limit the set to just five shapes.

Shapes revealed 👥

Learning objective

Code	Learning objective
5Gs4	Visualise 3D shapes from 2D drawings and nets, e.g. different nets of an open or closed cube.

Resources

pictures of a range of prisms and pyramids (per class).

What to do 📊

- Display: the **Nets tool** set to show a 3D shape, for example, a triangular prism.
- Cover the shape with a piece of paper and slowly reveal it to the class.
- Say: **Raise your hand as soon as you recognise the shape.**
- Ask the first learner who raises their hand what the name of the shape is. If correct they score a point.
- Repeat with a different 3D shape.

Variation

- Introduce pictures of a wider variety of prisms and pyramids. Ask learners to name the shape, using the shape of the base as a clue.

Geometry – Shapes and geometric reasoning

Shape-go-round 👥

Learning objective

Code	Learning objective
5Gs4	Visualise 3D shapes from 2D drawings and nets, e.g. different nets of an open or closed cube.

Resources

set of 3D shapes: cube, cuboid, triangular based pyramid, square based pyramid, triangular prism, pentagonal prism, hexagonal prism, tetrahedron, octahedron (per class)

What to do

- Ask learners to stand. Place a set of 3D shapes at the front of the class and hold one up.
- Go around the class, with each learner naming one of its properties then sitting down.
- The next learner states a different characteristic.
- If a learner is unable to name a property they are allowed to choose a new 3D shape.
- The game continues with learners naming properties for the new shape.
- Encourage learners to describe: shape and number of faces; number of vertices and edges.

Variation

- Reduce the set of prisms used.

Fact match 👥

Learning objective

Code	Learning objective
5Gs4	Visualise 3D shapes from 2D drawings and nets, e.g. different nets of an open or closed cube.

Resources

3D shapes: cube, cuboid, triangular based pyramid, square based pyramid, triangular prism, pentagonal prism, hexagonal prism, tetrahedron, octahedron (per class)

What to do

- Place the set of shapes on a table.
- Call out a number from 4 to 10, 12, 15 or 18 followed by one of the following terms: faces, vertices or edges.
- Ask learners to identify a shape that has that property, for example, if '7 faces' is stated then learners should identify a pentagonal prism.
- Repeat for different shapes.

Variation

- Remove prisms from the shape set and the numbers 7, 9, 10, 15 and 18.

Ruler angles

Learning objective

Code	Learning objective
5Gs6	Understand and use angle measure in degrees; measure angles to the nearest 5°; identify, describe and estimate the size of angles and classify them as acute, right or obtuse.

Resources

ruler (or other straight-edged object, such as a lolly stick) (per pair)

What to do

- Ask learners to use a ruler to construct three angles: a right angle; an angle greater then a right angle; an angle less than a right angle.
- They arrange the angles in order of size from smallest to largest.

Variation

- Ask learners to include two other angles: a straight angle (180°) and an angle greater than a straight angle.

Greater or less than a right angle

Learning objective

Code	Learning objective
5Gs6	Understand and use angle measure in degrees; measure angles to the nearest 5°; identify, describe and estimate the size of angles and classify them as acute, right or obtuse.

What to do

- Display the **Geometry set tool**.
- Select a shape and ask learners to say how many right angles the shape has.
- Invite volunteers to the board to identify the right angles and confirm that they are correct.
- Ask learners to identify angles of other sizes:
  ◊ How many angles does this shape have that are less than a right angle?
  ◊ How many angles does this shape have that are greater than a right angle?

Variation

- Ask learners to select an appropriate shape in response to an instruction. Say, for example: **Show me a shape that has four right angles. Show me a shape that has two angles that are greater than a right angle and two angles that are less than a right angle.**

Geometry – Shapes and geometric reasoning

Geometry – Shapes and geometric reasoning

Compass points 👥👥

Learning objective

Code	Learning objective
5Gs6	Understand and use angle measure in degrees; measure angles to the nearest 5°; identify, describe and estimate the size of angles and classify them as acute, right or obtuse.

Resources

large cards displaying the letters N, E, S and W (per class)

What to do

- Label the walls of the classroom N, E, S, W.
- Ask learners to stand and face north, then turn to face east, south, west, then back to north.
- Remind them that a clockwise turn is to the right, like the hands of a clock.
- Ask learners to make a quarter turn clockwise. Remind them that a quarter turn is a right-angled turn of 90 degrees.
- Ask: **Which card are you facing?** (east)
- Ask them to make other turns with questions involving the appropriate vocabulary, for example:
  ◊ **Turn clockwise through one/two/three right angles. Which card are you facing? How many degrees have you turned?**
  ◊ **Turn anticlockwise through one/two/three right angles. Which card are you facing?**
  ◊ **Complete a half turn clockwise. Which card are you facing?**

Variation

- Establish that half-way between north and west is called north-west. Repeat for NE, SE and SW. Give instructions for and ask questions about turns originating from NE, SE, SW and NW starting points.

Make a shape 👥👥

Learning objective

Code	Learning objective
5Gs7	Calculate angles in a straight line.

What to do 📊

- Display the **Geoboard tool**, displaying two triangles and one square reconstructed into a quadrilateral that has no right angles.
- Ask learners to come to the board to construct a shape with:
  ◊ one right angle
  ◊ two right angles
  ◊ four right angles
  ◊ three angles smaller than a right angle
  ◊ one right angle and two angles that are less than a right angle.
- Repeat for other combinations of right angles and angles that are greater/less than a right angle.

Variation

- Ask: **Who can show me how to place two shapes side by side so that the combined angles make a straight angle?** Expect learners to place two right angles side by side.

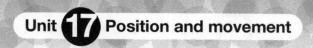

Four-in-a-row 👥

Learning objective

Code	Learning objective
5Gp1	Read and plot co-ordinates in the first quadrant.

Resource sheet 2: 0–100 number cards (per class)

What to do 📊

- Display the **Co-ordinates tool** set at the first quadrant.
- Place the number cards face down in two piles.
- Volunteer learners take turns to choose two cards. The number on the first card represents the co-ordinate on the x-axis; the number on the second card represents the co-ordinate on the y-axis.
- They plot the point on the grid.
- The first learner in the class to spot that the plotted point makes four points in a row, vertically or horizontally on the grid, wins the game.

Variation

- Make two sets of co-ordinates with the card digits each time, e.g. (7, 2) and (2, 7).

Find the treasure 👥

Learning objective

Code	Learning objective
5Gp1	Read and plot co-ordinates in the first quadrant.

What to do 📊

- Display the **Co-ordinates tool** set at the first quadrant.
- Secretly choose a point on the grid to be the location of a treasure chest and write it down on a slip of paper.
- Say: **I have hidden a treasure chest at a single point on the grid. Can you find its 'secret' location?**
- Learners take turns to call out a pair of co-ordinates and plot them on the grid.
- Call out 'treasure' and reveal the slip of paper, when a learner plots the point where the chest is located.

Variation

- Learners work in pairs to plot co-ordinates.

Lines of symmetry 🏃🏃

Learning objective

Code	Learning objective
5Gp2	Predict where a polygon will be after reflection where the mirror line is parallel to one of the sides, including where the line is oblique.

What to do 📊

- Display the **Symmetry tool**.
- Learners take turns to point to a shape and say how many lines of symmetry it has.
- They select the shape and click 'show lines' to reveal the number and position of the lines.
- They score points equivalent to the number of lines.
- The winner is the learner or group with the most points after an agreed amount of time/number of turns.

Variation

- Learners score an extra point for drawing lines of symmetry that match their actual position.

Reflecting shapes 🏃🏃

Learning objective

Code	Learning objective
5Gp2	Predict where a polygon will be after reflection where the mirror line is parallel to one of the sides, including where the line is oblique.

Resources

small stickers (per learner)

What to do 📊

- Display the **Symmetry tool** with the square positioned on the left of the vertical mirror line.
- Say: **I am going to reflect the square in the mirror line. Think where its image will be positioned.**
- Ask learners to each take a sticker and attach it to the whiteboard at a grid point that will be covered by the square after reflection.
- Click 'reflect' and confirm (and congratulate!) the learners that correctly positioned their sticker.

Variation

- Use a range of shapes. Learners must correctly position their sticker on a vertex of the shape.

Make a symmetrical shape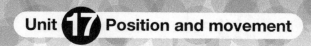

Learning objective

Code	Learning objective
5Gp2	Predict where a polygon will be after reflection where the mirror line is parallel to one of the sides, including where the line is oblique.

Resources

mirror (per class)

What to do

- Display the **Geoboard tool** with the square and two triangles selected.
- Learners take turns to come to the board to select a shape.
- They reposition the vertices of the shape to transform it into a shape with symmetry.
- Use a mirror to confirm that the shape is symmetrical.

Variation

- Use the **Geoboard tool** to make a shape. Learners have to say whether they think it is symmetrical or not. Confirm this using the mirror.

Country co-ordinates

Learning objective

Code	Learning objective
5Gp3	Understand translation as movement along a straight line, identify where polygons will be after a translation and give instructions for translating shapes.

What to do

- Display the **Co-ordinates tool** with the 'country map' background selected.
- Plot a point near to a feature on the map, for example, the tree or the duck pond.
- Learners take turns to state a direction and number, for example: 'up 4' or 'right 5'.
- Follow the instruction and reposition the grid point.
- If the point is re-plotted on a feature, for example, the café or swing, the learner scores a point.

Variation

- Learners must name the feature they predict the point will land on before they give directions.

Treasure map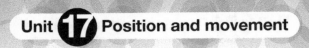

Learning objective

Code	Learning objective
5Gp3	Understand translation as movement along a straight line, identify where polygons will be after a translation and give instructions for translating shapes.

What to do

- Display the **Co-ordinates tool** with the 'treasure map' background selected.
- Plot a point anywhere on the map.
- Name a feature that shares the same *x*- or *y*-co-ordinate as the plotted point.
- Learners must say the direction and number of squares that will move the plotted point to the new location.
- Follow the instruction and re-plot the point.

Variation

- The teacher provides the direction; learners must provide the number of squares to move.

Around the shape

Learning objective

Code	Learning objective
5Gp3	Understand translation as movement along a straight line, identify where polygons will be after a translation and give instructions for translating shapes.

What to do

- Display the **Co-ordinates tool**.
- Plot a point in the middle of the grid.
- Divide the class into two teams.
- Name a shape, for example a square.
- The teams take turns to provide directions that will re-plot the point to form the vertices of a square.
- Repeat with a rectangle.

Variation

- The team must provide all the instructions to complete the square before re-plotting begins.

Estimation game

Learning objective

Code	Learning objective
5MI1	Read, choose, use and record standard units to estimate and measure length, [mass and capacity] to a suitable degree of accuracy.

Resources

ruler or tape measure (per group)

What to do

- Learners take turns to nominate a length or a height to measure, for example, 'height of a table' or 'width of the classroom', and a unit of length for this measurement. They each write an estimate, then measure with a ruler. The learner with the closest estimate scores a point.
- The game ends after five minutes or so, when several learners (or all learners if time allows) have nominated a length to measure. The winner is the learner with the highest score.

Variations

- Learners nominate lengths and heights that are less than one metre.
- Learners can choose to measure curved edges.

Counter flip

Learning objective

Code	Learning objective
5MI2	Convert larger to smaller metric units (decimals to one place), [e.g. change 2·6 kg to 2600 g.]

Resources

ruler or tape measure (per pair); counters (per pair); chalk or pen and large sheet of paper (per class)

What to do

- Learners take turns to flip a plastic counter from a starting line drawn on paper or chalked on the ground.
- They measure the distance travelled from the counter to the starting line in centimetres and convert this to millimetres.
- Continue for around five minutes until several different distances have been measured and converted.

Variations

- Learners use a ruler marked in millimetres to help them convert centimetres to millimetres.
- Learners use their fingers to flip a paper ball distances greater than one metre. They measure the distance in metres and centimetres, then convert the distance to millimetres.

Measure – Length, mass and capacity

Getting in order 👪

Learning objectives

Code	Learning objective
5MI3	Order measurements in mixed units.
5MI4	Round measurements to the nearest whole unit.

What to do

- Learners take turns to write down two single-digit numbers, between 1 and 9, one underneath the other and multiply the second digit by 1000. The first digit represents a measurement in metres; the second represents a measurement in millimetres.
- Learners convert their two measurements to the same units, add them together and compare. Who has the greatest measurement?

Variations

- The first digit represents a measurement in centimetres; the second represents a measurement in millimetres.
- Learners write three digits between 1 and 9. The first digit is a measurement in metres; the second is in centimetres; the third is in millimetres. They convert the units, add the three measurements together and compare.

Line lengths 👥

Learning objective

Code	Learning objective
5MI7	Draw and measure lines to the nearest centimetre and millimetre.

Resources

paper (per pair); 30 cm ruler (per pair); metre ruler (per pair)

What to do

- Each learner uses a 30 cm ruler to draw five straight lines of different lengths in various orientations on paper.
- Partners swap papers. They estimate the length of each line and record it.
- They measure the lines to the nearest millimetre and calculate the difference between their estimate and the measurement. The sum of the difference for all five lines is found. The learner with the lower number is the winner.

Variation

- Learners use a larger piece of paper and draw lines up to one metre in length.

Estimation game 👥

Learning objective

Code	Learning objective
5MI1	Read, choose, use and record standard units to estimate and measure [length,] mass [and capacity] to a suitable degree of accuracy.

Resources

1 kg weight, 'mystery' objects with mass between 0·5 kg and 1·5 kg, scales (per group)

What to do

- Learners take turns to close their eyes.
- They hold a 1 kg weight in one hand and a mystery object in the other. They estimate the mass of the object by comparing its mass with the known 1 kg weight.
- After estimating, learners use scales to measure the mass of the mystery object. They calculate the difference between the actual and the estimated mass.
- The winner of the game is the learner with the smallest difference.

Variation

- Groups select five different objects, only one of which has a mass of 1 kg. Groups swap objects and learners estimate which one has a mass of 1 kg. When all learners have estimated, they use scales to weigh the objects and confirm the object with a mass of 1 kg.

Watch the weight 👥

Learning objective

Code	Learning objective
5MI2	Convert larger to smaller metric units (decimals to one place), e.g. change 2·6 kg to 2600 g.

Resources

eight 'mystery' objects with mass between 0·1 kg and 1·5 kg (per group); kitchen scales (per pair)

What to do

- Learners take turns to select an item, place it on the scales and read the mass in kilograms.
- They score one point for a correct reading. Their partner must convert the measurement to grams to score two points.
- They play five rounds. The winner of the game is the learner with the higher score.

Variations

- Without their partner seeing, learners select an object and weigh it. They remove the item and place it back in the group of objects. They state the measurement and ask their partner to identify the object.
- Without their partner seeing, learners select two objects and weigh each one. They calculate the combined mass of both objects. They announce the combined mass in kilograms and the mass of one of the objects in grams. Their partner must calculate the mass of the second object, weighing it to confirm.

Measure – Length, mass and capacity

Mixed up masses 👥

Learning objectives

Code	Learning objective
5MI3	Order measurements in mixed units.
5MI4	Round measurements to the nearest whole unit.

Resources

two sets of four 'mystery' objects with mass between 0·1 kg and 1·5 kg (per group); kitchen scales (per pair)

What to do

- Without their partner seeing, a learner weighs the objects and places them in order of mass.
- They mix up the place order of the items and ask their partner to re-order them.
- Players swap roles using a new set of four items.

Variation

- Learners match rounded mass measurements to the items, then mix up the order. Partners have to order by mass and match each rounded measurement to the correct item.

Mystery measures 👥

Learning objectives

Code	Learning objective
5MI5	Interpret a reading that lies between two unnumbered divisions on a scale.
5MI6	Compare readings on different scales.

Resources

Resource sheet 12: Blank number lines (two per pair)

What to do

- Learners take turns to mark the divisions of two number lines/scales in different units, for example, divisions of 0·25 kg for one scale and divisions of 50 g for the other scale.
- They draw arrows to indicate two mass measurements.
- Partners must calculate the difference between the two measurements. If successful, they score a point.
- They play five rounds. The winner is the learner with the higher score.

Variations

- Only one scale/number line is used. Both arrows are drawn on the same scale.
- Arrows can be drawn between divisions (equidistant between divisions if possible).

Guess the capacity

Learning objective

Code	Learning objective
5MI1	Read, choose, use and record standard units to estimate and measure [length, mass and] capacity to a suitable degree of accuracy.

Resources

calibrated measuring jug: 1000 ml (per group); containers of various sizes (per group); funnel (per group), water (per group)

What to do

- Arrange the group in a circle with the containers in the middle. A learner is chosen to begin the game. They are the 'estimator'. They select a container and estimate its capacity.
- The learner to their right (the 'measurer') measures the capacity and awards one point to the 'estimator' if they are within 200 ml, two points within 100 ml and three points within 50 ml.
- The learner(s) with the highest score wins the game.

Variations

- Learners estimate in two categories: 'under 500 ml', 'over 500 ml'.
- Play in two teams. Each team has to estimate and order the containers by their capacity. They measure and check who has the order correct.

Capacities compared

Learning objective

Code	Learning objective
5MI2	Convert larger to smaller metric units (decimals to one place), e.g. change 2·6 kg to 2 600 g.

Resources

two packs of 0–9 number cards from Resource sheet 2: 0–100 number cards (per pair)

What to do

- Learners take turns to work in litres or millilitres.
- If working in litres, a learner selects two digit cards and places a decimal point between the digits. If working in millilitres, a learner selects four digit cards.
- Players compare capacities by converting litres to millilitres. The learner with the larger capacity scores a point.
- The learner with the larger score after three rounds is the winner.

Variations

- Play the game without decimal numbers. The learner playing as 'litres' selects only one card.
- Play game with litres to two or three decimal places, with the learner playing 'litres' selecting three or four digit cards.

Measure – Length, mass and capacity

Up or down 👥

Learning objectives

Code	Learning objective
5MI3	Order measurements in mixed units.
5MI4	Round measurements to the nearest whole unit.

Resources

0–9 number cards from Resource sheet 2: 0–100 number cards (per group); two set circles (per group)

What to do

- Arrange the group in a circle with cards in the middle. Arrange two set circles outside, one labelled 'up', the other 'down'.
- Learners take turns to draw four cards, laying them out to form a four-digit number.
- Learners go and stand in the set labelled 'up' or 'down' depending on how the number is rounded to the nearest 100 ml. Anyone standing in the wrong set has to sit out until the next game.

Variations

- Play the game with two digit cards to represent litres. Round to the nearest 10 *l*.
- Play the game with four digit cards to represent litres with two decimal places. Round to the nearest tenth.

Mark the scale 👥

Learning objectives

Code	Learning objective
5MI5	Interpret a reading that lies between two unnumbered divisions on a scale.
5MI6	Compare readings on different scales.

Resources

Resource sheet 12: Blank number lines (per pair); pencil (per pair)

What to do

- The number line acts as a capacity scale. Learners take turns to label two divisions that are an easily calculable number of divisions apart, for example: 300 ml and 400 ml (two, four, five or ten divisions apart).
- They draw an arrow that points to a division, or is equidistant between two unnumbered divisions.
- Their partner scores one point for identifying an arrow that points to a division and two points for an arrow between divisions.
- The learner with the higher score after four rounds is the winner.

Variations

- Arrows are drawn pointing directly to a single division, not between divisions.
- Learners use a range of scales up to 50 000 ml (50 ℓ).

Train times 👥

Learning objectives

Code	Learning objective
5Mt1	Recognise and use the units for time (seconds, minutes, hours, days, months and years).
5Mt2	Tell and compare the time using digital and analogue clocks using the 24-hour clock.

What to do

- Display the **Clock tool**. Set it to 24-hour digital clock, time: 10:25.
- Say: **A train arrives at a station every 15 minutes.**
- Ask: **What time will the clock show when the next train is due?** (10:40)
- Continue round the circle adding 15 minutes until 12:55 at least.
- Repeat for other starting times.

Variations

- Change the starting time to o'clock and increments to 5 or 10 minutes.
- Change increments to a number between 11 and 19, excluding 15.

Don't be late! 👥

Learning objectives

Code	Learning objective
5Mt1	Recognise and use the units for time (seconds, minutes, hours, days, months and years).
5Mt3	Read timetables using the 24-hour clock.

What to do 🖥

- Display **Slide R1**. Briefly talk through the train timetable.
- Learners ask each other questions in the form, **I have to be at station B at [time]. If I leave station A at [time] will I make it?** or **I missed the train at [time] by [x] minutes. How long will I have to wait until the next train?**

Variation

- Learners ask questions that require the gathering of single pieces of information, for example: **What time will the train leaving station A at [time] arrive at station B?**

Measure – Time

Time to time 👥

Learning objective

Code	Learning objective
5Mt4	Calculate time intervals in seconds, minutes and hours using digital or analogue formats.

What to do 📊

- Open two computer windows and display the **Clock tool** set at digital 12-hour in both.
- Set 'start' and 'end' times for a motor racing competition, no more than two hours apart.
- Ask: **How long did the event take?** The first pair to put up their hand and answer correctly scores a point.
- Play for five minutes. The winners are the pair of learners with the highest score.

Variation

- Display clocks in three computer windows and ask learners to calculate the total time as the sum of a two-part event.

Calendar count 👥

Learning objectives

Code	Learning objective
5Mt5	Use a calendar to calculate time intervals in days and weeks (using knowledge of days in calendar months).
5Mt6	Calculate time intervals in months or years.

Resources

two calendars (per pair); counters (per pair)

What to do

- Learners open two calendars showing pages for consecutive months.
- They take turns to flip a counter on each page and work out the time interval between the two dates covered by the counters.
- The learner with the greater time interval scores a point. The overall winner is the learner with the higher score after five rounds.

Variation

- Use one calendar page only.

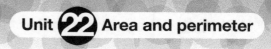

Geoboard shapes

Learning objective

Code	Learning objective
5Ma1	Measure and calculate the perimeter of regular and irregular polygons.

What to do

- Display the **Geoboard tool**.
- Construct a rectangle three units by one unit. Ask: **What is the perimeter?** (eight units)
- Ask: **Who can predict the perimeter of a rectangle three units by two units?** Construct and reveal the perimeter. (10 units)
- Convert a rectangle to a square, 3 × 3. Ask: **What is the perimeter?** (12 units)
- Construct rectangles 3 × 3, 4 × 3 and 5 × 3 and ask learners to predict and then find the perimeter.

Variation

- Ask learners to build squares 2 × 2, 3 × 3, 4 × 4 and so on, and find the perimeter of each. Ask: **What do you notice about the numbers? Is there a pattern?**

Measure perimeter

Learning objective

Code	Learning objective
5Ma1	Measure and calculate the perimeter of regular and irregular polygons.

What to do

- Display the **Geometry set tool**, showing the ruler.
- Select a scalene triangle. Ask: **Who can show us how to measure the perimeter of the shape?** A volunteer demonstrates the measurement.
- Repeat for a trapezium and a six-pointed star.
- Select the pink triangle. Say: **This is a right-angled isosceles triangle.** Ask: **Do we have to measure all three sides to work out the perimeter?** Accept answers and ask a volunteer to demonstrate that only the longer side and one of the shorter sides need to be measured as both shorter sides are equal.

Variation

- Ask learners to construct different scalene triangles with the same perimeter. How many different triangles can they make?

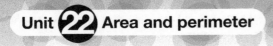

Measure – Area and perimeter

Count the squares

Learning objective

Code	Learning objective
5Ma2	Understand area measured in square centimetres (cm²).

What to do

- Display the **Geoboard tool**.
- Draw a rectangle on the board, 5 × 2. Say: **Each square is 1 cm long.** Ask: **By counting squares, who can tell me the area of the rectangle? What is the correct unit?** (10 cm²)
- Repeat for rectangles 5 × 3, 5 × 4 and 5 × 5. Ask: **What is happening to the area each time?** (increases in multiples of 5)
- Repeat for rectangles 4 and 6 squares long. Ask: **By what multiple is the area increasing?**

Variation

- Ask learners to investigate squares and how the area changes with side length.

Largest area

Learning objectives

Code	Learning objective
5Ma2	Understand area measured in square centimetres (cm²).
5Ma3	Use the formula for the area of a rectangle to calculate the rectangle's area.

Resources

set of rectangles (per group); ruler (per learner); 1 cm squared paper (per learner)

What to do

- Distribute the rectangles. The group select the two rectangles they estimate to have the largest area.
- They measure the dimensions of each rectangle and use the formula to calculate the area.
- The group with the shape with the largest area wins.

Variation

- Learners place the rectangles on 1 cm squared paper and use grid squares to find the area.

Tally tables 👥

Learning objective

Code	Learning objective
5Dh2	Draw and interpret frequency tables, pictograms and bar line charts, with the vertical axis labelled for example in twos, fives, tens, twenties or hundreds. Consider the effect of changing the scale on the vertical axis.

Resources

mini whiteboard and pen (per learner)

What to do

- Display **Slide R2**.
- Ask learners to copy the table on their mini whiteboards and complete the totals in the frequency column.
- Ask the class to answer questions of the type: 'Which sport was the most/least popular?'

Variations

- Ask learners to answer simpler questions that require reading of single pieces of information, for example: 'How many people voted for football?'
- Ask learners to compare two pieces of information, for example: 'How many more people voted for tennis than cricket?'

Pictogram goals 👥

Learning objective

Code	Learning objective
5Dh2	Draw and interpret frequency tables, pictograms and bar line charts, with the vertical axis labelled for example in twos, fives, tens, twenties or hundreds. Consider the effect of changing the scale on the vertical axis.

What to do 📊

- Before the lesson begins, use the **Pictogram tool** to construct a pictogram of number of goals scored by four football teams over a season. Use the scale 'one football equals two goals' and have the number of goals in the range 40 to 60, including odd and even numbers.
- Ask the class questions, such as: **How many goals were scored by Team A/Team B? Which team scored the most/fewest goals? How many goals in total were scored over the season?** Expect whole-class responses.

Variation

- Ask learners to compare two pieces of information, for example: Which team scored [x] more goals than Team B? Which team scored half the number of goals as Team C?

Plotting rainfall

Learning objective

Code	Learning objective
5Dh3	Construct simple line graphs, e.g. to show changes in temperature over time.

What to do

- Prior to the lesson, draw the table opposite on the board.
- Display the **Line grapher tool**. Select *y*-axis increments of 10.
- Ask volunteers to plot points on the graph.
- Ask: **Which month had the highest/lowest rainfall? How do you know?**
- Ask: **Between which months was there the largest drop in rainfall?**
- Ask learners to try to explain the rise and fall in rain measurements.

Variation

- Limit questions to asking learners to read single pieces of information from the graph, for example: **How much rain fell in the month of August?**

Month	Rainfall (mm)
Apr	25
May	40
June	50
July	30
August	25
September	20

Tide times

Learning objective

Code	Learning objective
5Dh4	Understand where intermediate points have and do not have meaning, e.g. comparing a line graph of temperature against time with a graph of class attendance for each day of the week.

What to do

- Prior to the lesson, draw the table opposite showing the sea levels for different times of the morning on the board.
- Display the **Line grapher tool**. Ask volunteers to help plot the data.
- Ask questions, such as: **At what times was the sea level above 50 cm?** (between 7 a.m. and 11 a.m.) **How far did the sea level fall between 9 a.m. and 11 a.m.?** (40 cm) **Why do you think this happened?** (change from high tide to low tide)

Variation

- Ask questions that involve reading data at intermediate points, for example: **Approximately, what is the sea level at 10.30 a.m.?** (60 cm)

Time	Height (cm)
6 a.m.	15
7 a.m.	50
8 a.m.	75
9 a.m.	90
10 a.m.	70
11 a.m.	50

Colour modes 🏫

Learning objective

Code	Learning objective
5Dh5	Find and interpret the mode of a set of data.

Resources

sticky labels (per learner)

What to do

- Learners write their favourite colour on the label and fix it to their top (clothing).
- When requested, learners organise themselves into sets of the same colour.
- Draw a frequency table on the board. Ask learners what colour headings they would give to the rows.
- Ask for volunteers to complete the frequency column.
- Ask: **Which is the largest group?**

Variation

- Ask learners what would be the lowest possible value for the frequency of the colour that is the mode of the set (given constant values for the other colours).

Favourite fruits 🏫

Learning objective

Code	Learning objective
5Dh1	Answer a set of related questions by collecting, selecting and organising relevant data; draw conclusions from their own and others' data and identify further questions to ask.

What to do 📊

- Say the following statement: **Given the choice of several fruits to eat – oranges, apples, bananas, pineapples and mangoes – most learners in the class would say they prefer bananas.**
- Ask: **How would I prove that this statement is true?** Accept comments and agree that a survey of favourite fruits would provide the information.
- Draw a frequency table of fruits with a tally column.
- Ask learners to make a tally mark next to their favourite fruit.
- Display the **Bar charter tool** and ask volunteers to help input data.
- Ask: **Does the data support the statement that 'most learners prefer bananas'. How do you know?**

Variation

- Ask learners to explain why the **Line grapher tool** would not have been a suitable graphing tool for this set of data.

Shoe size

Learning objective

Code	Learning objective
5Dh1	Answer a set of related questions by collecting, selecting and organising relevant data; draw conclusions from their own and others' data and identify further questions to ask.

What to do

- Say: **The most common shoe size in the class is 2/34.** (UK/EU)
- Ask: **How would you prove that this statement is true?** Accept comments and agree that a survey of shoe sizes would provide the information.
- Draw a frequency table and decide on the range of shoe sizes to survey.
- Ask learners to make a tally mark next to their shoe size.
- Display the **Bar charter tool** and ask volunteers to help input the data.
- Ask: **Does the data support the statement that 'most of the class have a shoe size of 2'. How do you know?**

Variation

- Ask learners to plot the data on two graphs with different scales.

Likely thumbs

Learning objective

Code	Learning objective
5Db1	Describe the occurrence of familiar events using the language of chance or likelihood.

Resources

bag (per class); coloured counters (blue and red) (per class)

What to do

- Tell learners that they are going to think about events and how likely they are to happen. They hold their thumb up if an event is likely to happen, and thumb down if it is unlikely to happen.
- Show learners the bag and coloured counters.
- Place several blue counters in the bag. Say: **I am going to pull out a counter.** Ask: **What are the chances it will be blue? And the chances it will be green?**
- Replace blue counters with red. Ask: **What are the chances it will be blue? And the chances it will be red?**
- Mix red and blue in different combinations. Ask learners to show likelihood by size of their angled thumb up or down.

Variation

- Ask learners to show an even chance with a sideways thumb and provide examples: half red, half blue counters and so on.

Unit 1: Whole numbers 1

Learning objectives

Code	Learning objectives
5Nn1	Count on and back in steps of constant size, extending beyond zero.
5Nn2	Know what each digit represents in five- and six-digit numbers.
5Nn3	Partition any number up to one million into thousands, hundreds, tens and units.
5Nn5	Multiply and divide any number from 1 to 10 000 by 10 or 100 and understand the effect. [Whole number answers]
5Nn6	Round four-digit numbers to the nearest 10, 100 or 1000.
5Nn8	Order and compare numbers up to a million using the > and < signs.
5Nn12	Recognise and extend number sequences.
5Nn13	Recognise odd and even numbers and multiples of 5, 10, 25, 50 and 100 up to 1000.
Strand 5: Problem solving	
5Ps3	Explore and solve number problems and puzzles, e.g. logic problems.
5Ps4	Deduce new information from existing information to solve problems.
5Ps5	Use ordered lists and tables to help to solve problems systematically.
5Ps6	Describe and continue number sequences, e.g. –30, –27, ☐☐ –18...; identify the relationships between numbers.
5Ps8	Investigate a simple general statement by finding examples which do or do not satisfy it, e.g. the sum of three consecutive whole numbers is always a multiple of three.
5Ps9	Explain methods and justify reasoning orally and in writing; make hypotheses and test them out.
5Ps10	Solve a larger problem by breaking it down into sub-problems or represent it using diagrams.

Number – Numbers and the number system

Unit overview

In this unit, learners investigate counting on and back in steps of constant size and find missing numbers, including decimal numbers. They extend their understanding of sequences by working out the term-to-term rule in number patterns. Learners work out the value of each digit in a five- or six-digit number by looking at its position within the number and they use this knowledge to partition numbers by place value. They investigate the effect of multiplying and dividing numbers by 10 and 100 and apply this to any given number from 1 to 10 000. They explore and apply rules and strategies for rounding and ordering numbers. The unit concludes with learners examining and identifying odd and even numbers.

Common difficulties and remediation

Some learners confuse numbers and digits. To avoid this, establish definitions of the terms before teaching the unit. Remind learners that a **digit** is a single numerical symbol, from 0 to 9, whereas a **number** is a string of one or more digits. Explain that numerals 0 to 9, for example 6, are both numbers and digits but 28 is a number because it is a string of two digits.

Immediate intervention is required for learners who believe that zero represents 'nothing' and so can be omitted from a number. They will need regular opportunities to work with zero as a placeholder.

Look out for learners who compare numbers based on the value of the digits, instead of place value, for example, writing 896 > 3005, because 8, 9 and 6 are 'larger' than 3 and 5. They may also make other errors, such as ordering numbers according to only the first digit in the number, or by the number of digits. These learners will require support to understand the concept that it is the **position and value** of the digits that is important when ordering numbers, not the **number of digits.**

Some learners round to 100 in stages. For example, rounding 3449 to the nearest 100 by: rounding 3449 to the nearest 10 (3450), 3450 to the nearest 100 (3500). Hence, the misconception leads to the incorrect answer, 3500. Encourage learners to use a blank number line as a practical visual aid, allowing them to see the choice of numbers to which they could round the given number.

Promoting and supporting language

If appropriate, when a new key word is introduced, ask learners to write a definition in their books, drawing a box around it for emphasis. Encourage them to write the definition in their own words, for example: 'An odd number is any number that cannot be divided by 2 to give another whole number.'

Encourage frequent use of mathematical language to help embed vocabulary. For example, when learners describe the next number in a sequence, encourage them to use the more mathematically correct phrase 'next term in the sequence'. This will help them make the transition to mathematically precise language.

Lesson 1: **Counting on and back (1)**

Learning objective

Code	Learning objective
5Nn1	Count on and back in steps of constant size [, extending beyond zero].

Strand 5: Problem solving

5Ps3	Explore and solve number problems and puzzles, e.g. logic problems.
5Ps4	Deduce new information from existing information to solve problems.
5Ps6	Describe and continue number sequences [, e.g. –30, –27, ☐☐ –18...]; identify the relationships between numbers.
5Ps9	Explain methods and justify reasoning orally and in writing; make hypotheses and test them out.

Prerequisites for learning

Learners need to:
- understand the place value system
- count forwards and backwards in multiples of any given number from and to zero.

Success criteria

Learners can:
- find missing numbers in a counting sequence of steps of constant size

- count on and back from any two-digit or three-digit number in steps of 2, 3, 4, 5, 9, 10 from and to zero.

Vocabulary

step, sequence, count on, count back

Resources

Resource sheet 1: Thermometer template; coloured pencils (per pair)

Refresh

- Choose the activity *Sequences*, from the Refresh activities.

Discover SB

- Direct learners to the Student's Book and ask them to look at the image. Say: **In your head, work out the total number of apples.** Ask: **How did you find the answer?** Ask learners to explain their strategies.

Teach and Learn SB 📊

- Discuss the text and example in the Learn section of the Student's Book. Ask learners to put their fingers on 43 and to count back along the number line in steps of three. Ask: **What other calculation resources could you use to help you to count on and back?** Take feedback. Learners might suggest using a number square, or a set of number cards, or using known number facts, such as times tables. Less able learners might suggest using manipulatives, such as counters.
- Display the **Number line tool** set at divisions of 1, scale: 0 to 50. Point to zero. Ask: **If you count on in steps of three, what would the fourth number be?** Take responses. Repeat for different starting points and steps, for example: start at 100, count back in steps of 10 or 25; start at 48, count back in steps of four or five. Learners demonstrate counting on and back.

Practise WB

- Workbook: Counting on and back page 2
- Refer to Activity 1 from the Additional practice activities.

Apply 👥

- Distribute Resource sheet 1: Thermometer template. Ask pairs of learners to draw a thermometer that goes from +40 to 0 degrees, with 1 degree divisions and labelled divisions of 5 degrees. Ask questions of the type: **If the temperature starts at 28 degrees and falls in steps of 2 degrees each day, what temperature will it be in nine days' time?**

Review 📊

- Display the **Number line tool**, set to 0 to 100. Divide the class into two teams. Say: **The scale measures the jumps of a superhero in metres.** Ask questions of the type: **A superhero jumps from four in steps of five. Where will he/she be in seven jumps?** Ask the learners to raise a hand to respond. If correct, they score a point for their team. Play ten rounds. The team that wins the most rounds is the winner.

Assessment for learning

- A man enters a lift on floor seven. He ascends in steps of four floors. What floor will he be on after six steps? (31) Explain your solution.
- The temperature outside fell from 16 °C to –4 °C in five equal steps. How many degrees are in each step? (3)

Lesson 2: **Number sequences**

Learning objective

Code	Learning objective
5Nn12	Recognise and extend number sequences.

Strand 5: Problem solving	
5Ps3	Explore and solve number problems and puzzles, e.g. logic problems.
5Ps4	Deduce new information from existing information to solve problems.
5Ps6	Describe and continue number sequences, e.g. −30, −27, □□ −18...; identify the relationships between numbers.
5Ps9	Explain methods and justify reasoning orally and in writing; make hypotheses and test them out.

Prerequisites for learning

Learners need to:

- find missing numbers in a number sequence, including negative and decimal numbers
- count on and back from any two- or three-digit number in steps of 2, 3, 4, 5, 9, 10 from and to zero.

Success criteria

Learners can:

- identify the rule by linking them to times table facts and knowledge of number.

Vocabulary

step, sequence, term, rule, common difference, value

Resources

sheet of paper and pencil (per learner); Resource sheet 2: 0–100 number cards (per class)

Refresh

- Choose the activity *Reverse*, from the Refresh activities.

Discover 〔SB〕

- Direct learners to the Student's Book and discuss the picture of the piano keys. Ask: **How are the black keys arranged?**

Teach and Learn 〔SB〕〔🖵〕

- Display **Slide 1**. Read the text with the class and discuss what each part of the sequence is called and how it is used. Ask: **What is the rule for this sequence?** (even numbers) **What is the value of the third term?** (6)
- Discuss the Example number sequence in the Student's Book. Ask: **Can you extend the number sequence beyond 51? What term comes next?** Learners continue the sequence for a few terms using the 'add 7' rule.
- On the board, write: 11, 17, 23, 29. Ask: **What pattern can you see in the numbers?** Accept comments. Ask: **What will the next term in the sequence be? How do you know?** Elicit that it will be 35, because the numbers go up by six each time.
- On the board, write: 5, 3, 1, −1, −3. Ask: **What is the third term in the sequence?** (1) **What will the next term be?** (−5) Elicit that the rule for this sequence is 'subtract 2'.
- Pairs discuss their solutions. Talk about the strategy of a **common difference** for finding the rule: subtracting

consecutive terms to check they increase/decrease by a constant value. Repeat for decimal sequences.

Practise 〔WB〕

- Workbook: Number sequences page 4
- Refer to Activity 1 from the Additional practice activities.

Apply 👥

- Give each group a target number and the steps to count on in. (Generate these by choosing number cards at random.) Each player writes a number between 0 and 100 on their own piece of paper. They pass their paper to the player on their left, who writes down the next term in the sequence. Learners continue writing and passing the paper on, until one player reaches or passes the target number. Repeat the game using subtraction.

Review

- Write on the board a whole number, decimal or negative as a sequence start number. Say: **Write a rule for a sequence that starts with this number and write five terms in the sequence. Keep the rule secret.** Ask for volunteers to write their sequences on the board. Ask: **Who can write the next three terms and say the rule?** Repeat for different sequences.

Assessment for learning

- What is the next term in the sequence: −25, −17, −9, −1? (7) How do you know?

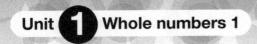
Unit **1** Whole numbers 1

Lesson 3: **Place value (1)**

Number – Numbers and the number system

Learning objectives

Code	Learning objective
5Nn2	Know what each digit represents in five- and six-digit numbers.
5Nn3	Partition any number up to one million into thousands, hundreds, tens and units.

Strand 5: Problem solving

5Ps3	Explore and solve number problems and puzzles, e.g. logic problems.
5Ps5	Use ordered lists and tables to help to solve problems systematically.
5Ps9	Explain methods and justify reasoning orally and in writing; make hypotheses and test them out.
5Ps10	Solve a larger problem by breaking it down into sub-problems or represent it using diagrams.

Prerequisites for learning

- Learners need to:
- understand what each digit represents in a three- or four-digit number.

Success criteria

Learners can:

- work out the value of each digit in a five- or six-digit number.

Vocabulary

place value, digit, hundred thousands, ten thousands

Resources

Resource sheet 3: Blank place value grid (per pair); mini whiteboard and pen or paper (per learner); pencil (per learner); coloured pencils (per learner)

Refresh

- Choose the activity *Dice digits*, from the Refresh activities.

Discover 〔SB〕

- Direct learners to the Student's Book and discuss the text and image. Ask: **Which suitcase has the greater mass? How do you know?** Accept comments. Say: **There is a maximum weight of 21 700 g per person allowed on board an aircraft.** After some discussion, ask: **Which suitcase will be allowed on board?** (21 409 g)

Teach and Learn 〔SB〕〔📊〕

- Remind learners that if there is a zero it acts as a place holder. For example, a zero in the hundreds place means there are no hundreds, so 58 902 is 50 000 + 8 000 + 900 + 2.
- Discuss the Learn text and example place value grid in the Student's Book. Then, on the board, draw a place value grid showing columns from Units to Hundred thousands. Write the numbers 9 997, 9 998 and 9 999 in the correct positions on the grid. Ask: **What is the next number in the sequence?** (10 000) Ask a volunteer to write this number. Agree that the first digit should go in the column headed 'Ten thousands'. Repeat for 99 997, 99 998, 99 999 and 100 000.
- Display the **Place value tool** showing 46 321. Ask: **What is the value of the digit 4?** Elicit that it is forty thousand and reveal the arrow card. Repeat for other digits.

- Distribute paper and ask learners to work in pairs. On the board, write: 374 592 and 903 408. Learners use Resource sheet 3: Blank place value grid and write the digits of the numbers in the correct positions. Say: **Choose a number and tell your partner the value of each digit.**

Practise 〔WB〕

- Workbook: Place value (1) page 6

Apply 〔👥〕〔🖥️〕

- Display **Slide 1** and read the word problem.
- Learners write the answer on their whiteboards. Check the answer as a class. Ask pairs to write their own five- or six-digit place value problem for another pair to solve. Share the best examples with the class.

Review 〔📊〕

- Display the **Place value tool**. Make 25 188. Point to the digits in random order and ask learners to say the value. Repeat for 420 709.

Assessment for learning

- What is the value of each seven in the number 777 777? How much bigger is the number 888 888? (one hundred and eleven thousand, one hundred and eleven, 111 111)
- The balance of a bank account is $939 939. What amount of money is $10 000 less than this?

Lesson 4: **Place value (2)**

Learning objectives

Code	Learning objective
5Nn2	Know what each digit represents in five- and six-digit numbers.
5Nn3	Partition any number up to one million into thousands, hundreds, tens and units.

Strand 5: Problem solving

5Ps3	Explore and solve number problems and puzzles, e.g. logic problems.
5Ps5	Use ordered lists and tables to help to solve problems systematically.
5Ps9	Explain methods and justify reasoning orally and in writing; make hypotheses and test them out.
5Ps10	Solve a larger problem by breaking it down into sub-problems or represent it using diagrams.

Prerequisites for learning

Learners need to:

- understand what each digit represents in a three- or four-digit number and partition into thousands, hundreds, tens and units.

Success criteria

Learners can:

- use a place value grid to partition a five or six-digit number.

• Vocabulary

place value, digit, partition

Resources

Resource sheet 3: Blank place value grid (per learner); Resource sheet 5: Arrow cards (per pair); paper or mini whiteboard and pen (per learner)

Refresh

- Choose the activity *Values* together, from the Refresh activities.

Discover [SB]

- Direct learners to the image in the Student's Book. Say: **Discuss a maths problem where you exchanged one value for another. What did you do? What is happening in the diagram?** One ten is exchanged for ten units.

Teach and Learn [SB] [📊]

- Discuss the text and arrow cards in the Student's Book. Model how arrow cards are used using Resource sheet 5: Arrow cards.
- Display the **Place value tool** showing 398 265 (set up before the lesson). Ask: **What is the value of the digit 3?** Separate the arrow card from the number and reveal 300 000. Repeat for the other digits.
- Distribute copies of Resource sheet 3: Blank place value grid. On the board, write: 432 987. Ask learners to write the digits of the number in the correct positions. Say: **The number 432 987 can be split by place value …** Use the **Place value tool** to model this in arrow form. Under the grid write: 432 987 = 400 000 + 30 000 + 2000 + 900 + 80 + 7. Repeat for 907 624. Demonstrate how each number can be recomposed.
- Show, in random order, the arrows for 8000, 4, 800 000, 600, 10 and 40 000. Say: **A number has been partitioned, into its separate place values. What is the number?** (848 614)

Practise [WB]

- Workbook: Place value (2) page 8
- Refer to Activity 2 from the Additional practice activities.

Apply [👥][💻]

- Display **Slide 1** and read the text. Say: **A charity receives two donations: $432 517 and $346 421. Use place value to find the total amount they received.** Encourage learners to solve the problem by using arrow cards to partition the numbers into their separate place values. Ask learners to explain how they found the sum.

Review [📊]

- Display the **Place value tool** showing 678 044. Point to digits in random order and ask learners to say the value. Reveal the arrow. Repeat for 942 311.
- Display five- and six-digit numbers. For each number, ask questions such as: **Write a number ten thousand more/three hundred thousand less.** Learners work in pairs and write the numbers on whiteboards.

Assessment for learning

- Here is the number 460 376. Show me how it can be split/partitioned into its separate place values, then recombined.
- Show separate arrows in random order for 4000, 2, 900 000, 500, 70 and 60 000. Say: **A number has been partitioned into its separate place values. What is the number?** (964 572)

Number – Numbers and the number system

65

Lesson 5: **Multiplying and dividing by 10 or 100**

Learning objective

Code	Learning objective
5Nn5	Multiply and divide any number from 1 to 10 000 by 10 or 100 and understand the effect.

Strand 5: Problem solving

5Ps3	Explore and solve number problems and puzzles, e.g. logic problems.
5Ps4	Deduce new information from existing information to solve problems.
5Ps10	Solve a larger problem by breaking it down into sub-problems or represent it using diagrams.

Prerequisites for learning

Learners need to:

- understand the effect of multiplying and dividing three-digit numbers by 10 or 100.

Success criteria

Learners can:

- multiply a number by 10 and 100

- divide a number by 10 and 100.

Vocabulary

place value, multiply, divide

Resources

mini whiteboard and pen (per learner); supermarket or fast-food flyers (per pair)

Refresh

- Choose the activity *Multiply or divide*, from the Refresh activities.

Discover [SB]

- Direct learners to the Student's Book and discuss the image. Ask: **What units of measurement are being converted?** (units of length) **What other units of measurement can be converted by multiplying, or dividing, by 10 or 100?** Take suggestions.

Teach and Learn [SB]

- Look at the text in the Student's Book to reinforce learners' understanding of multiplying and dividing by 10 and 100 and the effect it has on place value. Work through the examples as a class.

- Ask learners to each write one single digit on their whiteboard and keep it hidden. Place five chairs at the front of the class – one for each digit. Leave a larger space in between the second and third chairs (to represent the gap between the thousands and hundreds digits).

- Ask a learner to sit in the units place and reveal their number. Say: **Each new place to the left is worth ten times more than the last one. At the moment we have this amount of units, but what is ten times this amount?** The learner moves one place left to the tens chair. Ask: **What is this digit worth now? What is ten times this amount?**

- Point to the second chair from the left and ask: **What is this chair worth?** (thousands) Move the learner to sit in the thousands place and say the new value

of their digit. Ask: **If each new chair on the left is worth ten times more than the last one, what do you think ten times a thousand is?** (ten thousand) Move the learner to sit in the ten thousands place and say the new value of their digit. Choose different learners to sit in the chairs with their numbers and practise multiplying by 10 and 100.

- Repeat for division, (moving the correct number of places to the right when dividing by 10 or 100).

Practise [WB]

- Workbook: Multiplying and dividing by 10 or 100 page 10

Apply 👥

- Hand out some supermarket or fast-food flyers. Ask learners to make a list of prices of supermarket items. They then multiply to find the cost of 10 and 100 items.

Review

- On the board, write: 6741 × 10. Learners discuss the solution and solve it. Repeat for 2389 × 100.
- On the board, write: 9560 ÷ 10. Learners discuss the solution and solve it. Repeat for 7500 ÷ 100. Say: **A number multiplied by 100 is 3600. What is the number?** (360)

Assessment for learning

- Here is the number 4700. Show me a fast method for multiplying and dividing it by 100.

Lesson 6: **Rounding**

Learning objective

Code	Learning objective
5Nn6	Round four-digit numbers to the nearest 10, 100 or 1000.

Strand 5: Problem solving

5Ps3	Explore and solve number problems and puzzles, e.g. logic problems.
5Ps8	Investigate a simple general statement by finding examples which do or do not satisfy it, e.g. the sum of three consecutive whole numbers is always a multiple of three.

Prerequisites for learning

Learners need to:

- round three- and four-digit numbers to the nearest 10 or 100
- decide whether to round up or down.

Success criteria

Learners can:

- round a number by finding the rounding digit

- look at the digit to the right of the rounding digit and say that if its value is less than five, the rounding digit is left unchanged and if it is five or more, one is added to the rounding digit.

Vocabulary

round, rounding digit, nearest 10, nearest 100, nearest 1000

Resources

1–6 dice or Resource sheet 4: 1–6 spinner (per pair)

Refresh

- Choose the activity *Rounding*, from the Refresh activities.

Discover [SB]

- Read the text in the Discover section of the Student's Book together. Say: **Give your partner another example of where a newspaper report might round a figure.** Learners share suggestions. Ask: **Why might it not be a good idea to round the number of kilometres planned for a spaceship travelling to a distant planet?**

Teach and Learn [SB] [bar chart]

- Work through the example in the Student's Book. Review rounding three-digit numbers to nearest 10 and 100. Display the **Number line tool**, showing 610 to 620. Mark 614. Ask: **Between which two multiples of ten is 614?** (610, 620) **Which is it closer to?** (610) Repeat for 616 (620). Say: **What about 615?** Establish that 'halfway' numbers are rounded up.

- Say: **What rules do we know about when rounding?** Discuss that learners need to think about what multiples of 10, 100 or 1000 the number is between. They look at the digit to the right of the rounding digit (place position they are rounding to). If the digit is 5 or greater, round up. If it is less than 5, round down. Remember to include place holder zeros as needed.

- On the board, write: 6137, 6451, 6500, 6789. Display 6000–7000 on the **Number line tool**.

Demonstrate rounding to the nearest thousand, first with the number line, then by the rules. Elicit that the rounding digit is in the thousands place and the rules focus on the hundreds place.

Practise [WB]

- Workbook: Rounding page 12
- Refer to Activity 3 from the Additional practice activities.

Apply [two people icon]

- Distribute dice, which learners roll to create four four-digit numbers. These represent football match attendances. Say: **Write a newspaper report for the matches, rounding attendances to the nearest hundred.**

Review

- On the board, write: 3249 metres. Say: **Round this measurement to the nearest 10, 100 and 1000. Then, for each, write three other measurements that round to the same number.**

Assessment for learning

- Show me two methods for rounding 2751 to the nearest 10, 100 and 1000.
- A number with the digits 2, 3, 7, 4 rounded to the nearest hundred is 3500. What is the number? (3472)

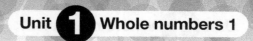

Number – Numbers and the number system

Lesson 7: **Comparing and ordering**

Number – Numbers and the number system

Learning objective

Code	Learning objective
5Nn8	Order and compare numbers up to a million using the > and < signs.

Strand 5: Problem solving

5Ps3	Explore and solve number problems and puzzles, e.g. logic problems.
5Ps8	Investigate a simple general statement by finding examples which do or do not satisfy it, e.g. the sum of three consecutive whole numbers is always a multiple of three.
5Ps9	Explain methods and justify reasoning orally and in writing; make hypotheses and test them out.

Prerequisites for learning

Learners need to:
• compare pairs of three- or four-digit numbers, using the < and > signs and find a number in between each pair.

Success criteria

Learners can:
• compare and order numbers by discussing the value of the digits and by considering the relative positions on a number line

Vocabulary

place value, digit, thousands, hundreds, tens, units

Resources

1–6 dice or Resource sheet 4: 1–6 spinner (per learner)

Refresh

• Choose the activity *Comparing numbers*, from the Refresh activities.

Discover ⬛⬜

• Direct learners to the Student's Book and display **Slide 1**. Ask: **When numbers become very large, such as the size of planets, are number lines a useful tool for ordering? Why/why not?** Accept comments and discuss.

Teach and Learn ⬛⬜⬜

• Display **Slide 2**. Say: **What do you know about comparing whole numbers with the same number of digits?** Read through the 'rules' with the class.
• Following a discussion of the text and Example in the Student's Book, on the board, write: 23 427 and 23 472. Ask: **Which is the larger number?** Accept suggestions. Establish that we compare numbers from the left. Say: **The ten thousands digits are most significant so compare them first.** Elicit that the digits are the same. Say: **If the digits are the same then compare the digits in next place to the right.** Elicit that these digits are also the same. Continue to the tens place and compare digits. Say: **When two numbers have different digits in the same place, the number with the larger digit is the higher number.** On the board, write: 23 472 > 23 427. Display the **Number line tool**, set 23 400 to 23 472 and confirm that the position of 23 472 is to

the right of 23 400 and therefore the greater number. On the board, write: 54 404, 54 640, 54 440, 54 604, 54 044. Say: **To order a set of numbers write them with their place values lined up.** Draw a place value grid, write the numbers in it and order them by working systematically across, applying the rules. (54 044, 54 404, 54 440, 54 604, 54 640)

Practise ⬛

• Workbook: Comparing and ordering page 14
• Refer to Activity 4 from the Additional practice activities.

Apply ⬛⬜

• Display **Slide 1** again. Ask pairs of learners to draw a diagram showing the planets in order of size.

Review

• On the board, write: 24 567, 24 765, 23 790, 23 709, 24 756. Ask learners to order the numbers, share answers and explain solutions. Now write number pairs with gaps between them, such as 673 281 and 673 199; 780 808 and 780 880. Ask learners to suggest numbers that come between the numbers and still keep the order.

Assessment for learning

• Arrange the four numbers 506 606, 560 660, 506 660, 560 066 in order. Explain the rules of ordering.

Lesson 8: **Odds, evens and multiples (1)**

Learning objective

Code	Learning objective
5Nn13	Recognise odd and even numbers and multiples of 5, 10, 25, 50 and 100 up to 1000.

Strand 5: Problem solving

5Ps3	Explore and solve number problems and puzzles, e.g. logic problems.
5Ps4	Deduce new information from existing information to solve problems.
5Ps6	Describe and continue number sequences [, e.g. –30, –27, ☐☐ –18...]; identify the relationships between numbers.
5Ps8	Investigate a simple general statement by finding examples which do or do not satisfy it, e.g. the sum of three consecutive whole numbers is always a multiple of three.
5Ps9	Explain methods and justify reasoning orally and in writing; make hypotheses and test them out.

Prerequisites for learning

Learners need to:

- count on and back from any number
- identify and explain simple number sequences.

Success criteria

Learners can:

- identify odd and even numbers by looking at the last digit

- confirm by halving that a number is odd or even
- identify a multiple of five by looking at the last digit.

Vocabulary

odd, even, multiple

Resources

counters (per pair); coloured pencils (per learner)

Refresh

- Choose the activity *Odd or even*, from the Refresh activities.

Discover 🔲SB

- Direct learners to the Student's Book and discuss the image. Ask: **Why is the number on the left door a multiple of 25 but not 50?** Accept and discuss suggestions

Teach and Learn 🔲SB 📊

- Challenge learners to recite multiples of 5/10/25/50/100 up to 1000.
- Display the **Number square tool**, highlighting zero. Count forward in steps of ten from zero. Highlight multiples of ten and confirm numbers ending zero are multiples of ten. Count forward in steps of 100 from zero and confirm numbers ending 00 are multiples of 100. Ask one half of the class to write the sequence of multiples of 25 from zero to 500 and the other half to write the sequence for multiples of 50 to 500. Ask each group: **What digits do the numbers in your sequence have in common?** Establish that multiples of 25 all end 25, 50, 75 or 00; multiples of 50 all end 50 or 00.
- Point to 1 on the **Number square tool**. Ask learners to count in twos from one, highlighting odd numbers. Say: **The numbers that end in one, three, five, seven or nine are odd numbers.** Ask: **If you were to use counters to make these numbers what would they look like?** Model and elicit that there is always an unpaired counter.

- Now work through the text and example in the Student's Book.

Practise 🔲WB

- Workbook: Odds, evens and multiples (1) page 16
- Refer to Activity 4 from the Additional practice activities.

Apply 👥

- Explain that the flats in a tower block are numbered from 1 to 1000. Learners draw a table with headings: flat number/multiple of 25/multiple of 50 and rows labelled in steps of 25 from 25 to 1000. They tick the boxes in the table and use this to work how many numbers between 1 and 1000 are multiples of 25 but not 50. (20)

Review

- On the board, write: 2340, 4567, 1852, 6355. Learners identify numbers as odd, even and multiples of five. Say: **Explain to your partner how to determine whether a number is odd or even and if it is a multiple of five.**

Assessment for learning

- How do you know if a number is odd? Even? A multiple of five? Give an example of each.
- Say a multiple of 25 that is between 3430 and 3470 (3450).

Additional practice activities

Activity 1

Learning objectives

- Create a number sequence using a known rule.
- Identify the rule for a number sequence and extend the terms.

Resources

paper or mini whiteboard and pen (per group)

What to do

- Arrange learners in circle groups of five or six.
- Learners suggest a rule for a sequence. One learner writes down a start number on the paper.
- The paper is passed to the next learner who writes the next term in the sequence, below the previous term. This continues until all members of the group have added a term.
- Groups swap papers. They write the next three terms in the sequence and determine the rule.

Variation

Challenge 1 Guide learners to use simple sequences with small increments between consecutive values.

Activity 2

Learning objectives

- Identify the value of each digit in a number.
- Use knowledge of place value to rearrange digits to make a number as close as possible to a given target number.

Resources

paper or mini whiteboard and pen (per learner); 1–6 dice or Resource sheet 4: 1–6 spinner (per pair)

What to do

- Pairs of learners choose a six-digit target number that uses all the digits 1 to 6, for example 435 126.
- They take turns to roll a dice six times and record the digits.
- Each learner rearranges their six digits to make a six-digit number as close as possible to the target number.
- The learner with the number closest to the target number scores a point.

Variation

Challenge 3 Learners roll the dice to make two six-digit numbers, one larger and one smaller than the target number. The closest 'larger' number and the closest 'smaller' number win a point.

Additional practice activities

Activity 3

Learning objectives

- Round a number to the nearest 100.
- Multiply a number by 10.

Resources

paper or mini whiteboard and pen (per pair); 1–6 dice or Resource sheet 4: 1–6 spinner (per pair)

What to do

- Learners take turns to roll a dice four times to create a four-digit number.
- They record the number and round it to the nearest 100.
- They multiply the rounded number by ten and record the number as their score.
- The first player to win 5 rounds is the overall winner.

Variations

Challenge 1 Learners roll the dice three times to make three-digit numbers. They round these to the nearest ten and multiply by ten.

Challenge 3 Learners roll the dice five times to make five-digit numbers. They round these to the nearest thousand and multiply by a hundred.

Activity 4

Learning objectives

- Order and compare five-digit numbers from smallest to largest.
- Recognise odd and even numbers.

Resources

0–9 number cards from Resource sheet 2: 0–100 number cards (multiple copies per pair)

What to do

- Learners shuffle the number cards and deal ten cards each. They place the remaining cards face down in a pile.
- They each place their ten cards face down in two rows of five cards.
- They turn over the cards to reveal two five-digit numbers.
- They place the four five-digit numbers in order from smallest to largest.
- The learner with the largest numbers scores a point. They also score one point if the number is even, two points for odd and five points for identifying any multiple of 5, 10, 25, 50 or 100.
- The learner with the higher score after six rounds is the winner.

Variation

Challenge 1 Learners work with three- or four-digit numbers.

Number – Numbers and the number system

Unit 2: Whole numbers 2

Learning objectives

Code	Learning objective
5Nn1*	Count on and back in steps of constant size, extending beyond zero.
5Nn3*	Partition any number up to one million into thousands, hundreds, tens and units.
5Nn9	Order and compare negative and positive numbers on a number line and temperature scale.
5Nn10	Calculate a rise or fall in temperature.
5Nn12*	Recognise and extend number sequences.
5Nn14	Make general statements about sums, differences and multiples of odd and even numbers.

*Learning objective revised and consolidated.

Strand 5: Problem solving

5Pt1 / 5Ps1	Understand everyday systems of measurement in [length, weight, capacity,] temperature [and time] and use these to perform simple calculations.
5Pt7	Consider whether an answer is reasonable in the context of a problem.
5Ps2	Choose an appropriate strategy for a calculation and explain how they worked out the answer.
5Ps3	Explore and solve number problems and puzzles, e.g. logic problems.
5Ps4	Deduce new information from existing information to solve problems.
5Ps5	Use ordered lists and tables to help to solve problems systematically.
5Ps6	Describe and continue number sequences, e.g. −30, −27, □□ −18...; identify the relationships between numbers.
5Ps8	Investigate a simple general statement by finding examples which do or do not satisfy it, e.g. the sum of three consecutive whole numbers is always a multiple of three.
5Ps9	Explain methods and justify reasoning orally and in writing; make hypotheses and test them out.
5Ps10	Solve a larger problem by breaking it down into sub-problems or represent it using diagrams.

Unit overview

The unit reviews skills and concepts covered in Unit 1, including place value, partitioning and number sequences. The use of thermometers to measure temperature is discussed and learners are introduced to the concept of a negative scale. They practise ordering and comparing negative and positive numbers on both number lines and temperature scales. Strategies for calculating temperature change are discussed and learners are given opportunities to solve simple problems involving rise or fall in temperature. The unit concludes with an investigation into the outcomes of addition, subtraction and multiplication of odd and even numbers.

Common difficulties and remediation

For some learners, negative numbers can cause confusion between magnitude and order, for example, assuming that −9 is greater than −3. Learners who make this error focus on absolute value and do not differentiate between positive and negative numbers. To remedy this, introduce the concept of negative integers with models. Number lines in particular should be used frequently throughout the unit.

To help learners think in terms of a scale beyond zero, teachers should include work with vertical scales, where values above zero are represented by positive numbers, with the negative numbers below zero. Alternative positive and negative scales, such as temperature, or pictures that illustrate heights above and below sea level, with zero at sea level, are appealing to learners and also provide extra contexts for discovering the importance of negative numbers.

Promoting and supporting language

Be aware of terms with more than one definition, particularly those with meanings outside mathematics, such as: negative, positive, scale.

Encourage frequent usage of mathematical vocabulary. For example, model the use of the question 'Which number is greater/smaller?' or 'Which is the higher/lower temperature?' when learners ask how much lower or higher a temperature is compared to another number on a vertical scale. This will help them make the transition to mathematically precise language.

Number – Numbers and the number system

Lesson 1: **Whole numbers**

Learning objectives

Code	Learning objective
5Nn1*	Count on and back in steps of constant size, extending beyond zero.
5Nn12*	Recognise and extend number sequences.

Strand 5: Problem solving

5Pt1 / 5Ps1	Understand everyday systems of measurement in [length, weight, capacity,] temperature [and time] and use these to perform simple calculations.
5Ps2	Choose an appropriate strategy for a calculation and explain how they worked out the answer.
5Ps3	Explore and solve number problems and puzzles, e.g. logic problems.
5Ps4	Deduce new information from existing information to solve problems.
5Ps5	Use ordered lists and tables to help to solve problems systematically.
5Ps6	Describe and continue number sequences, e.g. −30, −27, □□ −18...; identify the relationships between numbers.
5Ps9	Explain methods and justify reasoning orally and in writing; make hypotheses and test them out.

Prerequisites for learning

Learners need to:

- recognise and extend number sequences
- multiply and divide any number from 1 to 10 000 by 10 or 100 and understand the effect.

Success criteria

Learners can:

- find missing numbers in a number sequence, including sequences that extend into negative numbers

- investigate and work out the term-to-term rule that is applied in a number pattern, including sequences with negative numbers.

Vocabulary

negative number, positive number, step, sequence, count on, count back

Resources

Resource sheet 6: Squared paper (per pair); paper (per learner); ruler (per pair)

Refresh

- Choose the activity *Guess the steps*, from the Refresh activities.

Discover 🆂🅱 🖥

- Direct learners to the Student's Book and display **Slide 1**. Indicate the 10 on the temperature scale and count back with the class until you reach −5 °C. Continue to −10 °C.
- Discuss how temperature and money both use negative numbers.

Teach and Learn 🆂🅱 📊

- Discuss the text and example in the Student's Book. Display the **Number line tool**, scaled from −20 to 20. Point to 20. Ask learners to count back in fives. Stop at zero. Ask: **What happens to the sequence when we get to zero?** Accept comments. Establish that the numbers carry on beyond zero.
- Say: **A number less than zero is a negative number.** Continue the sequence: 0, −5, −10, −15, −20. Repeat for different start numbers and steps. Practise counting forwards from a negative number in steps of different size.
- On the board, write: 19, 13, 7, 1, □, □. Ask: **What are the next two terms in the sequence?** (−5, −11) Discuss solutions. Repeat for a forward sequence.

Practise 🆆🅱

- Workbook: Whole numbers page 18

Apply 👥 🖥

- Distribute Resource sheet 6: Squared paper. Display **Slide 2** and read the text to the class. Learners draw a section through the museum labelling the floors and the different displays on each floor. Ask questions such as: **How many levels are there altogether? Where are the books/ dinosaurs bones?**

Review 📊

- Display the **Thermometer tool**, set to 4 °C. Demonstrate readings below zero. Say: **Imagine that the temperature falls by two degrees every day. Write the temperature readings for a period of ten days.**

Assessment for learning

- A man enters a lift 13 floors underground. He ascends in steps of five floors. What floor will he be on after six steps? (17)

Number – Numbers and the number system

73

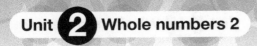

Lesson 2: **Positive and negative numbers**

Number – Numbers and the number system

Learning objectives

Code	Learning objective
5Nn9	Order and compare negative and positive numbers on a number line and temperature scale.

Strand 5: Problem solving

Code	Learning objective
5Pt1 / 5Ps1	Understand everyday systems of measurement in [length, weight, capacity,] temperature [and time] and use these to perform simple calculations.
5Pt7	Consider whether an answer is reasonable in the context of a problem.
5Ps2	Choose an appropriate strategy for a calculation and explain how they worked out the answer.
5Ps3	Explore and solve number problems and puzzles, e.g. logic problems.
5Ps6	Describe and continue number sequences, e.g. –30, –27, ☐☐ –18...; identify the relationships between numbers.

Prerequisites for learning

Learners need to:
- understand negative numbers
- mark a temperature on a thermometer.

Success criteria

Learners can:
- use a number line or temperature scale to order a set of numbers from –30 to 30.

Vocabulary

negative number, positive number, order, thermometer, temperature, degrees Celsius (°C)

Resources

mini whiteboard and pen (per learner); ruler (per pair)

Refresh

- Choose the activity *Up above, down below*, from the Refresh activities.

Discover [SB] 🖥

- Ask learners if they can think of any sports where the player with the lowest score wins. Take suggestions. Direct learners to the Student's Book and display **Slide 1**. Talk about how the scoring in a game of golf works.

Teach and Learn [SB] 📊

- Discuss the text and number line in the Student's Book. Talk about how the scale on a thermometer is like a vertical number line.
- Display the **Number line tool**, scaled from –20 to 20. Place pointers at –5 and 7. Ask: **Which of the two numbers is the greater?** (7) Establish that numbers on the left of the number line are smaller than numbers on the right. On the board, write: –5 < 7. Repeat for other positive and negative number pairs.
- On the board, write: 13, –3, 7, –11. Ask: **How would you order these numbers?** Elicit that they would place numbers on the number line and order them smallest to largest from left to right. Distribute paper. Ask learners to work in pairs. They draw a number line and order the numbers (–11, –3, 7, 13).

Practise [WB]

- Workbook: Positive and negative numbers page 20

Apply 👥 🖥

- Display **Slide 2**. Say: **Look at the average temperature in Helsinki, Finland over the past 12 months.** Ask learners to draw an appropriate number line on their whiteboards, mark the temperatures and establish the order. Invite learners to share their solutions.

Review 📊

- On the board, write a set of bank account balances: $27, $29, –$27, –$29. Ask learners to put them in order.
- Display the **Number line tool**, marked from –30 to 20. Highlight the numbers 16, –19, 19, –25, –8. Ask volunteers to put the numbers in order.

Assessment for learning

- Over a period of four days a bank account balance goes from –$25 to –$28. Is this good news for the account holder? Why?

Lesson 3: **Calculating temperature change**

Learning objectives

Code	Learning objective
5Nn9	Order and compare negative and positive numbers on a number line and temperature scale.
5Nn10	Calculate a rise or fall in temperature.

Strand 5: Problem solving

5Pt1 / 5Ps1	Understand everyday systems of measurement in [length, weight, capacity,] temperature [and time] and use these to perform simple calculations.
5Ps2	Choose an appropriate strategy for a calculation and explain how they worked out the answer.
5Ps3	Explore and solve number problems and puzzles, e.g. logic problems.
5Ps5	Use ordered lists and tables to help to solve problems systematically.
5Ps10	Solve a larger problem by breaking it down into sub-problems or represent it using diagrams.

Prerequisites for learning

Learners need to:

* use a number line or temperature scale to order a set of numbers from –30 to 30.

Success criteria

Learners can:

* select and use relevant number facts and mental strategies to work out differences between temperatures.

Vocabulary

negative number, positive number, thermometer, temperature, degrees Celsius (°C)

Resources

thermometer (per class); rulers (per pair)

Refresh

* Choose the activity *Temperature change*, from the Refresh activities.

Discover ⓢⓑ 🖥

* Show learners a thermometer. Ask: **Why do you think it has a negative scale?** Accept comments then display **Slide 1** and read the text in the Student's Book. Explain the function of different parts of the thermometer.

Teach and Learn ⓢⓑ 📊

* Display the **Thermometer tool**, with the scale set from –20 to 30 °C. Set the temperature at 11 °C. Say: **Tell your partner some important temperatures that you know and what happens when these temperatures are reached.** Ask: **Does anyone know the temperature at which water freezes?** (0 °C) **Water boils?** (100 °C) Say: **Temperature is measured in degrees Celsius. The scale is based on the freezing and boiling points of water, with 100 degrees between these temperatures.** Explain that when it is very cold temperatures fall below zero.
* Direct learners to the Student's Book and read the text. Work though the example as a class.
* Display two thermometers on the **Thermometer tool** set at –5 °C and 13 °C. Ask: **How would you work out the difference between these temperatures without counting along the scale?** Tell learners to think back to the example in the Student's Book. Complete the stages together: work out the difference between the positive value and 0 °C, then

the difference between 0 °C and the negative value, then add the differences together (18 degrees). Repeat for 25 °C and –19 °C (44 degrees).

* Set the temperature to –9 °C. Ask questions such as: **What would the temperature be if it rises/falls by [x] degrees? How much must it rise/fall to be [y] degrees?**

Practise 📒

* Workbook: Calculating temperature change page 22
* Refer to Activity 1 from the Additional practice activities.

Apply 👥 🖥

* Display **Slide 2** from Lesson 2. Learners write statements concerning the temperature difference between different months of the year.

Review 🖥

* Divide the class into two teams. Display **Slide 2.** Say: **This shows temperatures for different countries. If I travelled from country A to B/C to D ..., what would be the temperature change?** Learners must indicate rise or fall.

Assessment for learning

* Explain to your partner how you would work out the difference between a positive and a negative temperature reading.

Lesson 4: **Odds, evens and multiples (2)**

Learning objectives

Code	Learning objective
5Nn14	Make general statements about sums, differences and multiples of odd and even numbers.
Strand 5: Problem solving	
5Pt7	Consider whether an answer is reasonable in the context of a problem.
5Ps2	Choose an appropriate strategy for a calculation and explain how they worked out the answer.
5Ps3	Explore and solve number problems and puzzles, e.g. logic problems.
5Ps4	Deduce new information from existing information to solve problems.
5Ps8	Investigate a simple general statement by finding examples which do or do not satisfy it, e.g. the sum of three consecutive whole numbers is always a multiple of three.
5Ps9	Explain methods and justify reasoning orally and in writing; make hypotheses and test them out.

Prerequisites for learning

Learners need to:
• recognise odd and even numbers

Success criteria

Learners can:
• make general statements about the addition, subtraction and multiplication of odd and even numbers, based on evidence.

Vocabulary

odd, even

Resources

mini whiteboard and pen or paper (per pair);
15 cubes or counters (per pair)

Refresh

• Choose the activity *Odd and even guess*, from the Refresh activities.

Discover [SB]

• Direct learners to the Student's Book and discuss with learners what happens when you add odd and even numbers. Say: **What happens when you add 2 + 2? Is the answer an odd or even number? What about 3 + 3? Or 2 + 3?** Let learners work out some simple one-digit + or – one-digit questions. Ask: **Can you spot any patterns?**

Teach and Learn [SB]

• Ask learners to look at the example boxes in the Student's Book, which show all the possible combinations of the sum of odd and even numbers. Distribute paper to learners working in pairs. Ask them to test the combinations, writing and solving example sums. (More able learners may explore negative numbers.) Reconvene and discuss outcomes. Ask: **Are the results always the same?**

• Repeat for subtraction and multiplication combinations. Learners predict, test and discuss results.

Practise [WB]

• Workbook: Odd and even numbers page 24
• Refer to Activity 2 from the Additional practice activities.

Apply ▲▲

• Pairs predict, investigate and make general statements about sums of different combinations of three odd and even numbers, i.e. odd + odd + even; even + even + odd; odd + odd + odd and even + even + even.

Review

• On the board, draw the digits 1–9 at the top, with four boxes, connected by a circle, lower down. Led by the learners, select any four of the digits (all must be different) and put them in the boxes so the differences between connected boxes are odd. Ask: **What can you say about the sum of each pair of connected boxes?** (always odd) **What must you do to make the difference even?** (use even numbers)

Assessment for learning

• I multiply 1231 by 2437. Will the result be odd or even? How do you know? (odd)

Additional practice activities

Activity 1

Learning objective

• Count on and back in steps of a constant size and use this to work out the difference between 'start' and 'finish' numbers on a number line.

Resources

0–9 number cards from Resource sheet 2: 0–100 number cards (three sets per pair); number line (per pair)

What to do

• Learners shuffle and place sets of number cards face down in three piles.

• They take turns to turn over one card from each pile. The first card is the start number, the second is the step number and the third number is the number of steps.

• Learners count back from the start number in steps the given number of times.

• Learners work out the difference between the start and finish numbers, scoring an equivalent number of points. The winner is the learner with the higher score after five rounds.

Variation

Challenge 1 Use number cards 1–5 only.

Activity 2

Learning objective

• Calculate the difference of two temperatures, one positive and one negative, by breaking the problem into sub-problems to solve.

Resources

mini whiteboard and pen or paper (per pair); 1–6 dice or Resource sheet 4: 1–6 spinner (per pair)

What to do

• Draw a three row table with rows labelled 'Indoor temperature', Outdoor temperature' and 'Temperature difference' and column headings for 'Day 1' to 'Day 7'.

• Roll the dice three times and add the numbers rolled. Insert the number under Day 1 for 'Indoor temperature'. Continue to roll the dice and add the sum of three rolls to the table cells for Days 2–7.

• Do the same for the second row, Outdoor temperature, but making each number negative.

• Say: **Before calculating each temperature change, predict whether the difference will be odd or even. Then calculate and check your predictions.**

Variation

Challenge 3 Learners fill the table with the sums of five dice rolls.

Number – Numbers and the number system

Unit 3: Decimals 1

Learning objectives

Code	Learning objectives
5Nn4	Use decimal notation for tenths and hundredths and understand what each digit represents.
5Nn5	Multiply and divide any number from 1 to 10 000 by 10 or 100 and understand the effect.
5Nn16	Recognise equivalence between the decimal and fraction forms of halves, tenths and hundredths and use this to help order fractions, e.g. 0.6 is more than 50% and less than $\frac{7}{10}$.

Strand 5: Problem solving	
5Pt7	Consider whether an answer is reasonable in the context of a problem.
5Ps3	Explore and solve number problems and puzzles, e.g. logic problems.
5Ps4	Deduce new information from existing information to solve problems.
5Ps8	Investigate a simple general statement by finding examples which do or do not satisfy it [, e.g. the sum of three consecutive whole numbers is always a multiple of three].
5Ps9	Explain methods and justify reasoning orally and in writing; make hypotheses and test them out.
5Ps10	Solve a larger problem by breaking it down into sub-problems or represent it using diagrams.

Unit overview

In this unit, learners write decimals in tenths and hundredths and relate them to common fractions. They learn to construct a place value chart as a quick reference tool and correctly place a decimal number on a number line. They are expected to read a number with decimal place values up to hundredths. Reference is made to the relationship between decimals and dollars and cents. The unit then helps consolidate understanding of powers of ten and the magnitude of these. Learners investigate multiplication and division of any number from 1 to 10 by 10 or 100 and note the effect on the digits, moving them left or right a number of place positions. Learners apply their knowledge of multiplication and division by powers of ten to problems involving the measurement.

Common difficulties and remediation

Before teaching the unit, establish that learners have a secure understanding of the role of the decimal point in separating the whole number part from the fraction. It is equally important for learners to be comfortable with decimal notation and the language of tenths – reading 2·8 as 'two ones/units and eight tenths' and not just as the formulaic 'two point eight'.

Avoid telling learners to 'add a zero' when multiplying by 10 or 'add two zeros' when multiplying by 100. This could cause problems at Stage 6, when learners multiply and divide using decimals. Encourage learners to view zeros as placeholders, 'holding' a place when digits move.

Look out for learners who move digits in the wrong direction when multiplying or dividing. They are not yet secure in understanding that multiplication makes numbers bigger and division makes them smaller, and will need more experience of comparing and ordering numbers by place value.

Promoting and supporting language

At every stage, learners require mathematical vocabulary to access questions and problem-solving exercises. If appropriate, when a new key word is introduced, ask learners to write a definition in their books, drawing a box around it for emphasis. Encourage learners to write the definition in their own words, for example: 'A hundredth is what you get if you divide a tenth into ten equal parts.' Sketching a diagram alongside will help embed the definition.

Encourage frequent usage of mathematical language to help embed vocabulary. For example, model use of the question: 'How many times bigger/smaller do you expect the number to be after multiplying/dividing by 10/100?' when learners are considering movement of digits across place value positions for multiplication and division. This will help them make the transition to mathematically precise language.

Lesson 1: **Tenths**

Learning objective

Code	Learning objective
5Nn4	Use decimal notation for tenths [and hundredths] and understand what each digit represents.
5Nn16	Recognise equivalence between the decimal and fraction forms of halves, tenths and hundredths and use this to help order fractions, e.g. 0.6 is more than 50% and less than $\frac{7}{10}$.

Strand 5: Problem solving

5Ps3	Explore and solve number problems and puzzles, e.g. logic problems.
5Ps4	Deduce new information from existing information to solve problems.

Prerequisites for learning

Learners need to:

- understand fractions as equal parts of a whole
- understand the equivalence between decimals and fractions in tenths.

Success criteria

Learners can:

- write $\frac{1}{10}$ as 0·1

- recognise decimals and fractions that are equivalent, for example 0·3 and $\frac{3}{10}$
- use decimal notation to write numbers such as 1.1 and 2.3

Vocabulary

fraction, decimal fraction, tenths

Resources

Resource sheet 7: Equivalence snap (per pair)

Refresh

- Choose the activity *Fraction finder*, from the Refresh activities.

Discover 🖥 SB

- Discuss the image and the text in the Student's Book. On the board, write: 0.7 kg, 0.4 kg, 0.6 kg. Ask: **Which is the greatest mass?** Ask: **How do you know these masses are different from 7 kg, 4 kg, 6 kg?**

Teach and Learn 📊 SB

- Discuss the text and Example in the Student's Book. Ask learners to give examples of where they have seen tenths used in the real world. Examples include measurements of length/distance/height and mass.
- Display the **Number line tool**, set from 0 to 1, with ten divisions of 0·1 (no labels). Say: **The number line from zero to one has been divided into ten equal parts. What does each part represent?** Elicit that each part is a tenth. Ask volunteers to label divisions with corresponding fractions. Say: **Tenths can also be expressed in another way – as decimals.** Click to reveal decimal tenths labels.
- Explore similarities between decimal and fraction notation for tenths. Say: **Complete the following sentence: 'The number following the decimal point in a decimal tenth tells us …'** (how many tenths the value has)
- Remind learners that each place value position is ten times smaller than the one to its left. Say: **The value of the tenths position is ten times smaller than that of the units.** Display the **Fractions tool**, set to

ten parts. Colour three parts. Ask: **What fraction does the shaded part represent?** Elicit that it is three tenths. Write $\frac{3}{10}$ on the board. Ask a volunteer to write the decimal equivalent. Repeat for 0·7, 0·5, 0·9. Confirm answers by clicking 'show decimal'.

Practise 📒

- Workbook: Tenths page 26

Apply 👥 🖥

- Display **Slide 1**. Learners work together to write and order the given mass measurements as decimals.
- If time allows, learners can then play a game of 'equivalence snap' using the cards from Resource sheet 7.

Review 📊

- Display the **Fractions tool**, set to ten divisions (no fraction label). On the board, write $\frac{3}{10}$. Ask a volunteer to shade the correct number of parts (three tenths) using the digital tool and express the fraction in decimal notation. Repeat for five parts and nine parts shaded.
- Display the **Number line tool**, set from 0 to 2 in tenths (no labels). Set the pointer to 1·3. Ask: **What decimal number is indicated by the pointer?** Repeat for 1·8 and 1·5.

Assessment for learning

- Here are three decimal numbers: 0·3, 0·7, 1·4. What does the digit after the decimal point represent?

Unit 3 Decimals 1

Lesson 2: **Hundredths**

Learning objective

Code	Learning objective
5Nn4	Use decimal notation for tenths and hundredths and understand what each digit represents.

Strand 5: Problem solving

5Pt7	Consider whether an answer is reasonable in the context of a problem.
5Ps3	Explore and solve number problems and puzzles, e.g. logic problems.
5Ps4	Deduce new information from existing information to solve problems.
5Ps9	Explain methods and justify reasoning orally and in writing; make hypotheses and test them out.

Prerequisites for learning

Learners need to:
- have an understanding of fractions and the role of the denominator
- understand the equivalence between decimals and fractions in hundredths.

Success criteria

Learners can:
- write $\frac{3}{100}$ as 0·03

- use decimal notation to write numbers such as 0.47, 0.99 and 1.07
- count on in hundredths from any hundredths fraction.

Vocabulary

fraction, decimal fraction, hundredths, denominator

Resources

paper (per pair)

Refresh

- Choose an activity from Number – Numbers and the number system in the Refresh activities.

Discover [SB]

- Discuss the picture and the text in the Student's Book. Ask learners to try and remember the cost of the last thing they bought. Did it cost a whole number of dollars, or was it a mixture of dollars and cents?. On the board, write: $0.46, $0.64, $0.60. Ask: **Which is the highest price? How do you know?**

Teach and Learn [📊] [SB]

- Discuss the text and Example in the Student's Book. Display the **Number line tool**, set from 0 to 1 in increments of 0·1, sub-divide: 0·01. Say: **Each interval of 0·1 can be divided into ten parts. What do we call each part?** Elicit that the parts are 'hundredths' and comparable to money: a dollar is divided into 100 parts, called 'cents'.
- Display the **Fractions tool**, set to 100 parts, alongside the **Number line tool**. Shade one part, read the fraction: $\frac{1}{100}$ Show the decimal: 0·01. Remind learners that the first digit to the right of the decimal point is the tenths digit. Say: **The second digit to the right of the decimal point is the hundredths digit.**
- Point out that numbers one hundredth, two hundredths, and so on, up to nine hundredths, have no digits in the tenths place and are written with a

placeholder zero in the tenths place. Say: **Numbers greater than nine hundredths do not need a zero as a placeholder.** Count from 0.1 to 0·30 and highlight the decimal digits. Return to the number line. Point to 0·34. Ask a learner to write the number in this position in decimal notation. Repeat for other decimals.

Practise [WB]

- Workbook: Hundredths page 28
- Refer to Activity 1 from the Additional practice activities.

Apply [👥]

- Learners take turns to count from 9.01 to 9.99 in hundredths. Listen as they count, intervening if they have difficulty. Ask: **What happens to the count after 9·99?**

Review [📊]

- Display the **Fractions tool**, set to 100 divisions (no fraction label). On the board, write $\frac{7}{100}$. Ask a volunteer to shade the correct number of parts and express the fraction in decimal notation. Repeat for 17 parts and 23 parts shaded.

Assessment for learning

- Here are three decimal numbers: 0·03, 0·33, 1·61. What do the two digits after the decimal point represent?

Lesson 3: **Multiplying by 10 or 100**

Learning objective

Code	Learning objective
5Nn5	Multiply [and divide] any number from 1 to 10 000 by 10 or 100 and understand the effect.

Strand 5: Problem solving

5Pt7	Consider whether an answer is reasonable in the context of a problem.
5Ps3	Explore and solve number problems and puzzles, e.g. logic problems.
5Ps10	Solve a larger problem by breaking it down into sub-problems or represent it using diagrams.

Prerequisites for learning

Learners need to:

• understand the place value of decimals as tenths and hundredths.

Success criteria

Learners can:

• multiply a number by 10 by moving digits one place value to the left and multiply by 100 by moving digits two place values to the left.

Vocabulary

units, tenths, hundredths, multiply, place value

Resources

mini whiteboards and marker pens (per learner); coloured pencils: yellow, blue, green, red (per learner); spiked abacus (per group)

Refresh

• Choose an activity from Number – Numbers and the number system in the Refresh activities.

Discover SB

• Ask learners to look at the picture in the Student's Book. Discuss the problem. Ask: **How would you use multiplication to solve this problem?** Accept suggestions and discuss strategies.

Teach and Learn SB 📊 👥

• Discuss the text and Example in the Student's Book. Give groups of learners a spiked abacus and let them model the examples you are demonstrating in the lesson.

• Review multiplying by 10 or 100. On the board, write: 23 × 10 × 10, 23 × 100. Say: **Solve the multiplications.** Allow a short time for learners to respond. Ask: **Why do you think both questions give the same answer?** Elicit that multiplying by 100 is the same as multiplying by ten, then ten again. Check the answers. Ask: **What happens when a number is multiplied by ten?** Establish that when we multiply by ten, all the digits in the number move one place to the left and become ten times larger.

• Ask: **What happens when a number is multiplied by 100?** Draw a place value grid on the board. Establish that when we multiply by 100, all the digits in the number move two places to the left and become 100 times larger.

• Say: **Now let's think about multiplication of larger numbers.** Ask: **How would you use the grid to show 508 multiplied by ten or 100?** Ask a volunteer

to demonstrate movement of the digits one place left for multiplication by ten (5080) and two places left for multiplication by 100 (50 800). Confirm that the number becomes 10/100 times larger. Repeat for 7077 multiplied by ten or 100.

• Display the **Function machine tool**, set up with 4863 and '× 10'. Ask: **What will the output be?** (48 630) Repeat for '× 100' (486 300).

• On the board, draw a table with columns labelled '× 10' and '× 100' and rows labelled: 76, 564 and 3409. Ask learners to work in pairs to complete the table.

Practise WB

• Workbook: Multiplying by 10 or 100 page 30

Apply 🖥

• Display **Slide 1**. Ask: **How would you use your knowledge of multiplying by ten or 100 to convert these measurements?** Remind learners of the relationships between millimetres, centimetres and metres.

Review

• Review multiplying by ten and 100. Write 8041 × 10 on the board. Learners discuss the method and solve the problem. Repeat for 5733 × 100. Say: **A number divided by 100 is 9158. What is the number?** (915 800) Discuss solutions.

Assessment for learning

• Here is the number 3609. Show me how to multiply it by 100

Number – Numbers and the number system

Lesson 4: **Dividing by 10 or 100**

Number – Numbers and the number system

Learning objectives

Code	Learning objectives
5Nn4	Use decimal notation for tenths and hundredths and understand what each digit represents.
5Nn5	[Multiply and] divide any number from 1 to 10 000 by 10 or 100 and understand the effect.

Strand 5: Problem solving

5Pt7	Consider whether an answer is reasonable in the context of a problem.
5Ps3	Explore and solve number problems and puzzles, e.g. logic problems.
5Ps9	Explain methods and justify reasoning orally and in writing; make hypotheses and test them out.
5Ps10	Solve a larger problem by breaking it down into sub-problems or represent it using diagrams.

Prerequisites for learning

Learners need to:

• understand the place value of decimals as tenths and hundredths.

Success criteria

Learners can:

• divide by ten by moving digits one place value to the right

• divide by 100 by moving digits two place values to the right.

Vocabulary

units, tenths, hundredths, divide, place value

Resources

green and red coloured pencils (per learner); spiked abacus (per group)

Refresh

• Choose an activity from Number – Numbers and the number system in the Refresh activities.

Discover SB

• Discuss the problem in the Student's Book. Ask: **How would you use division to work this out?** Accept suggestions and discuss strategies. Work through the calculation with the learners.

Teach and Learn SB

• Read the text in the Student's Book with the class and check the Examples.

• Review dividing by ten or 100. Give groups of learners a spiked abacus and let them model the examples you are demonstrating in the lesson.

• On the board, write: 4400 ÷ 10 ÷ 10, 4400 ÷ 100. Say: **Solve the divisions.** Elicit that dividing by 100 is the same as dividing by ten, then ten again. Review the answers. Ask: **What happens when a number is divided by ten or 100?** Establish that the number becomes ten (or 100) times smaller and the digits move one (or two) places to right.

• Say: **We know 4400 divided by 100 is 44.** Ask: **What happens if we divide 44 by ten or 100?** Accept comments. On the board, draw a place value grid with columns through to decimals. Put 44 in the tens and units columns and, in the next row, move the digits one column to the right. Say: **44 has now become ten times smaller. It is now 4·4.** In the next row, move 4·4 one place to the right. Say: **44 has now become 0·44.** Ask: **How many times smaller**

is 0·44 than 44? (100) Write 6 in the grid. Say: **If we divide 6 by 100, it will move two places to the right.** Explain that 6 will move across the decimal point to the hundredths position, but it cannot be on its own; it needs a zero in the tenths position as a placeholder.

• Display **the Function machine tool**, set up with 301 and '÷ 10'. Ask: **What will the output be?** (30·1) Repeat for '÷ 100' (3·01). Repeat for 7707 and '÷ 10', '÷ 100'.

Practise WB

• Workbook: Dividing by 10 or 100 page 32

• Refer to Activity 2 from the Additional practice activities.

Apply 🖥

• Display **Slide 1**. Ask: **How would you use your knowledge of dividing by ten or 100 to convert these measurements?** Remind learners of the relationships between millimetres, centimetres and metres.

Review

• Write 457 ÷ 10 on the board. Learners discuss the method and solve the problem. Repeat for 9875 ÷ 100. Say: **A number multiplied by 100 is 24 300. What is the number?** (243)

Assessment for learning

• Here is the number 3609. Show me how to divide by 10 or 100.

Additional practice activities

Activity 1

Learning objective

• Use knowledge of decimal notation and place value to order a set of decimal numbers.

Resources

0–9 cards from Resource sheet 2: 0–100 number cards (two sets per pair); mini whiteboard and pen or paper (per pair)

What to do

• On the board, draw a place value grid with columns: units, tenths, hundredths. Mark the decimal point on the line between units and tenths. Write 0·86 and 0·85 in the grid, circle the hundredths and write 0·86 > 0·85 to the side of the grid.

• Say: **You can use a place value grid to order two-digit decimals. As with comparing whole numbers, we move left to right across the place value columns, comparing digits until we find digits of different values. The greater the value of the digit, the greater the number.**

• Learners copy the grid from the board and place a zero in the units column for five of the rows.

• They place the digit cards face down in two piles. They create five two-digit decimals less than one by picking cards: the first card is tenths, the second is hundredths. They write the digits in the grid, then order the decimals.

Variation

 Learners create five one-digit decimals and order the numbers.

Activity 2

Learning objective

• Use a combination of multiplication and division by 10 and 100 to determine the number of steps it takes to convert one number into another, for example: 78·4 to 784 000.

Resources

mini whiteboard and pen or paper (per pair)

What to do

• On the board, write: '7 to 70 000', '3000 to 0.3', '57 to 57 000', '900 to 0.09'.

• Ask: **Using only combinations of '× 10', '× 100' '÷ 10' and '÷ 100', what steps does it take to convert one number to the other?**

• Learners find solutions and discuss them with the class.

• They construct and write four new problems, swapping with another pair to solve.

Variations

 Learners work with '× 10' and '÷ 10' only.

Learners work with '× 100' and '÷ 100' only.

Number – Numbers and the number system

83

Unit 4: Decimals 2

Learning objectives

Code	Learning objective
5Nn4*	Use decimal notation for tenths and hundredths and understand what each digit represents.
5Nn7	Round a number with one or two decimal places to the nearest whole number.
5Nn11	Order numbers with one or two decimal places and compare using the > and < signs.

*Learning objective revised and consolidated.

Strand 5: Problem solving

5Pt7	Consider whether an answer is reasonable in the context of the problem.
5Ps3	Explore and solve number problems and puzzles, e.g. logic problems.
5Ps4	Deduce new information from existing information to solve problems.
5Ps8	Investigate a simple general statement by finding examples which do or do not satisfy it [, e.g. the sum of three consecutive whole numbers is always a multiple of three].
5Ps9	Explain methods and justify reasoning orally and in writing; make hypotheses and test them out.
5Ps10	Solve a larger problem by breaking it down into sub-problems or represent it using diagrams.

Unit overview

In this unit, learners review use of decimal notation for tenths and hundredths and the place value of digits in these positions. To consolidate understanding of place value, they practise writing numbers with one or two decimal places in expanded form. They use number lines and place value grids to compare the tenths and hundredths in pairs of numbers and use > and < signs to record comparisons. They extend this to ordering sets of numbers with one or two decimal places, adopting a systematic approach to find the greater decimal by comparing digits in different place value positions in the numbers, from left to right, starting with the largest place. The unit concludes with learners rounding decimals with one or two decimal places to the nearest whole number. Learners discuss similarities with strategies for rounding whole numbers.

Common difficulties and remediation

Learners may develop several misconceptions or difficulties about comparing and ordering decimal numbers. Some learners believe the longer the decimal, the larger the number. They treat the decimal part as a whole number, for example thinking that 4·13 > 4·2 because 13 > 2. Remediation exercises can be useful, including reading a decimal number and writing it in the expanded form, so learners are reminded of the importance of the position of the digits. Reading numbers from a number line helps learners establish links with graduated scales and prepares them for activities involving ordering of

decimals. Learners should be encouraged to read 4·7 as 'four units and seven tenths', not just as 'four point seven'. This gives a vocal indication of the delineation between whole numbers and fractional parts, in the same way that the decimal point separates the 'whole' from the 'parts' visually. Counting on and back in steps along decimal number lines helps familiarise learners with place value. It is important for learners to explore decimal numbers in context and compare the size of numbers. For example: Which is the greater amount, $0.85 or $0.90?

Promoting and supporting language

At every stage, learners require mathematical vocabulary to access questions and problem-solving exercises. If appropriate, when a new key word is introduced, ask learners to write a definition in their books, drawing a box around it for emphasis. Encourage learners to write the definition in their own words, for example: 'When rounding, you are finding the closest multiple of 10 (or 100, or other place value) to your number.' Sketching a diagram alongside will help embed the definition.

Encourage frequent use of mathematical language to help embed vocabulary. For example, model use of the question: 'How many decimal places do you think the number will have?' when learners are considering rounding one- or two-digit decimals. This will help them make the transition to mathematically precise language.

Lesson 1: **Tenths and hundredths**

Learning objectives

Code	Learning objective
5Nn4*	Use decimal notation for tenths and hundredths and understand what each digit represents.

Strand 5: Problem solving

5Pt7	Consider whether an answer is reasonable in the context of a problem.
5Ps3	Explore and solve number problems and puzzles, e.g. logic problems.
5Ps4	Deduce new information from existing information to solve problems.
5Ps10	Solve a larger problem by breaking it down into sub-problems or represent it using diagrams.

Prerequisites for learning

Learners need to:

- use decimal notation to write numbers, such as 1.1, 2.3, 0.47, 0.99, 1.07
- count up and down in tenths or hundredths.

Success criteria

Learners can:

- demonstrate their understanding of decimal place value by writing a decimal in expanded form.

Vocabulary

decimal fraction, tenth, hundredth, place value, expanded form

Resources

Resource sheet 8: Blank place value grid (decimals) (per learner); mini whiteboard and pen (per learner)

Refresh

- Choose an activity from Number – Numbers and the number system in the Refresh activities.

Discover [SB]

- Read the text in the Student's Book and discuss the problem with learners. Accept comments. Remind the learners of the similarity between decimal notation and money. Encourage them to think how the price of an item provides clues to the notes and coins they could use to pay for it.

Teach and Learn [SB] [graph]

- Read the text in the Student's Book with the class and work through the Examples to check they are correct.
- Display the **Number line tool**, set from 0 to 1 with increments 0·1, subdivisions 0·01. Say: **Look at the number 0·37.** Set the pointer to 0·37. On the board, draw a place value grid with units, tenths and hundredths. Write 0·37 on the grid. Say: **You can see from the chart that 0·37 has three tenths and seven hundredths.** Ask: **How would you write 0·37 as the sum of its separate place values – tenths and hundredths?**
- Give learners a copy of Resource sheet 8: Blank place value grid (decimals) for them to model the examples that you demonstrate in this lesson. Display the **Place value tool** and construct 0·37. Demonstrate the component place values by splitting the number into 0·3 and 0·07. On the board,

write: 0·37 = 0·3 + 0·07. Say: **This is called the expanded form, because it is written as a sum of the different parts that make up the number.**
- Repeat for 0·89 and 0·05. Point out the zero in 0·05, explaining that zero tenths is a placeholder to 'hold' the tenths position. On the board, write: 0·42 = ☐ + ☐ and 0·61 = ☐ + ☐. Ask learners to work in pairs to complete these expanded forms.

Practise [WB]

- Workbook: Tenths and hundredths page 34

Apply [icon]

- Display **Slide 1**. Learners write the prices in expanded form on their whiteboards.

Review [graph]

- Display the **Place value tool**. Without learners seeing, create some three-digit numbers with two decimal places, such as 2·37. Learners write the numbers in the expanded form. (2·37 = 2 + 0·3 + 0·07)
- Then, show some three-digit numbers with two decimal places expanded as separate place values. Learners write them recomposed as one number, using correct decimal notation.

Assessment for learning

- Here are three decimal numbers: 1·4, 99·47 and 2036·43. Show me that all three numbers have the same number of tenths.

Number – Numbers and the number system

85

Lesson 2: **Comparing decimals**

Learning objectives

Code	Learning objective
5Nn11	Order numbers with one or two decimal places and compare using the > and < signs.

Strand 5: Problem solving	
5Pt7	Consider whether an answer is reasonable in the context of a problem.
5Ps3	Explore and solve number problems and puzzles, e.g. logic problems.
5Ps10	Solve a larger problem by breaking it down into sub-problems or represent it using diagrams.

Prerequisites for learning

Learners need to:
- compare two 3- or 4-digit numbers, using the < and > signs.

Success criteria

Learners can:
- compare the tenths and hundredths in a number by placing them on a number line or place value grid

- decide which is the greater of two decimals that have equal whole numbers by comparing digits in each decimal place position.

Vocabulary

decimal, decimal place, place value

Resources

Resource sheet 8: Blank place value grid (decimals) (per learner)

Refresh

- Choose an activity from Number – Numbers and the number system in the Refresh activities.

Discover 〔SB〕

- Ask: **Given two decimals, how would you work out which was the smaller number?** Ask learners to share strategies. Look at the text and image in the Student's Book. Say: **Name some other real-world examples in which you are asked to compare decimal values.**

Teach and Learn 〔📊〕〔SB〕

- Display the **Number line tool**, from to 7 to 8, with increment 0·1. Position pointers at 7·3 and 7·6. Ask: **How do you know which of these decimal numbers is the greater?** Learners share their strategies. Say: **You can use the number line to compare the numbers, but another strategy is to use a place value grid.**
- Draw a place value grid on the board with columns for units, tenths and hundredths and give learners a copy of Resource sheet 8, so they can work through the examples you demonstrate during the lesson. Write 7·3 and 7·6 in the grid, aligning the place values in columns. Say: **To find the greater number, compare the different place values in the two numbers, from left to right.** Highlight the units column.
- Discuss the text and example in the Student's Book. Say: **The units have the greatest place value. The two numbers have the same amount of units so move to the tenths.** Highlight the tenths column. Say: **7·6 has more tenths than 7·3, so 7·6 is greater.**

- On the board, write: 7·6 > 7·3. Change the increments on the tool to 0·01. Position pointers at 7·83 and 7·81. Demonstrate the comparison with the place value grid, working from left to right.
- On the board, write 7·81 < 7·83 then repeat the above for another pair of decimals.

Practise 〔WB〕

- Workbook: Comparing decimals page 36
- Refer to Activity 1 from the Additional practice activities.

Apply 〔👥〕〔🖥️〕

- Display **Slide 1** and read the text with the the learners. Give learners a copy of Resource sheet 8. Allow learners to discuss the problem and strategies to solve it. Elicit that writing decimals in a place value grid will help establish that Shop B has the better offer.

Review

- On the board, write several number pairs: 0·08, 0·04; 2·75, 2·72; 43·26; 43·29. Ask learners to compare the numbers in each pair, writing number statements and using the correct symbol of comparison, < or >. They share answers and explain solutions.

Assessment for learning

- Here are two decimals: 4·34 and 4·32. Explain how you would find which is the greater number.

Lesson 3: **Ordering decimals**

Learning objectives

Code	Learning objective
5Nn11	Order numbers with one or two decimal places and compare using the > and < signs.

Strand 5: Problem solving

5Pt7	Consider whether an answer is reasonable in the context of a problem.
5Ps3	Explore and solve number problems and puzzles, e.g. logic problems.
5Ps10	Solve a larger problem by breaking it down into sub-problems or represent it using diagrams.

Prerequisites for learning

Learners need to:

• compare the tenths and hundredths in a number by placing them on a number line or place value grid.

Success criteria

Learners can:

• order and compare numbers with one or two decimal places.

Vocabulary

decimal, decimal place, place value

Resources

mini whiteboard and pen (per pair); Resource sheet 8: Blank place value grid (decimals) (per learner)

Refresh

• Choose an activity from Number – Numbers and the number system in the Refresh activities.

Discover [SB]

• Direct learners to the text and picture in the Student's Book. Ask: **Why is it important to be able to order decimal numbers?** Accept comments and suggestions. Ask: **How would you order a set of decimal numbers?**

Teach and Learn [chart] [SB]

• Discuss the text and Example in the Student's Book working through each numbered step as a class. Display the **Number line tool**, set from 4 to 5, with increment 0·1 and subdivisions 0·01. On the board, write: 4·77, 4·29, 4·73, 4·22, 4·75. Say: **Use the number line to order these numbers, from largest to smallest.** After a couple of minutes, ask learners to discuss and confirm the order.

• Give learners a copy of Resource sheet 8 and ask them to write the numbers in it, aligning the decimal points. Say: **You will remember how to use a grid to compare two numbers. You can also use it to compare and order larger sets.**

• Indicate the numbers on the board and say: **To find the largest decimal you compare different place values in the numbers, working from left to right.** Highlight the units column. Say: **The units have the largest place value. All these numbers have the same amount of units so move to the tenths.**

Highlight the tenths column. Say: **As 4·77, 4·73 and 4·75 have more tenths than 4·29 and 4·22, 4·29 and 4·22 are smaller, so focus on the larger numbers.** Explain that the larger numbers all have the same amount of tenths, so move to hundredths.

• Establish that 4·77 > 4·75 > 4·73 and record the order. Say: **Look again at 4·29 and 4·22. They have the same number of tenths so move to the hundredths.** Establish that 4·29 > 4·22. Record: 4·77, 4·75, 4·72, 4·29, 4·22.

Practise [WB]

• Workbook: Ordering decimals page 38

Apply [pair icon] [screen icon]

• Display **Slide 1**. Allow learners to discuss the problem and strategies to solve it. Ask: **Who won the race?** (B) **Who came last?** (A) Agree the order: B, D, E, C, A.

Review [screen icon]

• Display **Slide 2**. Say: **The table shows shop prices for the same digital camera.** Ask: **Which shop offers the best price?** (B) **Which is most expensive?** (C) **Which has the middle price?** (E)

Assessment for learning

• Here are five measurements in kilograms: 23·51, 22·87, 23·15, 22·78, 23·13. Explain how you would order them, from smallest to largest.

Number – Numbers and the number system

87

Unit **4** Decimals 2

Lesson 4: **Rounding decimals**

Number – Numbers and the number system

Learning objectives

Code	Learning objective
5Nn7	Round a number with one or two decimal places to the nearest whole number.

Strand 5: Problem solving

Code	Learning objective
5Ps3	Explore and solve number problems and puzzles, e.g. logic problems.
5Ps8	Investigate a simple general statement by finding examples which do or do not satisfy it, [e.g. the sum of three consecutive whole numbers is always a multiple of three].
5Ps9	Explain methods and justify reasoning orally and in writing; make hypotheses and test them out.
5Ps10	Solve a larger problem by breaking it down into sub-problems or represent it using diagrams.

Prerequisites for learning

Learners need to:
* round 3- and 4-digit numbers to the nearest ten or 100.

Success criteria

Learners can:
* round a number by finding the rounding digit and know that for rounding to the nearest whole number this is the units place

* look at the digit to the right of the rounding digit and say that if its value is less than five, the rounding digit is left unchanged and if it is five or more, one is added to the rounding digit.

Vocabulary

round, decimal, one decimal place, two decimal places, hundredths, tenths

Resources

mini whiteboard and pen (per pair); 1–6 dice, or Resource sheet 4: 1–6 spinners (per pair)

Refresh

* Choose an activity from Number – Numbers and the number system in the Refresh activities.

Discover [SB]

* Ask learners to look at the text and pictures in the Student's Book. Discuss the question. Ask: **What is the advantage of rounding numbers in this way?** Discuss the advantages. Ask: **Where else might rounding of decimals be useful?**

Teach and Learn [📊] [SB]

* Review rounding whole numbers to the nearest ten. On the board, write: 24, 25, 26. Ask: **How would you round each number to the nearest 10?** Display the **Number line tool**, from 20 to 30. Mark 24 on it. Ask: **Between which two multiples of ten is 24?** (20 and 30) **Which is it closer to?** (20) Say: **You round 24 to 20, as it is closer to this multiple of ten.** Repeat for 25 (30) and 26 (30).

* Say: **Rounding decimals to the nearest whole number is similar to rounding whole numbers.** Set the Number line tool from 2 to 3, with increments of 0·1. On the board, write: 2·4, 2·5, 2·6. Set the pointer to 2·4. Ask: **Which whole number is 2·4 nearest to?** (2) Say: **2·4 is rounded to 2 as this is the whole number it is closest to.** Repeat for 2·5 (3), 2·6 (3).

* Highlight the tenths digits of the numbers on the board. Say: **You will remember that when rounding off a whole number to the desired place value, the rounding digit is the digit at that place value. When rounding a decimal number to the nearest whole**

number, the rounding digit is in the units place. Look at the digit that is one place to the right of the rounding digit, the tenths, and follow the normal rounding rules. Remind learners of the 'round up or down' rules by referring to the Student's Book.

* Demonstrate more rounding using the rounding rules, for numbers such as 8·08, 8·49, 8·72. Stress that, because the rounding digit is in the units place, the focus remains on tenths; thus the hundredths digit is ignored when rounding to the nearest whole.

Practise [WB]

* Workbook: Rounding decimals page 40
* Refer to Activity 2 from the Additional practice activities.

Apply [👥] [🖥]

* Display **Slide 1**. Allow learners to discuss the problem. Agree that numbers can be rounded to make addition easier. Learners round and add the numbers. (29 m)

Review [👥]

* Distribute the dice and ask learners to use them to create four 4-digit numbers, each with two decimal places e.g. 32.46. Say that the numbers represent prices. Say: **Write a shopping list with items rounded to the nearest whole.**

Assessment for learning

* Here is the number 48·49. Show me two methods for rounding this decimal to the nearest whole number.

Additional practice activities

Activity 1

Learning objective
- Compare numbers with two decimal places in a shopping context.

Resources
price list of supermarket food items (per class); paper (per learner)

What to do
- Display a list of supermarket food items with prices in dollars and cents.
- Learners make their own list of items, writing two prices alongside each item: one is the original price; the other is the price plus multiples of 10 cents. The two prices should be labelled 'offer A' and 'offer B' and varied in order.
- They swap papers with another pair and find the best offer for each food item.
- Learners write their solutions in the form A > or < B.

Variation
Challenge 3 Learners calculate three prices per item by adding and subtracting a few cents to or from the original price.

Activity 2

Learning objectives
- Order and round numbers in the context of length measurements.
- Investigate whether rounding changes the numerical order of a set of measurements.

Resources
objects from around the classroom with straight edges (per class); mini whiteboard and pen or paper (per pair); metre ruler (per pair)

Variation
Challenge 1 Provide learners with a set of measurements in centimetres for them to work with.

What to do
- Learners measure and record five objects from around the classroom that they estimate to be between 1 cm and 1 metre (100 cm) in length.
- They divide each number by 100 to express measurements in metres, creating numbers with two decimal places.
- Say: **Order the five decimal measurements and then round each figure.**
- Ask: **Does the order change when the numbers are rounded? Why?**

Unit 5: Fractions

Learning objectives

Code	Learning objective
5Nn15	Recognise equivalence between: $\frac{1}{2}$, $\frac{1}{4}$ and $\frac{1}{8}$; $\frac{1}{3}$ and $\frac{1}{6}$; $\frac{1}{5}$ and $\frac{1}{10}$.
5Nn16	Recognise equivalence between the decimal and fraction forms of halves, tenths and hundredths and use this to help order fractions, e.g. 0·6 is more than 50% [half] and less than $\frac{7}{10}$.
5Nn17	Change an improper fraction to a mixed number, e.g. $\frac{7}{4}$ to $1\frac{3}{4}$; order mixed numbers and place between whole numbers on a number line.
5Nn18	Relate finding fractions to division and use to find simple fractions of quantities.

Strand 5: Problem solving	
5Pt2	Solve single [and multi-]step word problems (all four operations); represent them, e.g. with diagrams or a number line.
5Pt7	Consider whether an answer is reasonable in the context of a problem.
5Ps2	Choose an appropriate strategy for a calculation and explain how they worked out the answer.
5Ps3	Explore and solve number problems and puzzles, e.g. logic problems.
5Ps4	Deduce new information from existing information to solve problems.
5Ps8	Investigate a simple general statement by finding examples which do or do not satisfy it [, e.g. the sum of three consecutive whole numbers is always a multiple of three].
5Ps9	Explain methods and justify reasoning orally and in writing; make hypotheses and test them out.
5PS10	Solve a larger problem by breaking it down into sub-problems or represent it using diagrams.

Unit overview

In this unit, learners construct and use diagrams to identify equivalent fractions, learning to recognise equivalence between halves, quarters and eighths; thirds and sixths; and fifths and tenths. They extend their knowledge of equivalence by investigating the relationship between decimal and fraction forms of halves, tenths and hundredths. Learners apply their knowledge of this relationship when ordering fractions. Improper fractions and mixed numbers are introduced and learners are asked to identify them from mixed sets of numbers that include proper fractions. Visual models are used to help convert improper fractions to mixed numbers. The unit concludes with practical exercises that involve finding fractions of quantities that include money.

Common difficulties and remediation

Misconceptions can arise when learners try to apply their understanding of whole numbers to comparing fractions. For example, they may say: '$\frac{5}{9}$ is greater than $\frac{5}{8}$ because nine is greater than eight.' It is important to intervene where similar misunderstandings occur and question learners about their understanding of fractions. Practical exercises working with visual models will help to develop their thinking in this area.

It is important that learners have a solid understanding of what fractions represent before moving on to comparing and ordering them. In particular, they need to understand the relationship between denominator and numerator; that the **denominator** is the number of equal parts into which a whole is divided and the **numerator** is the number of parts being considered.

Some learners experience difficulty when naming fractional parts of a set of objects. It is important to remind them to count the objects in the set first, as this will help them identify the size of the equal parts.

As learners develop understanding of the part–whole model of fractions, provide opportunities for them to represent the fractions on number lines. This will help them to visualise fractions as numbers or values. A misconception arises when learners do not fully understand that fractions are unique numbers, as are decimals, each occupying a specific place on the number line.

Promoting and supporting language

Be aware of terms that have more than one meaning, particularly with meanings outside of mathematics. The term 'improper' implies that there is something incorrect about this type of fraction. Explain that there is nothing wrong about improper fractions – it's simply a word that helps differentiate them from proper fractions.

Encourage frequent usage of mathematical language to help embed vocabulary, for example consistently using the term 'numerator' for the 'top number' in a fraction. This will help learners make the transition to mathematically precise language.

Lesson 1: **Equivalent fractions**

Learning objectives

Code	Learning objective
5Nn15	Recognise equivalence between: $\frac{1}{2}$, $\frac{1}{4}$ and $\frac{1}{8}$; $\frac{1}{3}$ and $\frac{1}{6}$; $\frac{1}{5}$ and $\frac{1}{10}$.

Strand 5: Problem solving	
5Pt2	Solve single [and multi-]step word problems (all four operations); represent them, e.g. with diagrams or a number line.
5Pt7	Consider whether an answer is reasonable in the context of a problem.
5Ps2	Choose an appropriate strategy for a calculation and explain how they worked out the answer.
5Ps3	Explore and solve number problems and puzzles, e.g. logic problems.
5Ps9	Explain methods and justify reasoning orally and in writing; make hypotheses and test them out.

Prerequisites for learning

Learners need to:
- understand the key terms 'numerator' and 'denominator'
- understand equivalent fractions.

Success criteria

- recognise equivalent fractions, such as $\frac{1}{3}$ and $\frac{2}{6}$
- use and construct diagrams to identify equivalent fractions.

Vocabulary

equivalent, numerator, denominator, like fraction, unlike fraction

Resources

red, yellow and blue coloured pencils (per learner)

Refresh

- Choose an activity from Number – Numbers and the number system in the Refresh activities.

Discover SB 🖥

- Discuss the text and picture in the Student's Book. Ask: **What fraction of a dollar is 25 cents?** ($\frac{25}{100}$ or $\frac{1}{4}$).

Teach and Learn 📊 SB

- Read and discuss the text and Example in the Student's Book. Ask: **What are equivalent fractions?** Accept responses.
- Display the **Fractions tool** with four fraction bars (numbers 1 to 4). Divide fraction bar 1 in two halves (one half green; one half orange). Divide the fraction bar below it in four parts (make all parts orange). Ask: **Who can shade this bar green so that the green parts take up the same amount of space as the green parts in the bar above?** Expect volunteers to shade two parts. Say: **You can see that $\frac{2}{4}$ is equal to $\frac{1}{2}$. They are equivalent fractions; they take up the same amount of the whole.**
- Ask: **Are there other fractions equivalent to $\frac{1}{2}$ and $\frac{2}{4}$?** Learners draw fraction bars and investigate.
- Divide the third bar in eighths. Shade four parts. On the board, write: $\frac{1}{2} = \frac{2}{4} = \frac{4}{8}$.

- Repeat the process for thirds and sixths, writing on the board: $\frac{1}{3} = \frac{2}{6}$.
- Ask: **What other equivalent fractions can you find?** Focus on fifths. Establish that $\frac{1}{5} = \frac{2}{10}$.

Practise WB

- Workbook: Equivalent fractions page 42

Apply 👥 🖥

- Display **Slide 1**. Learners work together to solve the problem.

Review 📊

- Display the **Fractions tool**: four circular fractions with cake background modelling $\frac{1}{2}$, $\frac{1}{4}$, $\frac{2}{4}$, $\frac{2}{8}$. Say: **Explain to your partner which fractions are equivalent and why.** Repeat for $\frac{2}{10}$, $\frac{2}{6}$, $\frac{1}{5}$, $\frac{1}{3}$.

Assessment for learning

- What other fractions are equivalent to $\frac{2}{4}$? How can you prove this?

Number – Numbers and the number system

91

Lesson 2: **Fraction and decimal equivalents**

Number – Numbers and the number system

Learning objectives

Code	Learning objective
5Nn16	Recognise equivalence between the decimal and fraction forms of halves, tenths and hundredths and use this to help order fractions, e.g. 0·6 is more than 50% [half] and less than $\frac{7}{10}$.

Strand 5: Problem solving

5Pt7	Consider whether an answer is reasonable in the context of a problem.
5Ps2	Choose an appropriate strategy for a calculation and explain how they worked out the answer.
5Ps3	Explore and solve number problems and puzzles, e.g. logic problems.
5Ps9	Explain methods and justify reasoning orally and in writing; make hypotheses and test them out.

Prerequisites for learning

Learners need to:
- understand the equivalence between one-place decimals and fractions in tenths
- know the place value of decimal tenths and hundredths.

Success criteria

Learners can:
- change a decimal to the equivalent fraction
- use decimal and fraction equivalence to order a set of fractions.

Vocabulary

equivalent, mixed number, fraction, decimal, tenths, hundredths

Resources

calculator (per pair);

Refresh

- Choose an activity from Number – Numbers and the number system in the Refresh activities.

Discover [SB]

- Read and discuss the text and pictures in the Student's Book. Ask: **In what other situations might you convert from a decimal to a fraction, or vice versa?** Establish that recipes often use measures written in both fractions and decimals, for example: 0·75 litres of milk; $\frac{3}{4}$ cup of water.

Teach and Learn [SB] [📊]

- Discuss the text in the Student's Book. Ask: **What is the relationship between decimals and fractions?** Say: **The fraction shows part of a whole using the fraction bar and the decimal shows part of a whole using place value.**
- Distribute calculators. Write: $\frac{1}{2}$. Say: **A fraction is a short way of writing a division. For example, $\frac{1}{2}$ means one divided by two. Type one divided by two on your calculator.** Ask: **What appears on the screen?** (0·5) Say: **That is because $\frac{1}{2}$ is one divided by two.** Repeat for $\frac{1}{10}$, $\frac{3}{10}$, $\frac{7}{10}$ (0·1, 0·3, 0·7).
- Write: $\frac{1}{100}$. Ask: **How would you write this number as a decimal? Think about place value.** (0·01) Display the **Number square tool** and shade one square (0·01). Build up hundredths to 0·09 and record fraction–decimal equivalents. Shade 25 and 75 hundredths and show the equivalence: $\frac{25}{100} = \frac{1}{4}$; $\frac{75}{100} = \frac{3}{4}$.

- On the board, write: $\frac{7}{10}$, $\frac{1}{2}$, $\frac{75}{100}$. Ask: **How would you order these fractions?** Elicit that the fractions can be converted to decimals to compare and order them. Convert, with learners assisting: (0·7) $\frac{7}{10}$, (0·5) $\frac{1}{2}$, (0·75) $\frac{75}{100}$. Establish the decimal order: 0·5 < 0·7 < 0·75, so the fraction order is: $\frac{1}{2}$, $\frac{7}{10}$, $\frac{75}{100}$.

Practise [WB]

- Workbook: Fraction and decimal equivalents page 44
- Refer to Activity 1 from the Additional practice activities.

Apply 👥

- Display **Slide 1** and read the text. Learners convert between fractions and decimals on the chocolate cake recipe.

Review

- On the board, write: $\frac{3}{4}$, 0·3, $\frac{1}{4}$, 0·9, $\frac{2}{100}$, $\frac{1}{2}$. Say: **Order the numbers.** Learners share their solutions. $\left(\frac{2}{100}, \frac{1}{4}, 0·3, \frac{1}{2}, \frac{3}{4}, 0·9\right)$

Assessment for learning

- What decimals are equivalent to $\frac{3}{4}$ and $\frac{3}{10}$? How would you prove this?

Lesson 3: **Mixed numbers**

Learning objectives

Code	Learning objective
5Nn17	Change an improper fraction to a mixed number, e.g. $\frac{7}{4}$ to $1\frac{3}{4}$; order mixed numbers and place between whole numbers on a number line.

Strand 5: Problem solving

5Pt2	Solve single [and multi-]step word problems (all four operations); represent them, e.g. with diagrams or a number line.
5Pt7	Consider whether an answer is reasonable in the context of a problem.
5Ps2	Choose an appropriate strategy for a calculation and explain how they worked out the answer.
5Ps3	Explore and solve number problems and puzzles, e.g. logic problems.
5Ps4	Deduce new information from existing information to solve problems.

Prerequisites for learning

Learners need to:

- be able to order and compare two or more fractions with the same denominator (halves, quarters, thirds, fifths, eighths or tenths).

Success criteria

Learners can:

- recognise mixed numbers and improper fractions
- convert an improper fraction to a mixed number

- order a set of mixed numbers.

Vocabulary

numerator, denominator, improper fraction, proper fraction, mixed number

Resources

mini whiteboard and pen or paper (per learner); coloured pencils (per learner); rulers (per pair or group)

Refresh

- Choose an activity from Number – Numbers and the number system in the Refresh activities.

Discover SB

- Read the text and look at the pictures in the Student's Book. Ask: **Where might you find mixed numbers being used?** Explore the use of mixed numbers by asking learners their age, encouraging them to consider the number of months since their last birthday as a fraction of a year.

Teach and Learn SB 🖥

- Discuss the text in the Student's Book.
- Display **Slide 1**. Say: $\frac{3}{4}$ **is a fraction you see a lot. It is an example of a proper fraction: a fraction less than one. Its numerator is less than its denominator.** Point to $\frac{9}{5}$. Ask: **What kind of fraction is this?** Say: **This is an improper fraction, which is a fraction equal to or greater than one. The numerator is greater than or equal to the denominator.**
- Now define a mixed number as 'a number that is made up of a whole number and a fraction'. Display **Slide 2**. Ask learners to sort the fractions into three sets: mixed numbers, proper fractions and improper fractions.
- Indicate $\frac{9}{4}$. Say: **You can use a circle diagram to represent $\frac{9}{4}$. How many parts will be shaded?** Display **Slide 3**.

- Repeat for $\frac{13}{5}$ with circles drawn on the board and divided in fifths.
- On the board, draw a 0–5 number line, subdivided into fifths. Ask learners to count in fifths from zero to three. Stop at $\frac{13}{5}$. Establish that each whole number is equivalent to $\frac{5}{5}$ and show that:
$\frac{13}{5} = \frac{5}{5} + \frac{5}{5} + \frac{3}{5} = 2\frac{3}{5}$.
Write: $4\frac{4}{5}, 2\frac{4}{5}, 3\frac{3}{5}, 4\frac{2}{5}$.
Ask volunteers to insert the mixed numbers between whole numbers on the number line. Write the mixed numbers in order, from least to greatest:
$2\frac{4}{5}, 3\frac{3}{5}, 4\frac{2}{5}, 4\frac{4}{5}$.

Practise WB

- Workbook: Mixed numbers page 46.

Apply 👥

- Display **Slide 4**. Learners write down all the possible mystery fractions and convert them to mixed numbers. $\left(\frac{89}{11} \text{ to } \frac{98}{11}\right)$

Review 🖥

- Display **Slide 5** and give learners some paper and coloured pencils. Read the text with the class and ask learners to answer the questions on their own.

Assessment for learning

- $4\frac{2}{3}$ and $\frac{14}{3}$ are the same fraction. How can you prove this?

Lesson 4: **Fractions of quantities**

Number – Numbers and the number system

Learning objectives

Code	Learning objective
5Nn18	Relate finding fractions to division and use to find simple fractions of quantities.

Strand 5: Problem solving

5Pt2	Solve single [and multi-]step word problems (all four operations); represent them, e.g. with diagrams or a number line.
5Pt7	Consider whether an answer is reasonable in the context of a problem.
5Ps2	Choose an appropriate strategy for a calculation and explain how they worked out the answer.
5Ps3	Explore and solve number problems and puzzles, e.g. logic problems.
5Ps4	Deduce new information from existing information to solve problems.
5Ps8	Investigate a simple general statement by finding examples which do or do not satisfy it [, e.g. the sum of three consecutive whole numbers is always a multiple of three].
5Ps9	Explain methods and justify reasoning orally and in writing; make hypotheses and test them out.
5PS10	Solve a larger problem by breaking it down into sub-problems or represent it using diagrams.

Prerequisites for learning

Learners need to:

• understand non-unit fractions.

Success criteria

Learners can:

• determine a fraction of a quantity by first finding the unit fraction and then multiplying by the numerator.

Vocabulary

numerator, denominator, unit fraction, non-unit fraction

Resources

interlocking cubes (per class); mini whiteboard and pen (per learner)

Refresh

• Choose an activity from Number – Numbers and the number system in the Refresh activities.

Discover [SB]

• Refer learners to the Student's Book and discuss the problem. Ask: **Which offer would you choose?** Establish that to find a half or third, you divide the price by two or three.

Teach and Learn [chart] [SB]

• Discuss the text and Example in the Student's Book. Show learners a stick of 12 interlocking cubes. Ask: **How would I find three quarters of this set of cubes?** Write: $\frac{3}{4}$ of 12. Say: **When you think of fractions, you think of a number representing part of a whole – but the whole isn't always just one object; it may be a whole set of objects.** Point at $\frac{3}{4}$. Ask: **What does the denominator four tell you?** Establish that the whole is divided into four equal parts. Ask: **If you divide 12 cubes into four equal sets, how many cubes are in one set?** (3) Establish that this is the same as dividing by four. Write $\frac{1}{4}$ of 12 = 12 ÷ 4 = 3. Say: **This is the number of cubes in one part.** Ask: **How many parts must you think about?** Elicit that the numerator gives this information (three parts). Ask: **How can you find the number of cubes in three parts?** Elicit the solution: 'multiply by three'. On the board, write $\frac{3}{4}$ of 12 = 3 × 3 = 9.

• Now write: $\frac{2}{3}$ of $60. Say: **You can find a fraction of any quantity, for example, money.** Display the **Fractions tool**, divided into thirds. Write $60 above the bar, with $\frac{1}{3}$ in each part. Say: **You need to find a third of $60.** Write '$20' below each part. Ask: **How can you find $\frac{2}{3}$?** Elicit: 'multiply by the numerator'. Write: $20 × 2 = $40. Repeat the 'fraction bar' method for $\frac{4}{5}$ of 25 and $\frac{5}{8}$ of 40.

Practise [WB]

• Workbook: Fractions of quantities page 48
• Refer to Activity 2 from the Additional practice activities.

Apply [icon]

• Display **Slide 1**. Learners discuss and solve the first problem. They share solution and strategies. (offers a and c) They then discuss the statement: '$\frac{2}{5}$ of $100 is the same as $\frac{4}{10}$ and $\frac{40}{100}$ of $100.' Learners explain how they would show that the fractions are all equivalent.

Review

• On the board, write: Find $\frac{5}{6}$ of $48. Ask learners to solve the problem, reminding them that the 'fraction bar' model helps keep calculations organised. ($40)

Assessment for learning

• Draw a diagram to show me how to work out $\frac{3}{7}$ of 21 cm.

Additional practice activities

Activity 1 Challenge 2

Learning objective

- Construct diagrams that identify and illustrate equivalence between fractions and decimals.

Resources

mini whiteboard and pen or paper (per group); coloured pencils (per group); rulers (per group)

What to do

- Divide the class into seven groups and name them A to G.
- Inform each group that they will be responsible for a different fraction: $\frac{1}{2}$ (A), $\frac{1}{3}$ (B), $\frac{1}{4}$ (D), $\frac{1}{5}$ (E), $\frac{1}{10}$ (F), $\frac{3}{4}$ (G).
- Each group has to find as many different ways of representing their fraction as possible.
- Distribute paper, pencils and rulers. Say: **Draw diagrams on your paper, including fraction bars and hundred grids. Remember to include any equivalent fractions and decimals.**
- Share results and combine to form a fraction poster display.

Variation

Challenge 3 Learners provide instructions on how to order a set of mixed fractions and decimals.

Activity 2 Challenge 2

Learning objectives

- Find an improper fraction of an amount by applying knowledge of how to convert improper fractions to mixed numbers and how to find a fraction of an amount.
- Solve a multi-step word problem involving improper fractions, mixed numbers and fractions of amounts.

Resources

paper, or mini whiteboard and pen (per learner)

What to do

- On the board, write: $\frac{13}{3}$, $\frac{17}{5}$, $\frac{22}{6}$ and the amounts: £30 and £90.
- Ask learners find the fractions of each amount. Expect them to do this by converting improper fractions to mixed numbers and then to find the sum of multiple and fractions of each amount. For example: $\frac{13}{3} = 4\frac{1}{3}$; $4 \times 30 = 120$; $\frac{1}{3}$ of 30 = 10; 120 + 10 = 130)

Variation

Challenge 1 Replace the improper fraction with a simple mixed number, for example $1\frac{1}{5}$.

Unit 6: Percentages

Learning objectives

Code	Learning objective
5Nn19	Understand percentage as the number of parts in every 100 and find simple percentages of quantities.
5Nn20	Express halves, tenths and hundredths as percentages.
Strand 5: Problem solving	
5Pt2	Solve single and multi-step word problems (all four operations); represent them, e.g. with diagrams or a number line.
5Pt7	Consider whether an answer is reasonable in the context of a problem.
5Ps2	Choose an appropriate strategy for a calculation and explain how they worked out the answer.
5Ps3	Explore and solve number problems and puzzles, e.g. logic problems.
5Ps4	Deduce new information from existing information to solve problems.
5Ps8	Investigate a simple general statement by finding examples which do or do not satisfy it, [e.g. the sum of three consecutive whole numbers is always a multiple of three].
5Ps9	Explain methods and justify reasoning orally and in writing; make hypotheses and test them out.
5Ps10	Solve a larger problem by breaking it down into sub-problems or represent it using diagrams.

Unit overview

The unit introduces learners to the concept of 'per cent' and how it relates to the number of parts per hundred. Learners use 10 × 10 arrangements of counters and hundred grids to investigate the relationship between percentages and fractions. They are expected to know and recognise halves, tenths and hundredths as percentages and use this knowledge effectively to solve practical problems. Strategies for calculating simple percentages of quantities are modelled and learners apply these to working out percentages of numbers in real-life contexts.

Common difficulties and remediation

An important component of teaching percentages is to help learners acquire a sound understanding of the relationship between fractions and percentages. Learners need to understand that 50% is equivalent to one half, 25% is equivalent to one quarter, 10% is equivalent to one tenth and 1% is equivalent to one hundredth. Other percentages can be built up from these percentage 'units', for example 13% = 10% + (1% × 3). Some learners may experience difficulty understanding percentages in the early stages of the unit. These learners will need extended practical experience in working with fractional models, exploring the relationship between part of a set and part of a whole.

Look out for learners who view a percentage as a number, rather than part of an amount, or who have yet to make a sound link with fractions. Also, be aware of learners who have made a link with fractions, but use the value of the percentage as the denominator. For example, they think 15% is equivalent to $\frac{1}{15}$ and

subsequently find 15% of 200 by attempting to divide 200 by 15. To remediate these problems, provide models of percentages and fractions, saying 'out of a hundred' while drawing the fraction line. Similarly, provide models and examples of percentages as fractions. This will help learners make links with problem solving and finding the percentage of an amount, not just an amount out of a hundred.

Promoting and supporting language

At every stage, learners require mathematical vocabulary to access questions and problem-solving exercises. If appropriate, when a new key word is introduced, ask learners to write a definition in their books, drawing a box around it for emphasis. Encourage learners to write the definition in their own words, for example: 'A percentage is a way of expressing an amount as a fraction of 100.' Sketching a diagram alongside will help embed the definition.

When discussing percentage, it is important to say what the percentage relates to, stating the whole amount (what 100% is), for example '30% of learners travel to school by car' does not mean the same as the context-specific: 'In a school of 900 learners, 30% travel to school by car.'

Throughout the lesson, encourage learners to seek clarification and confirmation of the mathematical language. This may involve prompting students to call out and complete definitions, or praising students when they ask for terms to be defined.

Provide learners with frequent opportunities to experience percentages in real-life contexts – to give more meaning to their calculations and promote understanding.

Lesson 1: **Per cent symbol**

Learning objectives

Code	Learning objective
5Nn19	Understand percentage as the number of parts in every 100 [and find simple percentages of quantities].

Strand 5: Problem solving

Code	Learning objective
5Pt7	Consider whether an answer is reasonable in the context of a problem.
5Ps2	Choose an appropriate strategy for a calculation and explain how they worked out the answer.
5Ps3	Explore and solve number problems and puzzles, e.g. logic problems.
5Ps4	Deduce new information from existing information to solve problems.
5Ps8	Investigate a simple general statement by finding examples which do or do not satisfy it, [e.g. the sum of three consecutive whole numbers is always a multiple of three].
5Ps9	Explain methods and justify reasoning orally and in writing; make hypotheses and test them out.

Prerequisites for learning

Learners need to:

* understand and be able to use unit fractions, halves, quarters, fifths, tenths, and hundredths.

Success criteria

Learners can:

* explain percentage as 'the number of parts in every hundred'

* use a 100 grid or counters to model a percentage.

Vocabulary

per cent, percentage, hundredths

Resources

Resource sheet 9: 100 square (per class); coloured pencils (per learner)

Refresh

* Choose an activity from Number – Numbers and the number system in the Refresh activities.

Discover ⟦SB⟧

* Refer learners to the Student's Book and read the text. Ask: **In what other situation could you express something 'out of 100'?** Learners share their ideas.

Teach and Learn ⟦⟧ ⟦SB⟧

* Refer to the text and Example in the Student's Book. Display the **Number square tool** with numbers hidden and all squares shaded red. Shade three squares yellow. Ask: **How many squares out of the hundred are yellow?** (three)

* Repeat for 11 and 19 yellow squares out of 100. Say: **There is a special way of writing a number of parts out of 100 – it is called percentage.** Write: 3%. Point to the percentage symbol (%) and explain that it represents 'parts in every hundred'. Point to the 3. Explain that this is the number of parts.

* Now shade 15 squares yellow. Ask: **What percentage of the squares are yellow?** (15%) Repeat for 25 and 37 squares

* Write: 13%, 48% and 91%. Ask learners to model these percentages on Resource sheet 9.

* Say: **In a car park, 13 out of 100 cars are yellow. How would you represent this on the 100 grid? How would you write the percentage?** Ask a volunteer to demonstrate.

* Present other 'out of 100' stories and invite volunteers to model and write percentages.

Practise ⟦WB⟧

* Workbook: Per cent symbol page 50

Apply ⟦≗⟧

* On the board, write: 1%, 9% and 17%. Learners show these percentages with models and simple diagrams.

Review ⟦⟧

* Model two percentages that total 100% on the **Number square tool**. Learners invent a story for the models, for example: '100 birds sit in a tree. 54 of the birds have yellow feathers; 46 have red feathers.' Ask learners to share their stories.

Assessment for learning

* 16 out of 100 learners walk to school. How do you show this as a percentage?

Number – Numbers and the number system

97

Number – Numbers and the number system

Lesson 2: **Expressing fractions as percentages**

Learning objectives

Code	Learning objective
5Nn20	Express halves, tenths and hundredths as percentages.

Strand 5: Problem solving

5Pt2	Solve single and multi-step word problems (all four operations); represent them, e.g. with diagrams or a number line.
5Pt7	Consider whether an answer is reasonable in the context of a problem.
5Ps2	Choose an appropriate strategy for a calculation and explain how they worked out the answer.
5Ps3	Explore and solve number problems and puzzles, e.g. logic problems.
5Ps4	Deduce new information from existing information to solve problems.

Prerequisites for learning

Learners need to:

• understand fractions.

Success criteria

Learners can:

• express halves, tenths and hundredths as percentages

Vocabulary

per cent, percentage, hundredths, tenths, whole, half, equivalent

Resources

100 interlocking cubes or counters (per pair or group)

Refresh

• Choose an activity from Number – Numbers and the number system in the Refresh activities.

Discover SB

• Read the text in the Student's Book and count the flowers with the learners. Say: **Tell your partner what the term 'percentage' means.** Invite a volunteer to share their definition. Discuss the problem. Return to the question and establish that five out of ten flowers is a half. Ask: **What percentage is a half?** If 50% is mentioned, elicit that the gardener could have said '50% red and 50% yellow'.

Teach and Learn SB

• Discuss the text in the Student's Book. Give each group or pair of learners 100 interlocking cubes or counters. Let them use these to model the examples you are demonstrating throughout the lesson.

• Display the **Number square tool**, showing 1 to 100. Highlight all squares 1 to 100. Ask: **What percentage does each square represent?** (1%) Say: **All 100 squares are shaded. What fraction is this?** On the board, write: $\frac{100}{100}$. Say: **If each square represents one per cent, what percentage does the whole grid represent?** On the board, write: $\frac{100}{100}$ = 100%.

• Remove the highlight, then shade squares 1 to 50. Ask: **What fraction is shaded now?** Accept suggestions. If learners answer with hundredths only,

prompt them for a simplified fraction: **How do you usually write a half as a fraction?** $\left(\frac{1}{2}\right)$ On the board, write: $\frac{1}{2}$. Ask: **How many one per cent squares are shaded?** (50) Write on the board $\frac{1}{2}$ = 50%.

• Now shade squares 1 to 10. Ask: **What fraction is this?** On the board, write: $\frac{1}{10}$. Ask: **What percentage is equal to one tenth?** Accept suggestions and elicit the correct answer, ten per cent. Write $\frac{1}{10}$ = 10%.

• Repeat the above for other tenths.

Practise WB

• Workbook: Expressing fractions as percentages page 52

• Refer to Activity 1 from the Additional practice activities.

Apply 👥 🖥

• Display **Slide 1.** Ask learners to write the number of shaded squares as fractions, then as percentages.

Review

• On the board, write: $\frac{1}{100}, \frac{1}{10}, \frac{1}{2}, \frac{100}{100}$. Ask learners to suggest '...out of' sentence for each fraction, then convert it to a percentage, for example: 'One out of two cars is blue.' (50%)

Assessment for learning

• What percentages are equivalent to $\frac{1}{10}$ and $\frac{1}{100}$?

Lesson 3: **Percentages of quantities**

Learning objectives

Code	Learning objective
5Nn19	Understand percentage as the number of parts in every 100 and find simple percentages of quantities.

Strand 5: Problem solving

5Pt2	Solve single and multi-step word problems (all four operations); represent them, e.g. with diagrams or a number line.
5Pt7	Consider whether an answer is reasonable in the context of a problem.
5Ps2	Choose an appropriate strategy for a calculation and explain how they worked out the answer.
5Ps3	Explore and solve number problems and puzzles, e.g. logic problems.
5Ps9	Explain methods and justify reasoning orally and in writing; make hypotheses and test them out.
5Ps10	Solve a larger problem by breaking it down into sub-problems or represent it using diagrams.

Prerequisites for learning

Learners need to:

- understand the relationship between fractions and percentages.
- relate finding fractions to division
- find simple fractions of quantities.

Success criteria

Learners can:

- find simple percentages of quantities by finding one per cent, using a hundred grid, and multiplying

- find 50% of a quantity by halving, 25% of a quantity by quartering
- write any percentage as a fraction with a denominator of 100.

Vocabulary

per cent, fraction, divide, multiply

Resources

mini whiteboard and pen (per learner); paper (per learner)

Refresh

- Choose an activity from Number – Numbers and the number system in the Refresh activities.

Discover [SB]

- Read the text and look at the picture of the children in the Student's Book. Ask: **What other quantities in your classroom can you describe using percentages?** Take suggestions.

Teach and Learn [SB]

- Discuss the text and Example in the Student's Book. On the board, write: 1% of $2400. Ask: **How would you work out this percentage?** Invite volunteers to explain their strategies. Say: **You can use knowledge of percentage–fraction equivalents to help solve the problem.** Ask: **Which fraction is equivalent to one per cent?** $\left(\frac{1}{100}\right)$ Say: **One per cent is equivalent to one hundredth. When you want to work out 100% of any quantity you can simply divide by 100.** Invite learners to show how to divide the following quantities by 100 to find one per cent: 200 kg, 400 m, 800 ml. Establish that the quickest method is to move the digits two place value columns to the right. Agree that one per cent of $2400 is $24 and record it on the board.
- Write: 10% of $2400. Ask: **Which fraction is equivalent to ten per cent?** $\left(\frac{1}{10}\right)$ Say: **How can you use this to find ten per cent?** (divide by ten) Ask

learners to solve the problem, reminding them that the quickest way to divide by ten is to move the digits one place value column to the right ($240). Record this on the board. Repeat for 50% (divide by two) and 25% (divide by four). ($1200, $600)

- On the board, write: 1700 km. Ask learners to find 1%, 10%, 25% and 50% of this measurement. (17 km, 170 km, 425 km, 850 km)

Practise [WB]

- Workbook: Percentages of quantities page 54

Apply ▪▪

- Learners discuss and record a figure for their pocket money. They find 1%, 10%, 25% and 50% of this value. If learners do not get pocket money, suggest an amount for them to use.

Review ▪▪

- Ask learners to record a 4-digit starter number that is a multiple of 100, for example, 3200. Ask: **Using a combination of 50%, 25%, 10% and 1%, how far can you reduce the figure?** They must use each percentage at least once.

Assessment for learning

- How would you show that 50% of 280 is 140? 25% of 280 is 70?

Unit **6** **Percentages**

Lesson 4: **Percentage problems**

Number – Numbers and the number system

Learning objectives

Code	Learning objective
5Nn19	Understand percentage as the number of parts in every 100 and find simple percentages of quantities.

Strand 5: Problem solving	
5Pt2	Solve single and multi-step word problems (all four operations); represent them, e.g. with diagrams or a number line.
5Pt7	Consider whether an answer is reasonable in the context of a problem.
5Ps2	Choose an appropriate strategy for a calculation and explain how they worked out the answer.
5Ps3	Explore and solve number problems and puzzles, e.g. logic problems.
5Ps4	Deduce new information from existing information to solve problems.
5Ps9	Explain methods and justify reasoning orally and in writing; make hypotheses and test them out.
5Ps10	Solve a larger problem by breaking it down into sub-problems or represent it using diagrams.

Prerequisites for learning

Learners need to:

- understand the relationship between fractions and percentages.

Success criteria

Learners can:

- find simple percentages of quantities by finding one per cent, using a hundred grid and multiplying

- find 50% of a quantity by halving; 25% of a quantity by quartering
- write any percentage as a fraction with a denominator of 100.

Vocabulary

fraction, per cent, divide, multiply

Resources

paper or mini whiteboard and pen (per learner)

Refresh

- Choose an activity from Number – Numbers and the number system in the Refresh activities.

Discover 📘

- Refer learners to the Student's Book and discuss the text and diagram. Ask: **How can you use your knowledge of fractions and percentages to find 30% of $30?** Ask learners to share their strategies and discuss.

Teach and Learn 📘 🖥

- Display **Slide 1**. Ask: **What are the key facts in this problem?** Ask a volunteer to underline the key terms. Establish that '25% off', 'original price', '$800' and 'reduced' are key terms. Discuss the term 'reduced' and explain it means an amount is taken off.
- Say: Ask learners to help extract the problem. On the board, write: 25% of $800. Say: **This will give the amount the bicycle is reduced by.** Let the learners work this out to solve the problem. ($200)
- Display **Slide 2**. Ask a volunteer to underline the key terms. Elicit that the problem is '30% of 300'. Ask: **How can you find 30% of something?** Elicit that they will find 10% and multiply by 3. On the board, write: '10% of 300', '30% of 300'. Let the learners solve the problem. (90)

Practise 📒

- Workbook: Problems involving percentages page 56
- Refer to Activity 2 from the Additional practice activities.

Apply 👥 🖥

- Display **Slide 3**. Learners discuss and solve the problem. They share solutions and strategies. (75% of $120)

Review

- Learners choose any question from Workbook Lesson 4. Ask individual learners to share their answers and discuss the strategies they used. Ask: **Did anyone work out this answer a different way? Describe the strategy you used.**

Assessment for learning

- I need to find 40% of 2000 kilometres. How would I do this?

Additional practice activities

Activity 1 :: Challenge 2

Learning objective
• Identify known percentage equivalents of fractions.

Resources
Two spinners from Resource sheet 10: fraction spinner (per pair)

What to do
• Both learners spin their spinner at the same time.
• They identify and state the percentage equivalent of the fraction spun.
• The player with the higher percentage scores a point.
• Players continue spinning against each other for seven rounds. The player with the higher score wins the game.

Activity 2 :: Challenge 2

Learning objective
• Find simple percentages of quantities.

Resources
Number cards 1, 5, 10, 20, 25 and 50 from Resource sheet 2: 0–100 number cards (per pair)

What to do
• Learners choose a four-digit target number that is a multiple of 100, for example, 6700.
• They take turns to pick a digit card representing a percentage.
• They find this percentage of the four-digit number and add the amount to their score.
• The winner is the first learner to reach a score that is greater than the target number.

Variations
 Use the number cards 1, 10 and 50.

 Use other tenths and hundredths.

Number – Numbers and the number system

Unit 7: Ratio and proportion

Learning objectives

Code	Learning objective
5Nn21	Use fractions to describe and estimate a simple proportion, e.g. $\frac{1}{5}$ of the beads are yellow.
5Nn22	Use ratio to solve problems, e.g. to adapt a recipe for 6 people to one for 3 or 12 people.

Strand 5: Problem solving

Code	
5Pt2	Solve single and multi-step word problems (all four operations); represent them, e.g. with diagrams or a number line.
5Pt7	Consider whether an answer is reasonable in the context of a problem.
5Ps2	Choose an appropriate strategy for a calculation and explain how they worked out the answer.
5Ps3	Explore and solve number problems and puzzles, e.g. logic problems.
5Ps4	Deduce new information from existing information to solve problems.
5Ps8	Investigate a simple general statement by finding examples which do or do not satisfy it, [e.g. the sum of three consecutive whole numbers is always a multiple of three].
5Ps9	Explain methods and justify reasoning orally and in writing; make hypotheses and test them out.
5Ps10	Solve a larger problem by breaking it down into sub-problems or represent it using diagrams.

Unit overview

The unit introduces learners to the concepts of ratio and proportion. Proportion is introduced through the phrase 'in every' and modelled in repeating sequences. Learners identify proportions of a whole and use simple fractions to describe them. They use fractions to determine how the values of component parts change when the size of the 'whole' increases.

Learners practise identifying ratios from pictures, models and word problems. They are given the opportunity to model ratios themselves, using the phrase 'for every' when describing them. They are provided with real-life examples of using ratio to scale amounts up or down, for example modifying a recipe to calculate the amount of each ingredient needed for various numbers of servings.

Common difficulties and remediation

Ratio and proportion can be difficult concepts for learners to grasp. The subjects can be confusing because the terms are frequently used interchangeably in many aspects of daily life. The words 'ratio' and 'proportion' actually represent two quite different ways of looking at the same thing. It is important to make frequent use of familiar phrases that help support development of the concepts involved. Use of the phrase 'for every' helps to introduce ratio, for example: 'For every red bead there are three yellow beads.' Explain that ratio compares one part to another.

After the introduction of ratio, and having completed lessons on proportion, it is advisable to use practical examples to clarify the difference between the two terms. An effective way to illustrate the difference is to use beads on a string to model ratio and proportion. For example, show a necklace made of one red bead and three yellow beads. Say: **The ratio of red beads to yellow beads is one to three.** Explain that the ratio can be scaled for any number of necklaces. Make two necklaces and show that the ratio is now two red to six yellow. Say: **The two necklaces are in proportion because the ratio of red to yellow beads is the same, i.e. two to six simplified is one to three.**

Look out for learners who add rather than multiply when increasing in a ratio or proportion. For example, when given a flapjack recipe that includes two tablespoons of syrup for three people, and asked to work out how many tablespoons they would need for nine people, learners realise there are seven extra people, but add either seven tablespoons or 14 (two tablespoons each). They do not see that the recipe is for **three times as many** people, or recognise the ratio of 3 : 1, giving six tablespoons. Allow learners to take part in practical cooking activities and give them opportunities to practise calculations involving ratio and proportion.

Promoting and supporting language

At every stage, learners require mathematical vocabulary to access questions and problem-solving exercises. If appropriate, when a new key word is introduced, ask learners to write a definition in their books, drawing a box around it for emphasis. Encourage learners to write the definition in their own words, for example: 'proportion compares a part with the whole' and 'ratio compares a part with a part'. Sketching a diagram alongside will help embed the definition.

Lesson 1: **Proportion**

Learning objectives

Code	Learning objective
5Nn21	Use fractions to describe and estimate a simple proportion, e.g. $\frac{1}{5}$ of the beads are yellow.

Strand 5: Problem solving

5Ps2	Choose an appropriate strategy for a calculation and explain how they worked out the answer.
5Ps3	Explore and solve number problems and puzzles, e.g. logic problems.
5Ps4	Deduce new information from existing information to solve problems.
5Ps9	Explain methods and justify reasoning orally and in writing; make hypotheses and test them out.

Prerequisites for learning

Learners need to:

• understand and use equivalent fractions

Success criteria

Learners can:

• recognise approximate proportions of a whole and use simple fractions to describe them

• use the vocabulary of proportion to describe the relationship between two quantities, for example: three beads in every five beads are green.

Vocabulary

proportion, in every, whole, fraction

Resources

interlocking cubes (per class); squared paper (per pair); counters (per pair)

Refresh

• Choose an activity from Number – Numbers and the number system in the Refresh activities.

Discover [SB]

• Together, discuss the text and image in the Student's Book. Say: **Proportion compares parts to a whole.**

Teach and Learn [SB]

• Discuss the text and example in the Student's Book. Construct a tower from 16 cubes, using three green and one blue cube in every four. Establish the form of the pattern. Say: **In every four cubes, there are three green cubes and one blue cube.** Check that learners agree. Say: **When you hear the phrase 'in every', you know that you are talking about proportion – how the parts compare to the whole.** Reduce the tower to four cubes, comprising three green and one blue. Say: **Think of the whole as four cubes and the cubes as parts of the whole.** Ask: **What fraction of the whole are the green cubes? What fraction is the blue cube?** $(\frac{3}{4}, \frac{1}{4})$

• Say: **You can use 'proportion' instead of 'fraction'. The proportion of green cubes is $\frac{3}{4}$; the proportion of blue cubes is $\frac{1}{4}$.** Return the tower to its original length. Say: **The proportion is maintained along the whole tower: in every four cubes, there are three green and one blue.** Repeat for a tower of twenty cubes, consisting of three red and two yellow cubes in every five.

• Arrange two rows of three children, with one girl and two boys in each row. Ask: **What proportion of the learners are girls?** Elicit 'two out of six' or 'one out of three'. Record the fraction on the board. For $\frac{2}{6}$, show that both rows have the same proportion of red to blue counters and the fraction can be simplified to $\frac{1}{3}$.

Practise [WB]

• Workbook: Proportion page 58

Apply 👥 🖥

• Display **Slide 1**. Pairs of learners answer the questions and express the answers as fractions.

  i) 4 boxes? (4, 12) $(\frac{1}{4})$

  ii) 10 boxes? (10, 30) $(\frac{1}{4})$

Review 👥

• Give pairs of learners counters of two colours and ask them to make a repeating pattern. They swap their patterns with another pair and work out the proportions of the colours. Ask: **How do you work out the proportion in a pattern like this?** Discuss methods.

Assessment for learning

• In every eight ducks there are two blue and six green. What is the proportion of ducks of each colour?

Lesson 2: **Proportion problems**

Learning objectives

Code	Learning objective
5Nn21	Use fractions to describe and estimate a simple proportion, e.g. $\frac{1}{5}$ of the beads are yellow.

Strand 5: Problem solving

5Pt2	Solve single and multi-step word problems (all four operations); represent them, e.g. with diagrams or a number line.
5Pt7	Consider whether an answer is reasonable in the context of a problem.
5Ps2	Choose an appropriate strategy for a calculation and explain how they worked out the answer.
5Ps3	Explore and solve number problems and puzzles, e.g. logic problems.
5Ps4	Deduce new information from existing information to solve problems.
5Ps9	Explain methods and justify reasoning orally and in writing; make hypotheses and test them out.
5Ps10	Solve a larger problem by breaking it down into sub-problems or represent it using diagrams.

Prerequisites for learning

Learners need to:
- understand and use equivalent fractions
- understand simple ideas of proportion.

Success criteria

Learners can:
- recognise approximate proportions of a whole and use simple fractions to describe these

- use the vocabulary of proportion to describe the relationship between two quantities, for example: three beads in every five beads are green.

Vocabulary

proportion, in every, whole, fraction

Resources

beads (per pair); strings (per pair)

Refresh

- Choose an activity from Number – Numbers and the number system in the Refresh activities.

Discover [SB]

- Discuss the image in the Student's Book. Ask: **What tells you that the beads are in proportion?** Discuss responses. Ask: **If a string has 10 beads, what numbers of red and blue beads would keep the beads in the same proportion?** (6 red, 4 blue)

Teach and Learn [SB]

- Discuss the text and Example in the Student's Book. Ask all the learners to arrange themselves in a line, three girls to two boys. Ask learners to describe the arrangement, in terms of proportion. Elicit: 'In every five learners there are three girls and two boys.' Ask the first five learners standing in line to stay where they are, and the rest to return to their places. Ask: **What proportion of these are boys?** Agree $\frac{2}{5}$. Ask: **What proportion are girls?** ($\frac{3}{5}$). Ask: **If the number of learners in the line increased to 20, how many boys would you expect there to be if the numbers stay in the same proportion?** Discuss answers. Establish that if they know the proportion (fraction) of boys/girls then they can find the number of boys/girls in groups of different sizes. On the board, write $\frac{2}{5}$ of 20 = 8; $\frac{3}{5}$ of 20 = 12. Confirm the answers (8 boys, 12 girls) by bringing back groups of five learners (in the same proportion as before) to total

20, then counting the boys and the girls. Repeat for other groups of learners: three girls, five boys; three boys, seven girls, asking questions related to the numbers of boys/girls expected in larger sized groups.

Practise [WB]

- Workbook: Proportion problems page 60
- Refer to Activity 1 from the Additional practice activities.

Apply 👥

- Ask pairs of learners to make a short two-colour bead necklace. They ask their partner to tell them the number of beads of each colour for a larger necklace, assuming the colours are in the same proportion.

Review

- Write: In every six cars in a car park, one is black and five are white. Ask: **What is the proportion of black cars? White cars?** ($\frac{1}{6}$, $\frac{5}{6}$) Ask: **If the same proportion is maintained, how many black cars would you expect in a group of 30 cars?** (5)

Assessment for learning

- In every 12 flowers there are five roses and seven tulips. How many tulips are there in 24 flowers? How do you know?

Lesson 3: **Ratio**

Learning objectives

Code	Learning objective
5Nn22	Use ratio to solve problems, e.g. to adapt a recipe for 6 people to one for 3 or 12 people.

Strand 5: Problem solving

Code	Learning objective
5Pt2	Solve single and multi-step word problems (all four operations); represent them, e.g. with diagrams or a number line.
5Pt7	Consider whether an answer is reasonable in the context of a problem.
5Ps2	Choose an appropriate strategy for a calculation and explain how they worked out the answer.
5Ps3	Explore and solve number problems and puzzles, e.g. logic problems.
5Ps4	Deduce new information from existing information to solve problems.
5Ps9	Explain methods and justify reasoning orally and in writing; make hypotheses and test them out.
5Ps10	Solve a larger problem by breaking it down into sub-problems or represent it using diagrams.

Prerequisites for learning

Learners need to:

- understand and use equivalent fractions
- simplify fractions.

Success criteria

Learners can:

- express related amounts or parts of a whole as a ratio, using the correct symbol and simplifying where possible.

Vocabulary

ratio, for every, simplify, scale factor

Resources

interlocking cubes (per pair)

Refresh

- Choose an activity from the Number – Numbers and the number system in Refresh activities.

Discover ⌷SB⌷

- Discuss the text and image in the Discover section. Say: **This shows a ratio of four red cubes to two blue cubes.**

Teach and Learn ⌷SB⌷

- Discuss the text and Example in the Student's Book. Use interlocking cubes to construct a repeating pattern, with three yellow cubes for every one red cube. Ask: **What is the repeating pattern?** (three yellow, one red) Say: **This is an example of ratio: comparing a part to a part. In this pattern, the ratio of yellow cubes to red cubes is three to one.** In other words, there are three yellow cubes for every one red cube. Distribute interlocking cubes and ask learners to copy the pattern. Say: **If I used this ratio to make a pattern that had three red cubes, how many yellow cubes would I need?** Learners construct the pattern with the cubes. Elicit nine and ask learners to explain their solutions.
- Write the ratio: two yellow to three red. Ask: **How many red cubes would be needed for a pattern that had eight yellow cubes?** (12) Ask: **A pattern that has this ratio is made from 20 cubes. How would you work out the number of yellow cubes needed**

for this pattern? Discuss strategies. Ask learners to use cubes to construct the pattern and extend it to 20 cubes. Ask: **How many yellow cubes is that?** (8) Repeat for similar questions with different numbers of cubes, for example: **The pattern has 30 cubes. How many red cubes is that?** (18)

Practise ⌷WB⌷

- Workbook: Ratio page 62

Apply ⌷👥⌷ ⌷🖥⌷

- Display **Slide 1**. Ask pairs of learners to solve the problem by scaling up or down.

Review

- Say: **To make a small cake, two eggs are needed for every four small chocolate bars. Ask: How many bars of chocolate would be needed for a cake with eight eggs?** (16) **If the combined number of eggs and chocolate bars is 18, how many eggs is that?** (6)

Assessment for learning

- A triangle has sides 8 cm, 12 cm, 14 cm. What ratio would the sides of a triangle half the size have? (4 to 6 to 7)

Lesson 4: **Ratio problems**

Number – Numbers and the number system

Learning objectives

Code	Learning objective
5Nn22	Use ratio to solve problems, e.g. to adapt a recipe for 6 people to one for 3 or 12 people.

Strand 5: Problem solving	
5Pt2	Solve single and multi-step word problems (all four operations); represent them, e.g. with diagrams or a number line.
5Pt7	Consider whether an answer is reasonable in the context of a problem.
5Ps2	Choose an appropriate strategy for a calculation and explain how they worked out the answer.
5Ps3	Explore and solve number problems and puzzles, e.g. logic problems.
5Ps4	Deduce new information from existing information to solve problems.
5Ps9	Explain methods and justify reasoning orally and in writing; make hypotheses and test them out.
5Ps8	Investigate a simple general statement by finding examples which do or do not satisfy it, [e.g. the sum of three consecutive whole numbers is always a multiple of three].
5Ps10	Solve a larger problem by breaking it down into sub-problems or represent it using diagrams.

Prerequisites for learning

Learners need to:

- understand and use equivalent fractions
- simplify fractions.

Success criteria

Learners can:

- express related amounts or parts of a whole as a ratio, using the correct symbol and simplifying where possible.

Vocabulary

ratio, for every, simplify, scale factor

Resources

interlocking cubes (per class)

Refresh

- Choose an activity from Number – Numbers and the number system in the Refresh activities.

Discover [SB]

- Discuss the text and image in the Student's Book. Ask: **How have the quantities of the ingredients changed?** Ask: **What quantity of each ingredient would be needed to make cakes for 6 people?**

Teach and Learn [SB]

- Discuss the text and example in the Student's Book. Construct a tower of two red and five yellow cubes. Ask: **What is the ratio of the colours?** Establish the ratio and write it on the board: 2 to 5 (R to Y) Say: **You can increase or decrease the size of the tower by the same ratio – this is called scaling.** Draw a table with headings: '1 ×' and '2 ×' and rows 'yellow' and 'red'. Add figures for one tower (2 red, 5 yellow). Say: **For two towers, simply double the numbers of yellow and red cubes.** Ask: **What numbers will you put in the table?** (4 red, 10 yellow) Add columns: '4 ×', '8 ×', '10 ×'. Learners discuss the answers in pairs. Invite volunteer pairs to discuss their answers and methods.

- Ask pairs of learners to make a tower of two colours of cubes, then scale it up or down, keeping the ratio between the colours the same.

Practise [WB]

- Workbook: Ratio problems page 64
- Refer to Activity 2 from the Additional practice activities.

Apply 👥 🖥

- Display **Slide 1** and ask pairs of learners to solve the problem by scaling up or down.

Review

- On the board, write: Paint recipe: 4 litres red : 5 litres green : 9 litres yellow. Say: **If I wanted five times the amount of paint, how much yellow paint would I need? How much red? How much green?**

Assessment for learning

- Using any resource, show me how to scale a ratio of 2 in 4 up or down.

Additional practice activities

Activity 1

Learning objective

- Use proportional reasoning to solve a problem.

Resources

paper (per pair)

What to do

- On the board, draw a table with columns labelled: 'homework (min)', '1 day', '2 days', '4 days', '8 days' and rows labelled: maths, English, science.
- Add the following numbers to the first column under '1 day': 60 (maths), 50 (English), 40 (science)
- Ask learners to complete the table with the amount of time spent completing homework for different numbers of days. Inform learners that the numbers stay in proportion.
- Invite learners to share their answers and explain how they used proportion to calculate the missing values.

Variation

Challenge 1 Provide resources, such as counters, for learners to model problems.

Activity 2

Learning objective

- Use a scale factor to scale a ratio.

Resources

1–9 number cards from Resource sheet 2: 0–100 number cards (per pair); 1–6 dice, alternatively use Resource sheet 4: 1–6 spinners (per pair) paper (per pair)

What to do

- Learners shuffle the cards and place them face down in front of them.
- They take turns to pick three cards each and create a ratio, A : B : C.
- Say: **The ratio represents the number of containers of different flavours of ice cream carried by an ice cream van. The flavours are chocolate, strawberry and vanilla.**
- Learners roll the dice a fourth time. Say: **This number is a scale factor, the number that scales, or multiplies, the quantities.**
- They scale the flavours and the winner is the learner to score 5 points.

Variation

Challenge 1 Work with two ice cream flavours only.

Unit 8: Addition and subtraction 1

Learning objectives

Code	Learning objective
5Nc8	Count on or back in thousands, hundreds, tens and ones to add or subtract.
5Nc10	Use appropriate strategies to add or subtract pairs of two- and three-digit numbers [and numbers with one decimal place], using jottings where necessary.
5Nc18	Find the total of more than three two- or three-digit numbers using a written method.
Strand 5: Problem solving	
5Pt2	Solve single and multi-step word problems (all four operations); represent them, e.g. with diagrams or a number line.
5Pt3	Check with a different order when adding several numbers or by using the inverse when adding or subtracting a pair of numbers.
5Pt6	Estimate and approximate when calculating, e.g. using rounding, and check working.
5Pt7	Consider whether an answer is reasonable in the context of a problem.
5Ps2	Choose an appropriate strategy for a calculation and explain how they worked out the answer.
5Ps3	Explore and solve number problems and puzzles, e.g. logic problems.
5Ps4	Deduce new information from existing information to solve problems.
5Ps5	Use ordered lists and tables to help to solve problems systematically.
5Ps9	Explain methods and justify reasoning orally and in writing; make hypotheses and test them out.
5Ps10	Solve a larger problem by breaking it down into sub-problems or represent it using diagrams.

Unit overview

In this unit, learners focus on mental methods for addition and subtraction, using jottings to assist computation. They practise counting on or back in thousands, hundreds, tens and units and apply this to solving mental addition and subtraction problems involving multiples of ten, 100 and 1000. Learners meet a range of strategies for solving problems mentally, including partitioning, compensation and compatible numbers. Guided examples are provided, together with practical activities in which learners use an appropriate strategy to solve problems. In so doing, they learn that some strategies are more effective than others, depending on the calculation. The unit concludes with a review of written methods of addition, focusing on column addition. Learners are asked to find the sum of three or more two- or three-digit numbers.

Common difficulties and remediation

Some learners lack confidence in counting on or back in tens or hundreds and find crossing tens and hundreds boundaries difficult. Consequently they find addition and subtraction of multiples of ten and 100 challenging. They may not fully understand the relationship between 1, 10 and 100 and therefore do not recognise that to add or subtract 10 or 100 is no more difficult than to add or subtract 1. To remediate, play ordering games with digit cards, for example, 'tens race', in which learners order multiples of ten, forwards or back, as quickly as possible. The ability to count on or back in tens and hundreds from any

number is the foundation for successful mental computation; learners need to be secure in this in order to progress.

When number problems become more complex, but still achievable mentally, learners should adopt a strategy. Some struggle to use known number facts and place value to add or subtract mentally. These learners require guided examples, modelled on base-10 equipment, linking each practical step to a recorded step.

Look out for learners who experience difficulty with column addition, for example, inserting a column that should not exist, possibly because they do not fully grasp that 11 = 1 ten + 1 unit; failing to acknowledge that the hundreds column exists, when no digits reside in it initially; or forgetting to add the carrying digit. Without a sound knowledge of place value, learners will continue to make mistakes with column addition. These learners would benefit from reinforcing their knowledge of place value. Modelling composing numbers in multiple ways using base-10 blocks will help learners develop the prerequisite skills required for working with algorithms.

Promoting and supporting language

At every stage, learners require mathematical vocabulary to access questions and problem-solving exercises. If appropriate, when a new key word is introduced, ask learners to write a definition in their books, drawing a box around it for emphasis. Encourage learners to write the definition in their own words.

Adapt language when it becomes a barrier to learning, for example, using simpler terms to describe the strategy of compensation: 'take ten away and add one'.

Lesson 1: **Counting on or back (2)**

Learning objectives

Code	Learning objective
5Nc8	Count on or back in thousands, hundreds, tens and ones to add or subtract.

Strand 5: Problem solving

5Pt2	Solve single and multi-step word problems (all four operations); represent them, e.g. with diagrams or a number line.
5Pt7	Consider whether an answer is reasonable in the context of a problem.
5Ps2	Choose an appropriate strategy for a calculation and explain how they worked out the answer.
5Ps3	Explore and solve number problems and puzzles, e.g. logic problems.
5Ps5	Use ordered lists and tables to help to solve problems systematically.
5Ps9	Explain methods and justify reasoning orally and in writing; make hypotheses and test them out.

Prerequisites for learning

Learners need to:

• understand the place value of three-digit numbers
• be able to add ones to any three-digit number.

Success criteria

Learners can:

• count on in tens, hundreds and thousands using place value to know which digits will change
• count on in tens, hundreds and thousands using place value to know how digits will change when crossing the hundreds, thousands and ten thousands boundaries.

Vocabulary

units, tens, hundreds, thousands, multiples of, tens/hundreds/thousands boundary

Refresh

• Choose an activity from Number – Calculation: Mental strategies, Addition and subtraction in the Refresh activities.

Discover [SB]

• Discuss the text and image in the Student's Book. Ask: **How would you count on in hundreds to add $400 to $237? Would you have enough to buy the TV?** Discuss counting on strategies.

Teach and Learn [graph] [SB]

• Discuss the text and Example in the Student's Book. Display: the **Number line tool**, set from 267 to 277. Lead the class counting on in ones from 267 to 277. Ask: **How did the digits change during the count?** Highlight 269, 270, 271 on the screen. Establish that the units digit increases by one until it reaches nine. Say: **When the count crosses the tens boundary, in this case 270, the tens digit increases by one and the units digit resets to zero.**
• Lead learners as they count back in ones from 489 to 479. Highlight crossing the tens boundary in the opposite direction. Adjust the range, from 568 to 668. Lead learners as they count in tens from 568 to 668, pointing to numbers. Ask: **How did the digits change?** Establish that only the tens and hundreds digits changed. Say: **When counting in tens, the units digit remains the same.**
• Explain that when the count crosses the hundreds boundary, 600, the hundreds digit increases by one and the tens digit resets to zero.

• Set the tool from 2037 to 2137. Lead learners as they count back in tens from 2137 to 2037. Highlight crossing the hundreds boundary in the opposite direction. Write: 679 + 60, 3192 − 400. Set the tool to the appropriate range and show how number problems can be solved mentally by counting on or back through tens/hundred barriers, highlighting changing digits. Write 4786 + 40, 4186 − 400. Pairs of learners work these out. (4826, 3786) Repeat for addition and subtraction of multiples of 1000, highlighting that the hundreds, tens and units digits remain unchanged; the thousands digit increases by one until it reaches nine, then resets to zero; then the ten thousands digit increases by one.

Practise [WB]

• Workbook: Counting on or back (2) page 66

Apply [icons]

• Display **Slide 1** and ask pairs of learners to solve the problem, by counting on.

Review [icon]

• Display: **Slide 2.** Ask learners to complete the table.

Assessment for learning

• Which digits change when you count on or back with five-digit numbers in tens, hundreds and thousands? Explain.

Lesson 2: **Adding 2- and 3-digit numbers**

Learning objectives

Code	Learning objective
5Nc10	Use appropriate strategies to add [or subtract] pairs of two- and three-digit numbers [and numbers with one decimal place], using jottings where necessary.

Strand 5: Problem solving

5Pt2	Solve single and multi-step word problems (all four operations); represent them, e.g. with diagrams or a number line.
5Pt3	Check with a different order when adding several numbers or by using the inverse when adding or subtracting a pair of numbers.
5Ps2	Choose an appropriate strategy for a calculation and explain how they worked out the answer.
5Ps3	Explore and solve number problems and puzzles, e.g. logic problems.
5Ps5	Use ordered lists and tables to help to solve problems systematically.
5Ps9	Explain methods and justify reasoning orally and in writing; make hypotheses and test them out.
5Ps10	Solve a larger problem by breaking it down into sub-problems or represent it using diagrams.

Prerequisites for learning

Learners need to:
- count on in ones, tens, hundreds and thousands from any two- or three-digit number.

Success criteria

Learners can:
- support their mental calculations by use of jottings to record intermediate steps, for example using a blank number line

- choose an appropriate mental method from a bank of strategies including bridging, partitioning and compensation.

Vocabulary

addition, counting on, partitioning, compensation, compatible numbers

Resources

mini whiteboard and pen or paper (per pair); Resource sheet 13: Addition strategies

Refresh

- Choose an activity from Number – Calculation: Mental strategies, Addition and subtraction in the Refresh activities.

Discover [SB]

- Discuss the text in the Discover section. Say: **You can choose a mental strategy you feel confident with, to solve an addition problem.**

Teach and Learn [SB]

- Discuss the examples in the Student's Book. Say: **Adding large numbers in your head can be difficult unless you have a strategy to help you. Take a look at some of them.** Distribute Resource sheet 13: Addition strategies. Write: 79 + 60. Ask: **What strategy would you use to complete this calculation?** One strategy would be to count on in tens. Model it on an empty number line, counting on across the 100 boundary: 79, 89 ... 139.
- Discuss the use of addition facts as a mental method. Write: 467 + 500 and say: **I know four add five is nine, so I also know that 400 add 500 is 900, and 467 add 500 is 967.** Write: 480 + 268. Ask: **How can you complete this calculation?** Accept suggestions. Demonstrate the strategy of addition in stages by place value (partitioning): jump from 480 to 680 to add 200; add 60 by bridging to 700, then add 40: 740; finally add 8: 748.

- Write on the board 570 + 347. Pairs of learners complete the calculation. Write: 334 + 245. Ask: **How would you calculate this?** Accept suggestions. Demonstrate addition by both methods of partitioning shown on the Resource sheet. Discuss the concept of looking for compatible numbers in calculations, for example: 325 + 475 = 800 (25 + 75 = 100). Work through some guided examples for compensation, referring to the Resource sheet.

Practise [WB]

- Workbook: Adding 2- and 3-digit numbers page 68
- Refer to Activity 1 from the Additional practice activities.

Apply 👥 🖥

Display **Slide 1** and ask pairs of learners to solve the problem, then share their additions and the strategies they used.

Review

- Write the names of the various strategies on the board. Working in pairs, learners write two addition calculations that are suitable for each strategy. Accept suggestions and solve as a class.

Assessment for learning

- 476 + 398. What strategy would you use to solve this problem? Why?

Number – Calculation: Mental strategies, Addition and subtraction

Lesson 3: **Subtracting 2- and 3-digit numbers**

Learning objectives

Code	Learning objective
5Nc10	Use appropriate strategies to [add or] subtract pairs of two- and three-digit numbers [and numbers with one decimal place], using jottings where necessary.

Strand 5: Problem solving

5Pt2	Solve single and multi-step word problems (all four operations); represent them, e.g. with diagrams or a number line.
5Pt3	Check with a different order when adding several numbers or by using the inverse when adding or subtracting a pair of numbers.
5Ps2	Choose an appropriate strategy for a calculation and explain how they worked out the answer.
5Ps3	Explore and solve number problems and puzzles, e.g. logic problems.
5Ps5	Use ordered lists and tables to help to solve problems systematically.
5Ps9	Explain methods and justify reasoning orally and in writing; make hypotheses and test them out.
5Ps10	Solve a larger problem by breaking it down into sub-problems or represent it using diagrams.

Prerequisites for learning

Learners need to:

- count on and back in ones, tens, hundreds and thousands from any two- or three-digit number.

Success criteria

Learners can:

- support their mental calculations by use of jottings to record intermediate steps, for example, using a blank number line

- choose an appropriate mental method from a bank of strategies including counting back, partitioning and compensation.

Vocabulary

subtraction, counting back, partitioning, bridging compensation, compatible numbers

Resources

Resource sheet 14: Subtraction strategies

Refresh

- Choose an activity from Number – Calculation: Mental strategies, Addition and subtraction in the Refresh activities.

Discover SB

- Discuss the text and example in the Student's Book. Ask: **Has the person received the correct change?** Ask learners to help solve the problem. ($143 – $107 = $36. No)

Teach and Learn SB

- Discuss the text and Example in the Student's Book. Say: **Mental subtraction can be difficult, unless you have a strategy to help you.** Distribute Resource sheet 14: Subtraction strategies. Write: 597 – 340. Ask: **What strategy would you use to work this out?** Elicit that one strategy would be to count back in hundreds, then tens: 497, 397, 297, then 287, 277, 267, 257.

- Discuss the use of subtraction facts as a mental method. Say: **I know nine subtract six is three, so 97 subtract 60 is 37.** Write 826 – 554.

- Draw an empty number line ending in 826. Ask: **How can you work out this calculation?** Demonstrate subtraction in stages by place value (partitioning): jump from 826 to 326 to subtract 500; subtract 54 by bridging to 300 (subtract 26), then subtract 28: 274.

- Write 737 – 444. Ask learners to complete the calculation, using jottings and blank number lines.

- Write: 689 – 247. Ask: **How would you solve this?** Demonstrate partitioning. Discuss looking for compatible numbers in calculations, for example: 925 – 175 = 925 – 100 – 50 – 25. Subtract and rearrange: 825 – 25 – 50 = 750. Work through guided examples for compensation, referring to the resource sheet.

Practise WB

- Workbook: Subtracting 2- and 3-digit numbers page 70
- Refer to Activity 1 from the Additional practice activities.

Apply 👥 🖥

- Display **Slide 1** and ask pairs of learners to solve the problem and discuss their strategies.

Review

- Write the names of the strategies on the board. Working in pairs, learners write two subtraction calculations that are suitable for each strategy. Accept suggestions and solve as a class.

Assessment for learning

- 879 – 443. What strategy would you use to solve this problem? Why?

Lesson 4: **Adding more than two numbers**

Number – Calculation: Mental strategies, Addition and subtraction

Learning objectives

Code	Learning objective
5Nc18	Find the total of more than three two- or three-digit numbers using a written method.

Strand 5: Problem solving

5Pt2	Solve single and multi-step word problems (all four operations); represent them, e.g. with diagrams or a number line.
5Pt3	Check with a different order when adding several numbers or by using the inverse when adding or subtracting a pair of numbers.
5Pt6	Estimate and approximate when calculating, e.g. using rounding, and check working.
5Ps2	Choose an appropriate strategy for a calculation and explain how they worked out the answer.
5Ps3	Explore and solve number problems and puzzles, e.g. logic problems.
5Ps5	Use ordered lists and tables to help to solve problems systematically.
5Ps4	Deduce new information from existing information to solve problems.
5Ps9	Explain methods and justify reasoning orally and in writing; make hypotheses and test them out.
5Ps10	Solve a larger problem by breaking it down into sub-problems or represent it using diagrams.

Prerequisites for learning

Learners need to:

- understand the place value of two- and three-digit numbers
- use the formal written method and know how to carry units and tens.

Success criteria

Learners can:

- write the calculation vertically and make a sensible estimate

- use the formal written method, carrying units tens and hundreds when needed.

Vocabulary

place value, estimate, carry, palindrome

Resources

mini whiteboard and pen or paper (per pair)

Refresh

- Choose an activity from Number – Calculation: Mental strategies, Addition and subtraction in the Refresh activities.

Discover [SB]

- Discuss the Discover section. Ask: **How would you add the numbers in the leaf count?** Discuss strategies. Ask: **What other situation might require you to add more than two numbers?**

Teach and Learn [SB]

- Discuss the text and example in the Student's Book. On the board, write: 37, 64, 49, 57. Ask: **How would you find the sum of these numbers?** (mentally or by using a written method, such as column addition or the expanded written method). Ask: **What is your estimate for the answer?** Say: **You can use the column method in the same way for adding two numbers.** Remind learners to align the digits according to their place value. Say: **Remember, begin on the right and work to the left. If the sum of any column is nine or more, carry to the column to the left. Continue adding the digits in each column, remembering to add any carried digits.** Work as a class.

- Now work through the same calculation on the board, this time using the expanded written method,

partitioning, or the grid method. Emphasise that there are several methods the learners can use.

- Work through an example for a set of three-digit numbers: 478, 345, 289, 567. Show how the 16 hundreds in the final sum is written as 6 (for 600) in the hundreds column and 1 (for 1000) in the thousands column.

Practise [WB]

- Workbook: Adding more than two numbers page 72
- Refer to Activity 2 from the Additional practice activities.

Apply 👥 🖥

- Display **Slide 1** and ask pairs to solve the problem.

Review

- On the board, write a list of measurements: 547 cm, 248 cm, 619 cm, 488 cm. Say: **If these were lengths for pieces of rope, would the pieces combine to make a length of two metres?** Learners work in pairs to find the total. (1902 cm; no)

Assessment for learning

- Write down four three-digit numbers and add them. What happens when a column total is greater than nine?

Additional practice activities

Activity 1 Challenge 2

Learning objective

• Select and use an appropriate strategy to find the sum of two three-digit numbers.

Resources

mini whiteboard and pen or paper (per pair)

What to do

• Before starting the activity arrange the numbers 70, 99, 378, 125, 210, 575, 470, 347, 388, 429, 298, 330, 59, 102, 233, 477 as eight pairs, each on a separate piece of paper, and work out the sum in each case.

• On the board, write the same numbers, randomly arranged.

• Learners draw a two by four grid. They choose eight pairs of numbers from the board and find the sum of each pair. They write the answers in the grid.

• Read out the list of addition answers you made, one at a time. Learners cross through the numbers on their grid if they have them.

• The first learner to mark four numbers in a line calls out 'Lined up!' and wins the game.

Variations

Challenge 1 Provide a set of numbers in which half the numbers are multiples of ten.

Challenge 3 Learners make subtractions with pairs of numbers and write their answers on their grids.

Activity 2 Challenge 2

Learning objectives

• Select and use an appropriate strategy to subtract one three-digit number from another.
• Solve a number puzzle that involves the addition of four three-digit numbers.

Resources

mini whiteboard and pen or paper (per pair)

What to do 🖥

• Display: **Slide 1**.

• One learner in each pair secretly selects one number from each row. They construct a set of calculations using the numbers where the number in the row below is subtracted from the result of the previous subtraction in the row above, for example:

999 − 567 = 432
432 − 298 = 134
134 − 127 = 7

• They record the final answer and present it to their partner who must work out which set of four numbers (including the answer) would add together to make the original 999.

• Partners swap roles.

Variation

Challenge 1 Make two of the numbers in every row a multiple of ten.

Unit 9: Addition and subtraction 2

Learning objectives

Code	Learning objective
5Nc1	Know by heart pairs of one-place decimals with a total of 1, e.g. 0·8 + 0·2.
5Nc2	Derive quickly pairs of decimals with a total of 10, and with a total of 1.
5Nc8*	Count on or back in thousands, hundreds, tens and ones to add or subtract.
5Nc9	Add or subtract near multiples of 10 or 100, e.g. 4387 – 299.
5Nc10*	Use appropriate strategies to add or subtract pairs of two- and three-digit numbers [and numbers with one decimal place], using jottings where necessary.
5Nc11	Calculate differences between near multiples of 1000, e.g. 5026 – 4998, or near multiples of 1, e.g. 3·2 – 2·6.
5Nc18*	Find the total of more than three two- or three-digit numbers using a written method.
5Nc19	Add or subtract any pair of three- and/or four-digit numbers, with the same number of decimal places, including amounts of money.

*Learning objective revised and consolidated

Strand 5: Problem solving

5Pt2	Solve single and multi-step word problems (all four operations); represent them, e.g. with diagrams or a number line.
5Pt3	Check with a different order when adding several numbers or by using the inverse when adding or subtracting a pair of numbers.
5Pt6	Estimate and approximate when calculating, e.g. using rounding, and check working.
5Pt7	Consider whether an answer is reasonable in the context of a problem.
5Ps2	Choose an appropriate strategy for a calculation and explain how they worked out the answer.
5Ps3	Explore and solve number problems and puzzles, e.g. logic problems.
5Ps4	Deduce new information from existing information to solve problems.
5Ps5	Use ordered lists and tables to help to solve problems systematically.
5Ps8	Investigate a simple general statement by finding examples which do or do not satisfy it, e.g. the sum of three consecutive whole numbers is always a multiple of three.
5Ps9	Explain methods and justify reasoning orally and in writing; make hypotheses and test them out.
5Ps10	Solve a larger problem by breaking it down into sub-problems or represent it using diagrams.

Unit overview

In this unit, learners begin by focussing on mental methods for addition and subtraction. They use jottings to assist computation. They practise counting on or back in thousands, hundreds, tens and units and apply this to solving mental addition and subtraction problems involving multiples of ten. Learners encounter a range of strategies for solving problems mentally, including partitioning, compensation and compatible numbers. Guided examples are provided, together with practical activities that require learners to use appropriate strategies to solve problems. In so doing they learn that some strategies are more effective than others, depending on the calculation. The unit concludes with a review of written methods of addition and subtraction, focusing on adding and subtracting decimals in the context of money.

Common difficulties and remediation

Some learners lack confidence counting on or back in tens or hundreds, and find crossing tens and hundreds boundaries difficult. Consequently, they find addition and subtraction of multiples of ten and 100 challenging. Some learners might have difficulty adding and subtracting decimals in the context of money. They may not fully understand the relationship between 1, 10 and 100 and, therefore, do not recognise that adding or subtracting 10 or 100 is no more difficult than adding or subtracting 1.

When number problems become more complex, encourage learners to use a strategy. Some may still struggle to use known number facts and place value to add mentally. Provide these learners with guided examples, modelled with Base 10 equipment or other manipulatives, such as counters or place value arrow cards.

Link each practical step to a recorded step. Encourage learners to estimate to help find their answers, whilst acknowledging that estimating is a difficult skill to master – praise learners who are willing to try it.

For work with decimals, provide fraction tiles to allow learners to regroup tenths into wholes.

Promoting and supporting language

At every stage, learners require mathematical vocabulary to access questions and problem-solving exercises. If appropriate, when a new key word is introduced, ask learners to write a definition in their own words.

Lesson 1: **Near multiples of 10 or 100**

Learning objectives

Code	Learning objective
5Nc8	Count on or back in thousands, hundreds, tens and ones to add or subtract.
5Nc9	Add or subtract near multiples of 10 or 100, e.g. 4387 – 299.

Strand 5: Problem solving

5Pt2	Solve single and multi-step word problems (all four operations); represent them, e.g. with diagrams or a number line.
5Pt7	Consider whether an answer is reasonable in the context of a problem.
5Ps2	Choose an appropriate strategy for a calculation and explain how they worked out the answer.
5Ps3	Explore and solve number problems and puzzles, e.g. logic problems.
5Ps5	Use ordered lists and tables to help to solve problems systematically.
5Ps9	Explain methods and justify reasoning orally and in writing; make hypotheses and test them out.
5Ps10	Solve a larger problem by breaking it down into sub-problems or represent it using diagrams.

Prerequisites for learning

Learners need to:

- be able to count on or back in tens, hundreds and thousands, using place value to know which digits will change
- support mental calculations by use of jottings, for example, using a blank number line.

Success criteria

Learners can:

- add mentally near multiples of 10 or 100

- subtract mentally near multiples of 10 or 100

Vocabulary

compensation, compensate (adjust)

Resources

mini whiteboard and pen or paper (per learner); 1–6 dice, alternatively use Resource sheet 4: 1–6 spinners (per pair); counter (per learner); Resource sheet 12: Blank number lines (per learner)

Refresh

- Choose an activity from Number – Calculation: Mental strategies, Addition and subtraction in the Refresh activities.

Discover [SB]

- Discuss the text and images in the Student's Book. Say: **Tell your partner where you have seen prices with nines at the end.** Invite volunteers to share their experiences.

Teach and Learn 🖵[SB]

- Discuss the text and example in the Student's Book. Display: **Slide 1**. Ask: **What do the numbers on the right all have in common?** Elicit they are all 'near multiples' of 10 or 100.

- Look at the numbers and operations on the left. Ask: **Which strategy would you use to solve these addition and subtraction problems?** After discussion, take suggestions. Elicit that they could use the strategy of compensation (rounding and adjusting). Remind learners that one number is rounded, to simplify the calculation, then the answer is adjusted to compensate for the original change. Demonstrate through two examples.

- Say: **For 73 add 89, round 89 to 90. Solve 73 add 90 (163) then subtract one to adjust for adding**

one too many (162). Say: **For 237 – 161, round 161 to 160. Solve 237 subtract 160 (77) then subtract one for subtracting one too few (76).** Distribute Resource sheet 12: Blank number lines and ask pairs of learners to use them to help to solve the remaining problems.

Practise [WB]

- Workbook: Near multiples of 10 or 100 page 74

Apply 👥🖵

- Display **Slide 2** and ask learners to work in pairs to solve the problem, then discuss their solutions and strategies.

Review 🖵

- Display: **Slide 3**. Ask learners to copy and complete the table.

Assessment for learning

- When I subtract 237 from 526, I round 237 to 240, subtract and add three. Why do I add three?

Lesson 2: **Near multiples of 1000**

Number – Calculation: Mental strategies, Addition and subtraction

Learning objectives

Code	Learning objective
5Nc8	Count on or back in thousands, hundreds, tens and ones to add or subtract.
5Nc11	Calculate differences between near multiples of 1000, e.g. 5026 – 4998, [or near multiples of 1, e.g. 3·2 – 2·6].

Strand 5: Problem solving

5Pt2	Solve single and multi-step word problems (all four operations); represent them, e.g. with diagrams or a number line.
5Pt7	Consider whether an answer is reasonable in the context of a problem.
5Ps2	Choose an appropriate strategy for a calculation and explain how they worked out the answer.
5Ps3	Explore and solve number problems and puzzles, e.g. logic problems.
5Ps5	Use ordered lists and tables to help to solve problems systematically.
5Ps9	Explain methods and justify reasoning orally and in writing; make hypotheses and test them out.
5Ps10	Solve a larger problem by breaking it down into sub-problems or represent it using diagrams.

Prerequisites for learning

Learners need to:

- count on or back in tens, hundreds and thousands using place value to know which digits will change
- support mental calculations by use of jottings, for example, using a blank number line.

Success criteria

Learners can:

- subtract mentally near multiples of 1000 by using

a blank number line or jottings, bridging through the thousands barrier using number pairs

- subtract mentally by using the strategy of compensation (rounding and adjusting)

Vocabulary

difference, bridge, place value, jottings, compensation

Resources

counter (per learner); Resource sheet 15: 1–3 spinner (per pair)

Refresh

- Choose an activity from Number – Calculation: Mental strategies, Addition and subtraction in the Refresh activities.

Discover 〔SB〕

- Read the Discover text with the learners. Ask: **In what other situations might you need to find the difference between near multiples of 1000? How would you solve this problem?** Discuss possible strategies.

Teach and Learn 〔SB〕

- Discuss the text and Example in the Student's Book. On the board, write: 4013 – 1998. Ask: **How would you work this out?** Elicit rounding and adjusting and ask learners to use this method to calculate the answer (2015).
- Say: **Compensation is a good strategy but there are others.** Draw a number line on the board, labelled from 1998 to 4013. Say: **You can use a bridging strategy to find the answer.** Mark and label 2000 and 4000 on the line. Draw an arc from 1998 to 2000. Say: **First you bridge to the nearest thousand, a difference of two.** Write 2 above the line. Then bridge to the 4000, the nearest thousand to 4013. Draw an arc between 2000 and 4000. Say: **The difference between 4000 and 2000 is 2000.** Write 2000 above the line. Say: **Finally, bridge to 4013, a difference of 13.** Write 13 above the line.

- The sum of the numbers above the line gives the difference (2015).
- Display: the **Place value tool**. Demonstrate counting forwards from 1998 to 4013 in the same steps: + 2, + 2000, + 13. Write: 9026 – 4001. Pairs of learners use the bridging strategy to calculate the answer.

Practise 〔WB〕

- Workbook: Near multiples of 1000 page 76
- Refer to Activity 1 from the Additional practice activities.

Apply 〔icons〕

Display **Slide 1** and ask pairs of learners to find the difference and discuss their answers and solutions. (height of space above the cupboard = 1027 mm: yes there is enough space)

Review

- On the board, write: 7023 – 1997; 8001 – 3991; 9017 – 6992. Divide the class in half. Ask one half to work out the answers by rounding and adjusting, while the other half uses a bridging strategy. They solve the problems and then the groups swap methods.

Assessment for learning

- When I subtract 2998 from 5013, I calculate in three stages. What are the stages?

Lesson 3: Decimals that total 1 or 10

Learning objectives

Code	Learning objective
5Nc1	Know by heart pairs of one-place decimals with a total of 1, e.g. 0·8 + 0·2.
5Nc2	Derive quickly pairs of decimals with a total of 10, and with a total of 1.

Strand 5: Problem solving

5Pt2	Solve single and multi-step word problems (all four operations); represent them, e.g. with diagrams or a number line.
5Pt7	Consider whether an answer is reasonable in the context of a problem.
5Ps2	Choose an appropriate strategy for a calculation and explain how they worked out the answer.
5Ps3	Explore and solve number problems and puzzles, e.g. logic problems.
5Ps5	Use ordered lists and tables to help to solve problems systematically.
5Ps9	Explain methods and justify reasoning orally and in writing; make hypotheses and test them out.
5Ps10	Solve a larger problem by breaking it down into sub-problems or represent it using diagrams.

Prerequisites for learning

Learners need to:

- understand the place value of decimals to one place
- know number bonds to 10 and 100.

Success criteria

Learners can:

- quickly derive decimal complements to 1 by applying knowledge of number bonds to 10
- quickly derive decimal complements to 10 by applying knowledge of number bonds to 100.

Vocabulary

decimal pair

Resources

mini whiteboard and pen or paper (per learner): Resource sheet 12: Blank number lines

Refresh

- Choose an activity from Number – Calculation: Mental strategies, Addition and subtraction in the Refresh activities.

Discover [SB]

- Discuss the Discover image and text with the learners. Ask: **If a different length was sawn off, say 7·4m, how would this affect the answer?** Accept suggestions and discuss strategies.

Teach and Learn [SB]

- Discuss the text and Example in the Student's Book. Display: the **Number line tool**, set from 0 to 1 with increments of 0·1. Place the pointer above 0·3. Ask: **What do I need to add to 0·3 to make 1?** (0·7) Ask learners to say how they worked this out. Repeat for 0·4. (0·6) Say: **How many one-place decimal pairs that total one are there?** Ask learners to list them on their whiteboards or paper. Establish that there is no need to continue beyond five, as the order of adding two numbers does not matter. Ask: **How did you find all these pairs?** Ensure that learners realise digits in the tenths place value position of each pair are the number bonds to ten. Ask: **Why do you think this is?** Elicit that it is because the decimal bonds are the whole number bonds, but ten times smaller.
- Adjust the number line to range 0 to 10 and increments of 0·1. Place the pointer above 7·2.

Ask: **What do I need to add to 7·2 to make 10?** Establish that it is 2·8. Repeat for 3·9. (6·1) Say: **Use a number line to write ten decimal pairs that total ten.** Learners use Resource sheet 12, then discuss their findings. Say: **You may have used the number line to find decimal pairs but did anyone use a different strategy?** Praise learners who realised that digits in pairs correspond to number bonds to 100.

Practise [WB]

- Workbook: Decimals that total 1 or 10 page 78
- Refer to Activity 2 from the Additional practice activities.

Apply [icons]

- Display **Slide 1** and ask pairs to work out the length of the fourth piece of wood.

Review

- On the board, write: 0·2 + ? = 1; 4·4 + ? = 10; 7·1 + ? = 10. Ask learners to complete the number sentences. (0·8, 5·6, 2·9)

Assessment for learning

- I know that 3·3 add 6·7 makes 10 because 33 add 67 is 100. Explain my reasoning.

Unit 9 Addition and subtraction 2

Lesson 4: **Near multiples of 1**

Learning objectives

Code	Learning objective
5Nc11	Calculate differences between [near multiples of 1000, e.g. 5026 – 4998, or] near multiples of 1, e.g. 3·2 – 2·6.

Strand 5: Problem solving

5Pt2	Solve single and multi-step word problems (all four operations); represent them, e.g. with diagrams or a number line.
5Pt7	Consider whether an answer is reasonable in the context of a problem.
5Ps2	Choose an appropriate strategy for a calculation and explain how they worked out the answer.
5Ps3	Explore and solve number problems and puzzles, e.g. logic problems.
5Ps5	Use ordered lists and tables to help to solve problems systematically.
5Ps9	Explain methods and justify reasoning orally and in writing; make hypotheses and test them out.
5Ps10	Solve a larger problem by breaking it down into sub-problems or represent it using diagrams.

Prerequisites for learning

Learners need to:
- know one-place decimal pairs that total one.

Success criteria

Learners can:
- subtract mentally near multiples of one by using a blank number line or jottings to count up, bridging through the ones barrier and using knowledge of decimal complements of one

- subtract mentally by using the strategy of compensation (rounding to the nearest one and adjusting).

Vocabulary

decimal pair, bridging, compensation, compensate (adjust)

Resources

mini whiteboard and pen; 1–9 number cards from Resource sheet 2: 0–100 number cards (per learner)

Refresh

- Choose an activity from Number – Calculation: Mental strategies, Addition and subtraction in the Refresh activities.

Discover [SB]

- Discuss the Discover section text with the learners. Ask: **What strategies could you use to solve this problem? You will be solving it later in the lesson.**

Teach and Learn [SB]

- Discuss the text and Example in the Student's Book. On the board, write: 7·3 – 2·8. Ask: **How would you work this out?** Praise learners who suggest compensation. Agree that they could round 2·8 to 3, subtract, then adjust the answer by adding 0·2 to compensate for subtracting 0·2 too many. Ask learners to use this method to calculate the answer (4·5).

- Say: **Compensation is a good strategy, but you could also use bridging.** Draw a number line on the board, labelled from 2·8 to 7·3. Say: **You can calculate mentally, but use a blank number line to help picture the method.** Mark and label 3·0 and 7·0 on the number line. Draw an arc from 2·8 to 3·0. Say: **First you bridge to the nearest one, which is a difference of …?** Expect learners to say 0·2. Say: **This is where it is useful to know decimal number pairs to one. Knowing that 0·8 add 0·2 makes one really speeds up your mental calculation.** Write

+ 0·2 above line. Say: **Now bridge to the seven, the nearest one to 7·3.** Draw an arc between 3 and 7. Say: **The difference is four.** Write + 4 above line. Ask: **Finally, bridge to 7·3, which is a difference of …?** (0·3) Write + 0·3 above line. Show that the sum of the numbers above the line gives the difference (4·5).

- Display: the **Place value tool**. Demonstrate counting forwards from 2·8 to 7·3 in the same steps: + 0·2, + 4, + 0·3. On the board, write: 12·4 – 6·9. Learners use the bridging strategy to calculate the answer. Invite volunteers to explain their solution.

Practise [WB]

- Workbook: Near multiples of 1 page 80

Apply 👥 🖥

- Display **Slide 1** and ask learners to solve the problem and discuss solutions (1·7 m).

Review

- On the board, draw a table with columns headed 1·8, 2·7, 3·9 and rows labelled 8·3, 11·2, 15·1. Ask learners to copy and complete the table, by finding the difference between the numbers in the rows and those in the columns.

Assessment for learning

- When I subtract 2·8 from 9·3, I calculate in three stages. What are the stages?

Lesson 5: **Adding and subtracting 2- and 3-digit numbers**

Learning objectives

Code	Learning objective
5Nc10	Use appropriate strategies to add or subtract pairs of two- and three-digit numbers [and numbers with one decimal place], using jottings where necessary.

Strand 5: Problem solving

5Pt2	Solve single and multi-step word problems (all four operations); represent them, e.g. with diagrams or a number line.
5Pt6	Estimate and approximate when calculating, e.g. using rounding, and check working.
5Pt7	Consider whether an answer is reasonable in the context of a problem.
5Ps2	Choose an appropriate strategy for a calculation and explain how they worked out the answer.
5Ps3	Explore and solve number problems and puzzles, e.g. logic problems.
5Ps5	Use ordered lists and tables to help to solve problems systematically.
5Ps9	Explain methods and justify reasoning orally and in writing; make hypotheses and test them out.
5Ps10	Solve a larger problem by breaking it down into sub-problems or represent it using diagrams.

Prerequisites for learning

Learners need to:

- understand the place values of the digits in the numbers up to 1000
- be able to add and subtract mentally three-digit numbers and hundreds, tens and ones.

Success criteria

Learners can:

- support their mental calculations by use of jottings to record intermediate steps, for example using a blank number line

- choose an appropriate mental method from a bank of strategies, including bridging, partitioning and compensation.

Vocabulary

counting back, partitioning, bridging, compensation, compatible numbers

Resources

mini whiteboard and pen or paper (per pair); counters (per pair); Resource sheet 17: 0–9 spinner (per pair); Resource sheet 14: Subtraction strategies (per learner); Resource sheet 13: Addition strategies (per learner)

Refresh

- Choose an activity from Number – Calculation: Mental strategies, Addition and subtraction in the Refresh activities.

Discover [SB]

- Discuss the Discover section. Ask: **Where else might you need to calculate mentally the difference between two numbers?** Ask learners what strategies they would use to determine the difference.

Teach and Learn [SB]

- Discuss the text and Example in the Student's Book. Write: 673 – 250. Ask: **How would you work this out?** Discuss strategies, referring to Resource sheet 14: Subtraction strategies. Elicit that the simplest strategy is to apply knowledge of place value and think which digits will change and which will stay the same. Establish that only the hundreds and tens digits will change, as no units are being subtracted. Write the numbers in partitioned form: (600 + 70 + 3) – (200 + 50). Demonstrate subtraction of place values: 673 – 250 = (600 – 200) + (70 – 50) + 3 = 423. Repeat for 486 + 440. Write: 733 – 498, 675 + 425, 957 – 346, 841 + 376. Ask: **How do these problems differ from the previous one?** (all three digits will change). Discuss

strategies and ask learners to suggest the most appropriate. Choose one problem and model the use of a blank number line and jottings for support. Say: **Work out these calculations mentally, showing your working out.** Check answers.

Practise [WB]

- Workbook: Adding and subtracting 2- and 3-digit numbers page 82

Apply 👥 🖥

- Display **Slide 1**. Pairs establish the order (order: 219 ml, 248 ml, 267 ml, 283 ml; differences: 29 ml, 19 ml, 16 ml.) Discuss the strategies they used.

Review

- Write the names of the strategies on the board. Working in pairs, learners write three addition and three subtraction calculations that can be solved by each strategy. Accept suggestions, ask learners to estimate answers and then solve as a class.

Assessment for learning

- 813 + 498. What strategy would you use to solve this problem? Why?

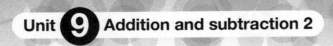

Unit **9** Addition and subtraction 2

Lesson 6: **Adding more than two numbers**

Learning objectives

Code	Learning objective
5Nc18*	Find the total of more than three two- or three-digit numbers using a written method.*

Strand 5: Problem solving	
5Pt2	Solve single and multi-step word problems (all four operations); represent them, e.g. with diagrams or a number line.
5Pt3	Check with a different order when adding several numbers or by using the inverse when adding or subtracting a pair of numbers.
5Pt6	Estimate and approximate when calculating, e.g. using rounding, and check working.
5Pt7	Consider whether an answer is reasonable in the context of a problem.
5Ps2	Choose an appropriate strategy for a calculation and explain how they worked out the answer.
5Ps3	Explore and solve number problems and puzzles, e.g. logic problems.
5Ps5	Use ordered lists and tables to help to solve problems systematically.
5Ps8	Investigate a simple general statement by finding examples which do or do not satisfy it, e.g. the sum of three consecutive whole numbers is always a multiple of three.
5Ps9	Explain methods and justify reasoning orally and in writing; make hypotheses and test them out.
5Ps10	Solve a larger problem by breaking it down into sub-problems or represent it using diagrams.

Prerequisites for learning

Learners need to:
• add using the formal written method

Success criteria

Learners can:
• use column addition to add several numbers, carrying ones, tens and hundreds when needed.

Vocabulary

budget, place value, estimate, carry

Resources

mini whiteboard and pen (per learner)

Refresh

• Choose an activity from Number – Calculation: Mental strategies, Addition and subtraction in the Refresh activities.

Discover 𝗦𝗕

• Discuss the text and image in the Student's Book. Ask: **How would you find the combined mass of the boxes of rice?** Use column addition and work through the problem. (843 g)

Teach and Learn 𝗦𝗕

• Discuss the example in the Student's Book. Ask: **If I went to a shop and bought four boxes of cereal weighing 125 g, 242 g, 356 g and 265 g, what mass of cereal would I have in total?** Remind learners they have added more than two numbers before. Ask: **Is this a problem to be calculated mentally?** Elicit it is easier to use a written method, such as column addition. Learners estimate the answer by rounding. Establish that the sum should be close to 1000 g (1 kg). Learners work in pairs to find the combined mass. Say: **Remember, begin on the right and work left. If the sum of any column is ten or more, carry to the column to**

the left. **Continue adding digits in each column, not forgetting to add in any carried digits.** Share solutions. (988 g) Establish that learners are confident in using column addition, for example: **Why was a one placed in the tens column under the line? Why is the sum of the digits in the tens column 18; I make it 17.**

Practise 𝗪𝗕

• Workbook: Adding more than two numbers page 84
• Refer to Activity 3 from the Additional practice activities.

Apply 👥 🖥

• Display **Slide 1** and ask pairs of learners to work through the problem.

Review

• On the board, write a set of volumes: 683 ml, 437 ml, 885 ml, 727 ml. Learners find the total and write it in mixed units. (2 ℓ 732 ml)

Assessment for learning

• When I add 433, 679, 378 and 895 I get 2395. Is that right? (no; 2385)

Lesson 7: **Adding decimals**

Learning objectives

Code	Learning objective
5Nc19	Add or subtract any pair of three- and/or four-digit numbers, with the same number of decimal places, including amounts of money.

Strand 5: Problem solving

5Pt2	Solve single and multi-step word problems (all four operations); represent them, e.g. with diagrams or a number line.
5Pt6	Estimate and approximate when calculating, e.g. using rounding, and check working.
5Pt7	Consider whether an answer is reasonable in the context of a problem.
5Ps2	Choose an appropriate strategy for a calculation and explain how they worked out the answer.
5Ps3	Explore and solve number problems and puzzles, e.g. logic problems.
5Ps5	Use ordered lists and tables to help to solve problems systematically.
5Ps8	Investigate a simple general statement by finding examples which do or do not satisfy it, e.g. the sum of three consecutive whole numbers is always a multiple of three.
5Ps9	Explain methods and justify reasoning orally and in writing; make hypotheses and test them out.
5Ps10	Solve a larger problem by breaking it down into sub-problems or represent it using diagrams.

Prerequisites for learning

Learners need to:

• understand the formal written method for addition with whole numbers.

Success criteria

Learners can:

• use column addition to add decimals with one or two places, aligning digits and decimal points correctly.

Vocabulary

place value, estimate, carry

Resources

mini whiteboard and pen or paper (per pair)

Refresh

• Choose an activity from Number – Calculation: Mental strategies, Addition and subtraction in the Refresh activities.

Discover [SB]

• Discuss the image in the Discover section. Ask: **How would you find the sum of three items on the menu? Does the decimal point present a problem?** Elicit that learners can use column addition, remembering to put the decimal point in the answer.

Teach and Learn [📊] [SB]

• Discuss the text and example in the Student's Book. Display: the **Place value tool** showing 3·47. Say: **Tell me a number that is three tenths more than this number.** (3·77) **Tell me a number that is two hundredths more.** (3·49) **Tell me a number that is seven tenths less.** (2·77) **Tell me a number that is 9 hundredths less.** (3·38) Check learners' answers by asking them to show their whiteboards. Ask similar questions for other decimals: 2·6, 42·9, 58·64.

• Ask: **When might you use decimal numbers?** Accept suggestions. Prompt discussion of money and other measures. Say: **It is important to be able to add and subtract decimals just as you do with whole numbers. Whether you do this mentally or with written methods depends on the decimals.** Write: $24.56 + $37.39. Demonstrate column addition, with learners assisting. Emphasise the importance

of aligning the decimal points and arranging the digits by their place value columns, and then writing the decimal point in the correct place in the answer. Say: **Six hundredths and nine hundredths add up to 15 hundredths, which have the same value as one tenth and five hundredths. Write the five hundredths in the hundredths column and carry the one to the tenths column.** Model clearly writing the decimal point in the answer line ($62.05). Write: $46.57 + $38.39. Ask learners to work in pairs to work it out. Discuss any problems. ($84.96)

Practise [WB]

• Workbook: Adding decimals page 86

Apply [👥][🖥]

• Display **Slide 1** and ask pairs of learners to calculate the answers.

Review

• On the board, write: $29.48 + $45.37; $58.77 + $33.66. Learners discuss their solutions and working. ($74.85, $92.43)

Assessment for learning

• When I use column addition to add $47.26 to $35.49, I write a one in the tenths column below the answer line. Why?

Number – Calculation: Mental strategies, Addition and subtraction

121

Lesson 8: **Subtracting decimals**

Number – Calculation: Mental strategies, Addition and subtraction

Learning objectives

Code	Learning objective
5Nc19	Add or subtract any pair of three- and/or four-digit numbers, with the same number of decimal places, including amounts of money.

Strand 5: Problem solving

Code	Learning objective
5Pt2	Solve single and multi-step word problems (all four operations); represent them, e.g. with diagrams or a number line.
5Pt6	Estimate and approximate when calculating, e.g. using rounding, and check working.
5Pt7	Consider whether an answer is reasonable in the context of a problem.
5Ps2	Choose an appropriate strategy for a calculation and explain how they worked out the answer.
5Ps3	Explore and solve number problems and puzzles, e.g. logic problems.
5Ps4	Deduce new information from existing information to solve problems.
5Ps5	Use ordered lists and tables to help to solve problems systematically.
5Ps8	Investigate a simple general statement by finding examples which do or do not satisfy it, e.g. the sum of three consecutive whole numbers is always a multiple of three.
5Ps9	Explain methods and justify reasoning orally and in writing; make hypotheses and test them out.
5Ps10	Solve a larger problem by breaking it down into sub-problems or represent it using diagrams.

Prerequisites for learning

Learners need to:
- understand the formal written method for subtraction with whole numbers.

Success criteria

Learners can:
- use column subtraction to subtract decimals with one or two places, aligning digits and decimal points correctly.

Vocabulary

place value, estimate, rename (decompose)

Resources

mini whiteboard and pen or paper (per pair)

Refresh

- Choose an activity from Number – Calculation: Mental strategies, Addition and subtraction in the Refresh activities.

Discover [SB]

- Discuss the text and images in the Student's Book. Ask: **How would you calculate the reductions?** Share possible strategies.

Teach and Learn [tool] [SB]

- Discuss the text and example in the Student's Book. Display the **Base 10 tool**, showing two rows of blocks representing 63·86 and 37·59
- Write: $63.86 – $37.59. Demonstrate column subtraction, modelling each stage on the tool. Say: **As you cannot subtract nine hundredths from six hundredths, you need to change the six hundredths. Look to the eight tenths and rename this value as seven tenths and ten hundredths.** Cross out the 8 and write 7 above it. Explain that the ten hundredths are added to the six in the hundredths, making 16 hundredths. Write a 1 next to the six hundredths (or cross out the 6 and write 16 above it). Write: 16 – 9 = 7. Write 7 in the answer line in the hundredths column. Say: **Seven tenths subtract five tenths is two tenths.** Write 2 in the answer line then write the decimal point to the left of the 2.

Point to the 3 in the ones column. Say: **As you can't subtract seven ones from three ones, change the three ones. To do this, look to the six tens and rename this value as five tens and ten ones.** Cross out the 6 and write 5 above it. Explain that ten ones are added to the three in the ones, making thirteen ones. Write a 1 next to the three ones. Show how the subtraction proceeds. ($26.27)

Practise [WB]

- Workbook: Subtracting decimals page 88
- Refer to Activity 4 from the Additional practice activities.

Apply [pair] [screen]

- Display **Slide 1**. Pairs solve the problem then share their working.

Review

- On the board, write: $67.44 – $38.29; $96.96 – $48.39. Learners discuss their solutions and working.

Assessment for learning

- When I use column subtraction to subtract $24.93 from $38.78, I cross through 8 in the ones column, writing 7 next to it and a 1 next to the 7 in the tenths column. Why?

Additional practice activities

Activity 1

Learning objectives

- Use a range of strategies to mentally add or subtract near multiples of 10, 100 and 1000.
- Discuss and evaluate the relative strengths of different strategies.

Resources

mini whiteboard and pen or paper (per pair)

What to do

- On the board, write these addition and subtraction problems: 79 + 59; 93 – 41; 756 + 598; 832 – 403; 4887 + 2402; 7383 – 3598.
- Divide the class into three groups: 'the count-on-ers', 'the compensators' and 'the bridgers'.
- Say: **Work in pairs within your group. Use your given strategy to solve all six problems. Share your solutions with your group to confirm that you are all in agreement with the answer.**
- Time the groups to see how long they take to finish. Learners share their answers and confirm that all groups have correct answers. Invite them to discuss and evaluate the strengths of the strategy they used.

Variation

Challenge 1 Help learners by displaying number lines on the board with ranges appropriate to each problem.

Activity 2

Learning objective

- Use decimal pairs that total 1 or 10 to calculate the difference between decimals to one place.

Resources

1–6 dice, alternatively use Resource sheet 4: 1–6 spinners (per pair)

What to do

- On the board, write these numbers: 1·8, 2·7, 3·9.
- Learners roll a dice three times to create a target number, a three-digit number with one decimal place, for example, 47·3.
- They choose a number from the board. One learner repeatedly subtracts this number from the target decimal, stopping before the answer becomes negative. The other learner counts on in steps of the board number until they reach a number equal to, or greater than, the target number. Before they do, learners estimate the number at which the two sequences will cross.
- The learner closer to the crossing point after the sequences are revealed is the winner.

Variation

Challenge 1 Narrow the range to 1 to 10 and use these numbers on the board: 0·7, 0·8 and 0·9.

Additional practice activities

Activity 3

Learning objective
- Subtract a three-digit number from the sum of three three-digit numbers as part of a game involving addition and subtraction.

Resources
mini whiteboard and pen or paper (per pair); Resource sheet 16: 0–3 spinner and Resource sheet 17: 0–9 spinner (per pair)

What to do
- Learners decide to be either the 'adder' or the 'subtractor'. The adder begins the game with 1000 points.
- Players take turns to spin. The adder spins the 0–3 spinner nine times to create three three-digit numbers, then uses column addition to find the sum of the three numbers. The sum is added to their 1000 points and this is their starting score.
- The subtractor spins the 0–9 spinner three times to create a three-digit number. They use column subtraction to subtract this number from the adder's total.
- Each player has ten rounds of spins. If the total points drop to zero the subtractor wins; if the points remain positive then the adder wins.

Variation
Challenge 1 Play the game with two-digit numbers. The adder begins on 100.

Activity 4

Learning objective
- Add or subtract prices in dollars and cents as part of a shopping game.

Resources
card labels for prices (per class); classroom objects to be used as items in a shop (per class); play money - optional (per class)

What to do
- Ask learners to set up three or four tables to be used as stalls displaying items for sale. On each stall, there should be at least ten items for sale. Learners place a price label next to each item and write a price, using correct dollar and cent notation.
- Choose two learners to work as 'shopkeepers' on each stall. Other learners are the shoppers. They each begin with $100 and record on their whiteboards or paper what they have spent.
- 'Shoppers' take turns to buy items, calculating the money they have left. Use 'play money' if available; if not learners should simply pretend to hand over money.
- At any point shopkeepers may advertise details of a price increase or discount. They display this in the shop and shoppers must calculate the new price before buying.

Variation
Challenge 3 Shopkeepers use percentages to indicate price discounts or rises, for example: '10% off'.

Unit 10: Addition and subtraction 3

Learning objectives

Code	Learning objective
5Nc9*	Add or subtract near multiples of 10 or 100, e.g. 4387 − 299.
5Nc10	Use appropriate strategies to add or subtract pairs of [two- and three-digit numbers and] numbers with one decimal place, using jottings where necessary.
5Nc11*	Calculate differences between near multiples of 1000, e.g. 5026 − 4998, or near multiples of 1, e.g. 3·2 − 2·6.
5Nc18*	Find the total of more than three two- or three-digit numbers using a written method.
5Nc19*	Add or subtract any pair of three- and/or four-digit numbers, with the same number of decimal places, including amounts of money.

*Learning objective revised and consolidated

Strand 5: Problem solving

5Pt2	Solve single and multi-step word problems (all four operations); represent them, e.g. with diagrams or a number line.
5Pt3	Check with a different order when adding several numbers or by using the inverse when adding or subtracting a pair of numbers.
5Pt6	Estimate and approximate when calculating, e.g. using rounding, and check working.
5Pt7	Consider whether an answer is reasonable in the context of a problem.
5Ps2	Choose an appropriate strategy for a calculation and explain how they worked out the answer.
5Ps3	Explore and solve number problems and puzzles, e.g. logic problems.
5Ps4	Deduce new information from existing information to solve problems.
5Ps5	Use ordered lists and tables to help to solve problems systematically.
5Ps8	Investigate a simple general statement by finding examples which do or do not satisfy it, e.g. the sum of three consecutive whole numbers is always a multiple of three.
5Ps9	Explain methods and justify reasoning orally and in writing; make hypotheses and test them out.
5Ps10	Solve a larger problem by breaking it down into sub-problems or represent it using diagrams.

Unit overview

This unit reviews content covered in Units 8 and 9. Learners are reminded of mental strategies for adding and subtracting decimals and near multiples of 10, 100 or 1000. Compensation and 'find the difference' strategies are modelled and learners use these methods to solve problems. They revisit the use of the column method for addition of three or more numbers, and addition and subtraction of decimals.

Common difficulties and remediation

In preparation for the final lesson in the unit, spend time discussing the monetary system with learners. It is important to point out that in recording money and metric measures the convention is to use two decimal places, even when the second place is occupied by a zero. Although this is not a significant figure, it is recorded and may cause learners confusion if they try to solve money calculations with a calculator. For example, after the calculation to find the cost of six small balls costing $1.75 each, the calculator would display 10·5. Look out for learners who interpret the figure after the decimal point as – 10·5 could be incorrectly written as $10.05.

Money is an example of a non-proportional model and should not be used to introduce place value work. Only use it when learners have acquired a sound conceptual understanding of place value, for additional reinforcement.

Promoting and supporting language

At every stage, learners require mathematical vocabulary to access questions and problem-solving exercises. If appropriate, when a new key word is introduced, ask learners to write a definition in their books, drawing a box around it for emphasis. Encourage learners to write the definition in their own words. Sketching a diagram alongside will help embed the definition.

It is also important to adapt language when it becomes a barrier to learning, for example, using simpler terms to describe the strategy of compensation, such as 'take ten away and add one'.

In addition, be aware of terms that have more than one definition, inside and outside of mathematics, for example, 'degrees' (in angles and in temperature) and 'point' (as in decimal point and a point in space), as well as terms in financial language such as 'rise', 'fall' and 'change'. Some learners may not understand that the words 'money' and 'change' can be interpreted to mean the same thing in the context of mathematical questions or real-life transactions. This can be remedied by setting up a pretend shop in the classroom and giving learners regular opportunities to practise buying and selling items, enabling them to develop an understanding of the various meanings.

Lesson 1: **Adding and subtracting decimals mentally**

Number – Calculation: Mental strategies, Addition and subtraction

Learning objectives

Code	Learning objective
5Nc10	Use appropriate strategies to add or subtract pairs of [two- and three-digit numbers and] numbers with one decimal place, using jottings where necessary.
5Nc11*	Calculate differences between [near multiples of 1000, e.g. 5026 – 4998, or] near multiples of 1, e.g. 3·2 – 2·6.

Strand 5: Problem solving

5Pt2	Solve single and multi-step word problems (all four operations); represent them, e.g. with diagrams or a number line.
5Pt3	Check with a different order when adding several numbers or by using the inverse when adding or subtracting a pair of numbers.
5Pt6	Estimate and approximate when calculating, e.g. using rounding, and check working.
5Pt7	Consider whether an answer is reasonable in the context of a problem.
5Ps2	Choose an appropriate strategy for a calculation and explain how they worked out the answer.
5Ps3	Explore and solve number problems and puzzles, e.g. logic problems.
5Ps4	Deduce new information from existing information to solve problems.
5Ps9	Explain methods and justify reasoning orally and in writing; make hypotheses and test them out.
5Ps10	Solve a larger problem by breaking it down into sub-problems or represent it using diagrams.

Prerequisites for learning

Learners need to:
- know one-place decimal pairs that total 1 and 10
- understand the place value of decimals.

Success criteria

Learners can:
- use a range of strategies to add or subtract

decimals mentally including partitioning, 'find the difference' and compensation.

Vocabulary

partition, compensation, 'find the difference'

Resources

mini whiteboard and pen or paper (per pair)

Refresh

- Choose an activity from Number – Calculation: Mental strategies, Addition and subtraction in the Refresh activities.

Discover [SB]

- Together, discuss the Discover image. Ask: **How would you solve the problems?**

Teach and Learn [SB]

- Discuss the examples in the Student's Book. On the board, write 23·3 + 3·4. Ask: **How would you work this out? You can use jottings to solve the problem mentally.** Learners write their answers and share solutions. Establish that the easiest strategy is to partition and add by place value: 23·3 + 3·4 = (23 + 3) + (0·3 + 0·4) = 26·7. Write 48·6 + 11·9. Elicit that if they know 0·6 + 0·9 is 1·5 then learners can solve this mentally: (48 + 11) + (0·6 + 0·9) = 60·5. Review mental addition of near multiples of tens and ones and the application of place value to deduce new facts from known facts: 6 + 9 = 15, so 0·6 + 0·9 = 1·5. Some learners may use compensation. Model the use of jottings: round 11·9 to 12, add 48·6 and 12 then subtract 0·1 to adjust (60·5).
- Write 36·7 – 12·5. Learners discuss possible strategies, including partitioning the second number.

Say: **36·7 subtract 12 is 24·7, then subtract 0·5 to give 24·2.** Model the 'find the difference' strategy on the board, using number line jumps from 12·5 to 13 (0·5), to 36 (23) then to 36·7 (0·7), giving 24·2.
- Write 48·2 – 16·8. Ask learners to find the difference. Review alternative strategies, including compensation. Model: 48·2 – 16·8 = (48·2 – 17) + 0·2 = 31·4.
- Write 17·7 + 12·2, 39·9 + 27·8, 85·8– 21·3, 75·2 – 42·9 for learners to solve in pairs.

Practise [WB]

- Workbook: Adding and subtracting decimals mentally page 90

Apply 👥 🖥

- Display **Slide 1** and ask learners to complete the calculation, then share solutions. ($76.50)

Review 🖥

- Display: **Slide 2.** Ask learners to copy and complete the table.

Assessment for learning

- When I subtract 25·9 from 98·2, I round one of the numbers, subtract, then adjust the figure. Show me this calculation on paper.

Lesson 2: Adding and subtracting near multiples mentally

Learning objectives

Code	Learning objective
5Nc9*	Add or subtract near multiples of 10 or 100, e.g. 4387 – 299.
5Nc11*	Calculate differences between near multiples of 1000, e.g. 5026 – 4998, [or near multiples of 1, e.g. 3·2 – 2·6].

Strand 5: Problem solving	
5Pt2	Solve single and multi-step word problems (all four operations); represent them, e.g. with diagrams or a number line.
5Pt3	Check with a different order when adding several numbers or by using the inverse when adding or subtracting a pair of numbers.
5Pt6	Estimate and approximate when calculating, e.g. using rounding, and check working.
5Pt7	Consider whether an answer is reasonable in the context of a problem.
5Ps2	Choose an appropriate strategy for a calculation and explain how they worked out the answer.
5Ps3	Explore and solve number problems and puzzles, e.g. logic problems.
5Ps5	Use ordered lists and tables to help to solve problems systematically.
5Ps9	Explain methods and justify reasoning orally and in writing; make hypotheses and test them out.
5Ps10	Solve a larger problem by breaking it down into sub-problems or represent it using diagrams.

Prerequisites for learning

Learners need to:

- add and subtract mentally near multiples of 10 or 100
- support mental calculations by use of jottings, for example, using a blank number line.

Success criteria

Learners can:

- add mentally near multiples of 10, 100 or 1000

- subtract mentally near multiples of 100 or 1000

Vocabulary

compensation, compensate (adjust), bridging, jottings

Resources

mini whiteboard and pen or paper (per pair)

Refresh

- Choose an activity from Number – Calculation: Mental strategies, Addition and subtraction in the Refresh activities.

Discover [SB]

- Discuss the text and image in the Student's Book. Say: **Look at the problems in the thought bubble. How would you solve them?** Discuss possible strategies.

Teach and Learn [SB]

- Discuss the examples in the Student's Book. Write: 44 + 39, 456 + 191, 2463 + 4999. Ask: **What strategy would you use to solve each problem? Why?** Elicit that the best choice is compensation because the numbers are near multiples of 10, 100 or 1000. Ask a volunteer to model the strategy for 44 + 39 (round 39 to 40, add 44 and adjust by subtracting 1 (83)). Learners solve the remaining problems, on their whiteboards. Discuss solutions, focusing on rounding and adjusting. Ask: **What other strategy could you have used?** Elicit counting on.
- Draw a blank number line and use it to solve 63 + 28. Mark 63 and count on 20 to 83, then add 8 (91).
- Write 98 – 29, 776 – 498, 6823 – 3997. Ask: **What strategy would you use to solve these problems? Why?** Ask a volunteer to model the compensation

strategy for 98 – 29 (round 29 to 30, subtract from 98 and adjust by adding 1 (69)). Learners solve the remaining problems. Discuss rounding and adjusting. Elicit that they could also have counted on from the smaller number, adding in stages to 'find the difference'.

Practise [WB]

- Workbook: Adding and subtracting near multiples mentally page 92
- Refer to Activity 1 from the Additional practice activities.

Apply 👥 🖥

- Display **Slide 1** and ask pairs to solve the problem. Say: **In Round 12 the team loses 91 points. What would their score be if they continued to lose 91 points every round for the next 10 rounds?**

Review 👥

- Say: **Work in pairs. Begin with the number 563. One learner adds 399 and subtracts 191; the other adds 297 and subtracts 492. What do you notice?**

Assessment for learning

- When I subtract 4998 from 7634, I round 5000, subtract from 7634, then I add two. Why do I add two?

Lesson 3: **Adding more than two numbers**

Number – Calculation: Mental strategies, Addition and subtraction

Learning objectives

Code	Learning objective
5Nc18*	Find the total of more than three two- or three-digit numbers using a written method.

Strand 5: Problem solving

5Pt6	Estimate and approximate when calculating, e.g. using rounding, and check working.
5Pt7	Consider whether an answer is reasonable in the context of a problem.
5Ps2	Choose an appropriate strategy for a calculation and explain how they worked out the answer.
5Ps3	Explore and solve number problems and puzzles, e.g. logic problems.
5Ps4	Deduce new information from existing information to solve problems.
5Ps8	Investigate a simple general statement by finding examples which do or do not satisfy it, e.g. the sum of three consecutive whole numbers is always a multiple of three.
5Ps10	Solve a larger problem by breaking it down into sub-problems or represent it using diagrams.

Prerequisites for learning

Learners need to:
- use column addition to add several numbers.

Success criteria

Learners can:
- use column addition to add several numbers, carrying units, tens and hundreds when needed.

Vocabulary

place value, estimate, carry

Resources

mini whiteboard and pen or paper (per pair); 1–6 dice, alternately use Resource sheet 4: 1–6 spinners (per pair)

Refresh

- Choose an activity from Number – Calculation: Mental strategies, Addition and subtraction in the Refresh activities.

Discover [SB]

- Discuss the text and menu in the Student's Book. Ask: **How would you add a set of one-, two- and three-digit numbers together?** Discuss how this relates to column addition.

Teach and Learn [SB]

- Say: **In this lesson you will arrange numbers by row so that they are as easy as possible to add.** Write: 238, 456, 552, 376. With reference to the example in the Student's Book, ask: **Using the column addition layout, how would you order these numbers?** Accept suggestions. Say: **Look for doubles and numbers that make ten, and group the numbers to make adding easier.** Agree that 238 and 552 should be grouped (8 + 2 = 10), and 446 and 386 (double 6 is 12). Say: **You could also have grouped 238 and 376, as three tens and seven tens makes ten tens. The same with 456 and 552.** Explain that any rearrangement that makes addition easier is helpful.
- Write 387, 388, 389. Ask: **Is the sum of these numbers odd or even?** Elicit it is odd. Say: **The numbers 387, 388 and 389 follow each other in the number line.** Ask: **Is the sum of any three consecutive numbers always odd?** Learners investigate. (yes)

- Write: 777, 249, 557, 663, 701. Ask learners to use column addition, estimating first by rounding. Take solutions and discuss how they decided the row-by-row arrangement of numbers. (2947)
- Write: 467, 92, 383, 437, 998, 63. Ask: **How do you add numbers with different amounts of digits?** Establish that the numbers must be aligned by place value. Say: **Align the digits from the right, so that all the units digits are correctly lined up.** Learners discuss their solutions and strategies. (2440)

Practise [WB]

- Workbook: Adding more than two numbers page 94
- Refer to Activity 2 from the Additional practice activities.

Apply 👥 🖥

- Display **Slide 1.** Learners find the total of the bill. ($599) Then say: **Unfortunately, the waiter forgot to add three items costing $247, $311 and $153.** Ask: **How does this affect the bill?** ($1310)

Review 👥

- In pairs, learners roll a dice to create five three-digit numbers, then add them using column addition. The learner with the higher total scores a point. They play five rounds. The learner with the higher score wins.

Assessment for learning

- Show me how you use column addition to add 137, 442, 273, 678 and 49.

Lesson 4: **Adding and subtracting decimals**

Learning objectives

Code	Learning objective
5Nc19*	Add or subtract any pair of three- and/or four-digit numbers, with the same number of decimal places, including amounts of money.

Strand 5: Problem solving

5Pt6	Estimate and approximate when calculating, e.g. using rounding, and check working.
5Pt7	Consider whether an answer is reasonable in the context of a problem.
5Ps2	Choose an appropriate strategy for a calculation and explain how they worked out the answer.
5Ps3	Explore and solve number problems and puzzles, e.g. logic problems.
5Ps4	Deduce new information from existing information to solve problems.
5Ps5	Use ordered lists and tables to help to solve problems systematically.
5Ps8	Investigate a simple general statement by finding examples which do or do not satisfy it, e.g. the sum of three consecutive whole numbers is always a multiple of three.
5Ps9	Explain methods and justify reasoning orally and in writing; make hypotheses and test them out.
5Ps10	Solve a larger problem by breaking it down into sub-problems or represent it using diagrams.

Prerequisites for learning

Learners need to:

- use column addition and subtraction for decimals with one or two places

Success criteria

Learners can:

- use effective, efficient and appropriate strategies when adding and subtracting decimals

- use column addition and subtraction for decimals with one or two places, remembering to align the numbers according to their place value columns and to line up the decimal points

Vocabulary

place value, estimate, carry, rename (decompose)

Resources

mini whiteboard and pen or paper (per pair); dice (per pair)

Refresh

- Choose an activity from Number – Calculation: Mental strategies, Addition and subtraction in the Refresh activities.

Discover 〔SB〕

- Discuss the text and image in the Discover section. Learners calculate the difference: 3·5 degrees. Ask: **What other scales give readings in decimals?** (millimetre divisions of a ruler, digital scales)

Teach and Learn 〔.ıl〕〔SB〕

- Discuss the text and example in the Student's Book. Say: **Money and temperature are real life examples of using decimals.** Display the **Thermometer tool**, with both scales 0 to 100. Cover '100' and rename the scale '10'. Set one scale to read 8·7 °C, the other 3·4 °C. Ask: **How do you work out the difference between the readings on the scales?** (counting on, mental or written subtraction strategies). Learners use a preferred method to solve the problem. (5·3 °C). Set two new temperatures: 7·3 °C and 4·8 °C. Learners find the difference. (2·5 °C)

- Write: 8·64 °C, +/– 6·37 degrees. Say: **Imagine the thermometer gives temperatures to two decimal places and reads 8·64 °C. At one point it rises by**

6·37 degrees; at another point it falls the same amount. Work out the new temperature each time. Discuss solutions, ensuring that learners carry and rename correctly. (15·01 °C, 2·27 °C)

- Write: $45.45 – $37.28. Ask learners to solve this problem. Check that they rename correctly. ($8.17)

Practise 〔WB〕

- Workbook: Adding and subtracting decimals page 96
- Refer to Activity 2 from the Additional practice activities.

Apply 〔••〕〔▭〕

- Display **Slide 1** and ask learners to complete the calculation and share solutions.

Review

- On the board, write: 34·99. Ask learners to call out a four-digit number with two decimal places, which they subtract from the total, using column subtraction. They continue until the total falls below 10.

Assessment for learning

- Show me how to use column addition to add 3·68 and 4·79.

Additional practice activities

Activity 1 :: Challenge 2

Learning objectives

- Add and subtract decimals and near multiples of 10, 100 or 1000 mentally.
- Solve a number problem using appropriate mental strategies with jottings.

Resources

mini whiteboard and pen or paper (per pair)

What to do

- On the board, write: $198, $9.70, $345, $101, $146, $7.40, $299, $3.70, $5.90, $4.60.
- Say: **Using mental addition and jottings, find four amounts that add to make $355.10.** ($198, $146, $7.40, $3.70)
- Learners discuss their solutions to confirm they are correct.
- Say: **Using mental subtraction and jottings, find four amounts that give a total that can be deducted from $500 to leave $85.70.** ($299, $101, $9.70, $4.60)
- Learners discuss their solutions to confirm they are correct.

Variation

Challenge 1 Separate the numbers for addition and subtraction problems. Addition: give a choice of: $198, $122, $146, $7.40, $2.80, $3.70; subtraction: give a choice of: $299, $167, $101, $9.70, $8.80, $4.60.

Activity 2 :: Challenge 2

Learning objectives

- Use knowledge of adding more than two numbers to add more than two decimal numbers.
- Use an appropriate written strategy to solve a number problem.

Resources

mini whiteboard and pen or paper (per pair)

What to do

- On the board, write: 'Order A: $23.40, $37.70 and $48.60', 'Order B: $19.50, $42.30 and $38.90' and 'Order C: $28.80, $32.60 and $49.10'.
- Say: **Imagine that you are working for a company that has received three sales orders, A, B and C. Your manager asks you to arrange the total costs of the sale orders from smallest to largest. You will need to use column addition to add the numbers.**
- Learners calculate the sum of each sales order, then arrange them in ascending order.
- They discuss their solutions to confirm they are correct. (A: $109.70; B: $100.70; C: $110.50; order: B, A, C)

Variation

Challenge 3 Include an extra two amounts per sales order: A ($237.44 and $166.53); B ($248.39 and $156.94) and C: ($228.39 and $154.58).

Unit 11: Multiplication and division 1

Learning objectives

Code	Learning objective
5Nc3	Know multiplication and division facts for the 2× to 10× tables.
5Nc4	Know and apply tests of divisibility by 2, 5, 10 and 100.
5Nc5	Recognise multiples of 6, 7, 8 and 9 up to the 10th multiple.
5Nc7	Find factors of two-digit numbers.
5Nc20	Multiply or divide three-digit numbers by single-digit numbers.
5Nc21	Multiply two-digit numbers by two-digit numbers.
5Nc23	Divide three-digit numbers by single-digit numbers, including those with a remainder (answers no greater than 30).
5Nc25	Decide whether to group (using multiplication facts and multiples of the divisor) or to share (halving and quartering) to solve divisions.

Strand 5: Problem solving

Code	Learning objective
5Pt2	Solve single and multi-step word problems (all four operations); represent them, e.g. with diagrams or a number line.
5Pt4	Use multiplication to check the result of a division [, e.g. multiply 3.7 × 8 to check 29.6 ÷ 8].
5Pt6	Estimate and approximate when calculating, e.g. using rounding, and check working.
5Pt7	Consider whether an answer is reasonable in the context of a problem.
5Ps2	Choose an appropriate strategy for a calculation and explain how they worked out the answer.
5Ps3	Explore and solve number problems and puzzles, e.g. logic problems.
5Ps4	Deduce new information from existing information to solve problems.
5Ps5	Use ordered lists and tables to help to solve problems systematically.
5Ps9	Explain methods and justify reasoning orally and in writing; make hypotheses and test them out.
5Ps10	Solve a larger problem by breaking it down into sub-problems or represent it using diagrams.

Unit overview

In this unit, learners review knowledge of multiplication and division facts for the two to ten times tables and factors of two-digit numbers. They recognise and find multiples for 6, 7, 8 and 9 up to the tenth multiple and recall squares of numbers up to 10 × 10. They use this knowledge to answer simple problems. Factors are introduced and learners identify pairs of factors of two-digit numbers. Learners extend their understanding of written methods to include three-digit by one-digit and two-digit by two-digit multiplication, including estimating and checking the answer to a calculation. Mental strategies are discussed, together with three written methods of multiplication: partitioning, the grid method and the expanded written method. Division is extended to include three-digit divided by one-digit problems. Mental strategies are discussed, together with two written methods: partitioning and the expanded written method. Learners are encouraged to decide whether a problem involves sharing or grouping.

Common difficulties and remediation

Difficulties with times tables often stem from an inability to recall addition and subtraction bonds to ten. Learners need to understand the relationship between addition facts and subtraction facts, and between multiplication facts and division facts.

To help develop recall of number facts, use rules that give facts based on those already known.

Problems recalling the 6× table may stem from inability with quick recall of the 2×, 3× and 5× tables. Problems recalling answers to 7× and 9× tables may result from a lack of confidence with other tables. Return regularly to the other tables to develop confidence, playing step-counting games and solving missing-number problems.

Learners who make consistent errors with written strategies for multiplication and division may need to consolidate their knowledge of informal methods before proceeding. Errors and misconceptions that learners experience when developing understanding of the varied methodologies and concepts of multiplication and division can, in most cases, be dealt with by providing opportunities to practise related, more fundamental, strategies first.

Promoting and supporting language

Learners must learn many terms related to the concepts of multiplication and division, for example: multiply – times, product, repeated addition; division – sharing, divided by, factors, repeated subtraction. Some of these words, such as 'factor' and 'square' have more than one meaning and some are used imprecisely outside of mathematics. It is important that learners experience a variety of terms and that the terms are used accurately.

Lesson 1: **Multiples**

Learning objectives

Code	Learning objective
5Nc4	Know and apply tests of divisibility by 2, 5, 10 and 100.
5Nc5	Recognise multiples of 6, 7, 8 and 9 up to the 10th multiple.

Strand 5: Problem solving

5Pt4	Use multiplication to check the result of a division [, e.g. multiply 3.7 × 8 to check 29.6 ÷ 8].
5Ps2	Choose an appropriate strategy for a calculation and explain how they worked out the answer.
5Ps3	Explore and solve number problems and puzzles, e.g. logic problems.
5Ps5	Use ordered lists and tables to help to solve problems systematically.
5Ps9	Explain methods and justify reasoning orally and in writing; make hypotheses and test them out.

Prerequisites for learning

Learners need to:
- count on and back in steps of 6, 7, 8 and 9.

Success criteria

Learners can:
- recognise the multiples of 6, 7, 8 and 9

- apply tests of divisibility by 2, 5, 10 and 100.

Vocabulary

multiple, divisibility rule

Resources

mini whiteboard and pen or paper (per pair)

Refresh

- Choose an activity from Number – Calculation: Mental strategies, Multiplication and division in the Refresh activities.

Discover [SB]

- Refer learners to the Student's Book. Ask: **Can you answer these problems?** Accept comments.

Teach and Learn [SB] [📊]

- Discuss the text and Examples in the Student's Book. Remind learners that a 'multiple' is the result of multiplying one number by another.
- Say: **Times tables are made up of multiples.** Ask learners to list multiples of three in a row and multiples of six in a row below. Ask: **What do you notice?** Establish that multiples of six are also multiples of three, and one set of multiples can be used to work out the other. Elicit that multiples of six are the **even** multiples of 3.
- Display the **Number square tool**. Skip count in sixes, highlighting numbers on the square. Ask: **Is there a pattern to the multiples?** Accept suggestions. Ask questions: **What is the eighth multiple of six?** (48) **What is nine times six?** (54) Show how multiples of 4 lead to multiples of 8.
- Write: 7, 14, 21, 28. Say: **These numbers are multiples of a number. Which one?** (7) Highlight the multiples on the tool. Repeat for multiples of 9.
- Write: 18, 25, 36, 80, 102, 260, 335, 400, 2505, 3000. Ask: **Which of the numbers is divisible by 2?** Establish that all even numbers are divisible by 2 and end with a 0, 2, 4, 6 or 8.

- Lead learners in a count of fives up to 50. Ask: **What do you notice about the numbers?** Establish that the units digits is either a five or a zero. Say: **Numbers divisible by five end with zero or five.** Ask them to identify the numbers from the list that are divisible by five (25, 80, 260, 335, 400, 2505, 3000).
- Discuss divisibility rules for ten (numbers that end with a zero) and 100 (numbers that end with 00).

Practise [WB]

- Workbook: Multiples page 98
- Refer to Activity 1 from the Additional practice activities.

Apply [👥]

- Display **Slide 1**. Accept answers and discuss solutions. (2 sets: 8788; 5 sets: 2785; 10 sets: 6340; 100 sets: 4200) Say: **Jar B 4200 could be divided by 2,5,10 and 100; but to solve the whole problem, you needed to divide it by 100.**

Review

- Write: 36, 54, 28, 63, 40, 72, 27, 81. Ask learners to sort the numbers into four groups, depending on whether they are multiples of 6, 7, 8 or 9.

Assessment for learning

- I know that 3 456 780 is divisible by 2, 5 and 10 without needing to divide. Explain.

Lesson 2: **Factors**

Learning objectives

Code	Learning objective
5Nc7	Find factors of two-digit numbers.

Strand 5: Problem solving	
5Pt2	Solve single and multi-step word problems (all four operations); represent them, e.g. with diagrams or a number line.
5Pt4	Use multiplication to check the result of a division [, e.g. multiply 3.7 × 8 to check 29.6 ÷ 8].
5Ps2	Choose an appropriate strategy for a calculation and explain how they worked out the answer.
5Ps3	Explore and solve number problems and puzzles, e.g. logic problems.
5Ps5	Use ordered lists and tables to help to solve problems systematically.
5Ps9	Explain methods and justify reasoning orally and in writing; make hypotheses and test them out.
5Ps10	Solve a larger problem by breaking it down into sub-problems or represent it using diagrams.

Prerequisites for learning

Learners need to:

- recall the multiplication tables up to 10 × 10.

Success criteria

Learners can:

- use a trial and improvement method of skip-counting to find factors

- use their knowledge of multiplication tables to find factors.

Vocabulary

multiple, key fact, divided by, product, factor

Resources

mini whiteboard and pen or paper (per pair)

Refresh

- Choose an activity from Number – Calculation: Mental strategies, Multiplication and division in the Refresh activities.

Discover [SB]

- Ask learners to look at the Student's Book page. Ask: **How many different groups are possible?** Ask learners to work in pairs to answer. (1 × 24, 2 × 12, 3 × 8, 4 × 6, 6 × 4, 8 × 3, 12 × 2, 24 × 1)

Teach and Learn [SB]

- Discuss the text and Example in the Student's Book. On the board, draw a 'factor rainbow' (see diagram opposite). Ask: **How many pairs of numbers can you find that, when multiplied, make 36?** Accept suggestions. Explain that one method is to jump in units twos, threes, fours, … to see whether the skip-multiple goes to the target number in equal steps. Say: **The multiple and the number of steps make the number pair.** Ask learners to find number pairs. Record their results on the board, in linked pairs.

- Say: **These numbers are called 'factors'.** Explain that a factor is a whole number that divides exactly into another whole number. On the board, write 'factor' and its definition. Point to factor pairs for 36. Say: **A factor rainbow is a way of writing factors for numbers. It uses a series of arcs. It helps to make sure you have all the factors, by listing them in consecutive order.** Explain that learners

should check every number, from two to half of the number, to see if it goes evenly into the number.

- On the board, draw factor rainbow templates for 32 (three arcs), 42 (four arcs), 80 (5 arcs). Ask learners to work in pairs to complete the diagrams. Encourage them to make their own rainbow diagrams for other two-digit numbers. Invite learners to discuss their factor rainbows with the class.

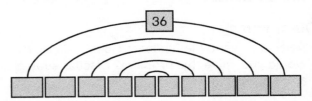

Practise [WB]

- Workbook: Factors page 100

Apply 👥🖥

- Display **Slide 1**. Ask learners to discuss their solutions for finding perfect numbers. (6, 28)

Review

- On the board, write: 12, 28, 48. Ask learners to find factor pairs for these numbers.

Assessment for learning

- I know that 18 and 30 have four factors in common. What are they?

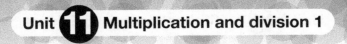

Lesson 3: **Multiples and factors**

Learning objectives

Code	Learning objective
5Nc3	Know multiplication and division facts for the 2× to 10× tables.
5Nc4*	Know and apply tests of divisibility by 2, 5, 10 and 100.
5Nc5	Recognise multiples of 6, 7, 8 and 9 up to the 10th multiple.
5Nc7*	Find factors of two-digit numbers.

Strand 5: Problem solving

5Pt4	Use multiplication to check the result of a division [, e.g. multiply 3·7 × 8 to check 29.6 ÷ 8].
5Pt6	Estimate and approximate when calculating, e.g. using rounding, and check working.
5Ps2	Choose an appropriate strategy for a calculation and explain how they worked out the answer.
5Ps3	Explore and solve number problems and puzzles, e.g. logic problems.
5Ps5	Use ordered lists and tables to help to solve problems systematically.
5Ps9	Explain methods and justify reasoning orally and in writing; make hypotheses and test them out.
5Ps10	Solve a larger problem by breaking it down into sub-problems or represent it using diagrams.

Prerequisites for learning

Learners need to:

- recall multiplication and division facts to 10 × 10
- recognise the multiples of 2, 6, 8 and 9

Success criteria

Learners can:

- recognise that a whole number is a multiple of each of its factors
- recognise and use factor pairs and commutativity

Vocabulary

number fact, multiple, factor, factor pair, array

Resources

squared paper (per learner)

Refresh

- Choose an activity from Number – Calculation: Mental strategies, Multiplication and division in the Refresh activities.

Discover [SB]

- Refer learners to the Student's Book. Ask: **How would you solve this problem?** Discuss strategies and solve the problem.

Teach and Learn [SB] [chart]

- Discuss the text and examples in the Student's Book. Revise the terms 'multiple' and 'factor'. Remind learners that factors are what they multiply to get a number and a multiple is the product of two or more factors.

- Say: **Tell me some multiples of 7.** Display the **Number square tool** set to 'colour squares'; 'hide all'. Colour a six by four grid. Say: **You can think of multiplication as an array – a set of rows and columns. This array shows the number 24.** The grid models the number fact 6 × 4 = 24 and factor pair 4, 6. Demonstrate removing and adding rows to model different multiples of six. Ask: **Does the array model any other number facts? What about division?** Prompt the learners: **How many groups of six can you divide 24 into?** Elicit four.

- On the board, write: 24 ÷ 4 = 6. Ask: **What multiplication fact does the array model now? What division fact?** Agree and record: 4 × 6 = 24, 24 ÷ 6 = 4. Set a challenge: **How many different arrays can we make with 24 squares?** Create other arrays using the Number square tool and discuss results. Expect; 1 × 24, 2 × 12, 3 × 8, Discuss the relationship to factor pairs of 24: (1, 24), (2, 12) (3, 8) and (4, 6).

Practise [WB]

- Workbook: Multiples and factors page 102. Give learners some squared paper for this activity.

Apply [group][screen]

- Display **Slide 1**. Invite learners to discuss their solutions. Answers (rows; seats): (1, 45); (3, 15); (5, 9); (9, 5); (15, 3); (45, 1).

Review [group]

- Learners take turns to state the number of people requiring seating at a cinema. Their partner describes all possible arrangements.

Assessment for learning

- How many different ways are there to arrange 64 eggs in rows and columns?

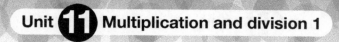
Lesson 4: **Multiplying a 3-digit number by a single-digit number (1)**

Learning objectives

Code	Learning objective
5Nc20	Multiply [or divide] three-digit numbers by single-digit numbers.

Strand 5: Problem solving

5Pt2	Solve single and multi-step word problems (all four operations); represent them, e.g. with diagrams or a number line.
5Pt6	Estimate and approximate when calculating, e.g. using rounding, and check working.
5Pt7	Consider whether an answer is reasonable in the context of a problem.
5Ps2	Choose an appropriate strategy for a calculation and explain how they worked out the answer.
5Ps3	Explore and solve number problems and puzzles, e.g. logic problems.
5Ps9	Explain methods and justify reasoning orally and in writing; make hypotheses and test them out.
5Ps10	Solve a larger problem by breaking it down into sub-problems or represent it using diagrams.

Prerequisites for learning

Learners need to:

- add mentally a two-digit number to a two- or three-digit multiple of ten.

- partition a three-digit number in order to multiply by a single-digit number.

Success criteria

Learners can:

- make a reasonable estimate for the answer

Vocabulary

multiple, key fact, estimate, partition

Resources

mini whiteboard and pen or paper (per pair)

Refresh

- Choose an activity from Number – Calculation: Mental strategies, Multiplication and division in the Refresh activities.

Discover [SB]

- Read the text in the Student's Book with the class and discuss the problem. Remind learners of the mental and written strategies they have used so far.

Teach and Learn [SB] [☐]

- Discuss the examples in the Student's Book.
- On the board, write: 343 × 3 = ☐. Ask: **How would you estimate the answer?** Accept suggestions. Explain the 'estimation method': round the three-digit number to the nearest 100 and multiply the numbers together (900).
- Ask: **How would you partition 343?** (300, 40 and 3) Demonstrate multiplication by partitioning: 343 × 3 = (300 × 3) + (40 × 3) + (3 × 3) = 900 + 120 + 9 = 1029. Say: **The estimate, 900, shows that the answer is likely to be correct.**
- On the board, write: 649 × 6 = ☐. Display **Slide 1** and demonstrate how to find the answer by the grid method.
- Refer to **Slide 1**. Say: **Now look at the expanded method.** Explain that the method works vertically

and begin with the units column. Ask: **What is six multiplied by six?** (36) Write 36 in the correct place value columns. Continue for tens and hundreds. Provide further guided examples to confirm that learners understand the method.

Practise [WB]

- Workbook: Multiplying a 3-digit number by a single-digit number (1) page 104
- Refer to Activity 2 from the Additional practice activities.

Apply [☐☐] [☐]

- Display **Slide 2**. Learners share their solutions. (583 × 3 = 1749 g; 249 × 7 = 1743 g. Three boxes of 583 g gives more cereal)

Review [☐☐☐]

- On the board, write: 464 × 8 = ☐. Split the class into three groups: Partitioning, Grid and Expanded. Learners use the method allocated to work this out, then swap twice to use the other methods. (3712)

Assessment for learning

- Show me two different methods for calculating 747 × 7.

Number – Calculation: Mental strategies, Multiplication and division

Lesson 5: **Multiplying a 2-digit number by a 2-digit number (1)**

Learning objectives

Code	Learning objective
5Nc21	Multiply two-digit numbers by two-digit numbers.

Strand 5: Problem solving

5Pt2	Solve single and multi-step word problems (all four operations); represent them, e.g. with diagrams or a number line.
5Pt6	Estimate and approximate when calculating, e.g. using rounding, and check working.
5Pt7	Consider whether an answer is reasonable in the context of a problem.
5Ps2	Choose an appropriate strategy for a calculation and explain how they worked out the answer.
5Ps3	Explore and solve number problems and puzzles, e.g. logic problems.
5Ps9	Explain methods and justify reasoning orally and in writing; make hypotheses and test them out.
5Ps10	Solve a larger problem by breaking it down into sub-problems or represent it using diagrams.

Prerequisites for learning

Learners need to:

- multiply a number by a multiple of ten
- partition numbers into tens and units

Success criteria

Learners can:

- make a reasonable estimate for the answer
- multiply two two-digit numbers by two-digit numbers.

Vocabulary

multiple, key fact, estimate, partition

Resources

mini whiteboard and pen or paper (per pair)

Refresh

- Choose an activity from Number – Calculation: Mental strategies, Multiplication and division in the Refresh activities.

Discover 🆂🅱

- Read the text in the Student's Book. Ask: **What amount must the restaurant owner spend?** Discuss strategies and ask learners to solve. ($1564)

Teach and Learn 🆂🅱 🖥

- Discuss the examples in the Student's Book. Write 48 × 33 =. Ask: **How would you estimate the answer?** Explain the estimation method: round both numbers to the nearest ten and multiply together. (1500) Say: **You can use the estimate to check the accuracy of your answer.** Demonstrate the partitioning method: 48 × 33 = (48 × 30) + (48 × 3) = 1440 + 144 = 1584. The estimate, 1500, shows the answer is likely to be correct.
- Write 74 × 67 = ☐. Learners work in pairs to estimate, then use partitioning to work out the answer. (4958). Display **Slide 1**. Demonstrate the grid method.
- Look at the expanded method. Say: **You can choose to multiply first by the most or least significant digit, which is two (20) or seven. We will choose**

the least significant digit. Remind learners that the method works vertically and begin at the units column. Ask: **What is 59 multiplied by 7?** (413). Write 413 in the correct place-value columns. Then multiply 59 × 20 (1180), finally adding to get the total (1593). Provide further guided examples to confirm that learners understand the method.

Practise 🆆🅱

- Workbook: Multiplying a 2-digit number by a 2-digit number (1) page 106

Apply 👥

- Display **Slide 2**. Ask learners to discuss their solutions. (47 × 86) + (33 × 96) = 7210 ml

Review

- On the board, write 89 × 89 = ☐. Split the class into three groups: Partitioning, Grid and Expanded. Learners use their method to work this out, then swap twice to use the other methods. (7921)

Assessment for learning

- Show me two different methods for calculating 93 × 93.

Lesson 6: **Multiplying a 2-digit number by a 2-digit number (2)**

Learning objectives

Code	Learning objective
5Nc21	Multiply two-digit numbers by two-digit numbers.

Strand 5: Problem solving

5Pt2	Solve single and multi-step word problems (all four operations); represent them, e.g. with diagrams or a number line.
5Pt6	Estimate and approximate when calculating, e.g. using rounding, and check working.
5Pt7	Consider whether an answer is reasonable in the context of a problem.
5Ps2	Choose an appropriate strategy for a calculation and explain how they worked out the answer.
5Ps3	Explore and solve number problems and puzzles, e.g. logic problems.
5Ps9	Explain methods and justify reasoning orally and in writing; make hypotheses and test them out.
5Ps10	Solve a larger problem by breaking it down into sub-problems or represent it using diagrams.

Prerequisites for learning

Learners need to:

- recall the multiplication tables up to 12 × 12 and associated facts involving multiples of ten
- understand the terms factor and multiple
- be able to double and halve whole numbers.

Success criteria

Learners can:

- multiply a two-digit number by two-digit numbers.

Vocabulary

doubling, halving, factor

Refresh

- Choose an activity from Number – Calculation: Mental strategies, Multiplication and division in the Refresh activities.

Discover [SB]

- Refer learners to the Student's Book. Say: **Doubling and halving can be really useful to help you solve multiplication problems.**

Teach and Learn [SB]

- Discuss the text and Example in the Student's Book. Write 36 × 16 = ☐. Ask: **How would you solve this?** Say: **The multiplication looks difficult but it can be simplified.**
- Demonstrate using a doubling and halving strategy. Write: 36 × 16 = 72 × 8. Say: **You can double one number and halve the other. You continue halving and doubling until it's not difficult anymore.** Continue: … = 114 × 4 = 228 × 2. Ask: **Who can solve the problem now?** Agree that as they are multiplying by 2, they only need to double 228 (456).
- Explain that if you multiply one number by two and divide the other by two the product does not change. Say: **Doubling and halving may not always the best strategy to use.** Write: 18 × 24, 14 × 25 and ask the learners to work them out. (432, 350)
- Say: **Another strategy involves breaking one of the numbers into its factors and then multiplying.**

Demonstrate: 18 × 16 = 18 × 2 × 8 = 18 × 2 × 2 × 4 = 36 × 2 × 4 = 72 × 4 = 288. Say: **Continue to split numbers into factors until the problem is easier to solve.**

- Write 24 × 12 and ask learners to work it out, in pairs. (288)

Practise [WB]

- Workbook: Multiplying a 2-digit number by a 2-digit number (2) page 108
- Refer to Activity 3 from the Additional practice activities.

Apply 👥 🖥

- Display **Slide 1**. Ask learners to discuss their solutions. (25 × 32 = 50 × 16 = 100 × 8 = 800)

Review

- On the board, write: 36 × 24. Split the class into two groups: a Halving and doubling group and a Factors group. Learners use their method to work this out, then swap. (864)

Assessment for learning

- Show me two methods for calculating 25 × 16.

Lesson 7: **Dividing a 3-digit number by a single-digit number (1)**

Learning objectives

Code	Learning objective
5Nc20	[Multiply or] divide three-digit numbers by single-digit numbers.
5Nc23	Divide three-digit numbers by single-digit numbers, including those with a remainder (answers no greater than 30).
5Nc25	Decide whether to group (using multiplication facts and multiples of the divisor) or to share (halving and quartering) to solve divisions.

Strand 5: Problem solving

5Pt2	Solve single and multi-step word problems (all four operations); represent them, e.g. with diagrams or a number line.
5Pt4	Use multiplication to check the result of a division [, e.g. multiply 3.7×8 to check $29.6 \div 8$].
5Ps2	Choose an appropriate strategy for a calculation and explain how they worked out the answer.
5Ps9	Explain methods and justify reasoning orally and in writing; make hypotheses and test them out.
5Ps10	Solve a larger problem by breaking it down into sub-problems or represent it using diagrams.

Prerequisites for learning

Learners need to:
- multiply and divide multiples of 10 and 100.

Success criteria

Learners can:
- partition three-digit numbers

- divide ten times a multiple of a number by a one-digit number.

Vocabulary

key fact, divide, divisible by, estimate, partition

Resources

mini whiteboard and pen or paper (per pair)

Refresh

- Choose an activity from Number – Calculation: Mental strategies, Multiplication and division in the Refresh activities.

Discover [SB]

- Refer learners to the Student's Book. Ask: **How do you know if a 'three-for' offer is good value for money?**

Teach and Learn [SB]

- Discuss the text and Example in the Student's Book. Write $320 \div 4 = \square$. Say: **You can solve such problems by using knowledge of division facts.** Write: $32 \div 4 = \square$. Ask: **What is the answer?** Elicit it is 8. Say: **320 is ten times larger than 32. You know 32 divided by four is eight. Therefore you know that 320 divided by four is ten times larger, 80.** Repeat for $420 \div 7 = \square$. (60)
- Say: **There are other numbers easily divisible by one-digit numbers.** On the board, write $666 \div 3$, $444 \div 2$, $888 \div 4$. Ask: **How would you find the answers to these calculations?** Discuss possible strategies. Elicit that divisions can be solved by considering multiples, for example, 444 can be split into $400 + 40 + 4$, all multiples of two.
- Say: **You can partition such numbers in your head and divide. Look at this division.** Write

$444 \div 2 = (400 \div 2) + (40 \div 2) + (4 \div 2) = 222$. Write $496 \div 8$. Ask: **How would you split 496?** Say: **Look for a known fact. You know $8 \times 6 = 48$, so you also know $8 \times 60 = 480$.** Say: **If you split 496 into 480 and 16 you can divide both parts by eight.**

- Demonstrate: $496 \div 8 = (480 \div 8) + (16 \div 8) = 60 + 2 = 62$. Write $294 \div 7$ for learners to solve.

Practise [WB]

- Workbook: Dividing a 3-digit number by a single-digit number (1) page 110

Apply 👥 [SB]

- Learners decide if the offer in the Discover section of the Student's Book is reasonable. (no; $177 \div 3$ is 59. Single cans only cost 49 cents)

Review

- Say: **A baker divides 272 cakes between eight trays. How many cakes are there per tray?** (34)

Assessment for learning

- I calculate $552 \div 6$ by splitting 552 into two numbers and dividing each by six. What are the numbers?

Lesson 8: **Dividing a 3-digit number by a single-digit number (2)**

Learning objectives

Code	Learning objective
5Nc20	[Multiply or] divide three-digit numbers by single-digit numbers.
5Nc23	Divide three-digit numbers by single-digit numbers, including those with a remainder (answers no greater than 30).

Strand 5: Problem solving

5Pt2	Solve single and multi-step word problems (all four operations); represent them, e.g. with diagrams or a number line.
5Pt4	Use multiplication to check the result of a division [, e.g. multiply 3.7 × 8 to check 29.6 ÷ 8].
5Ps2	Choose an appropriate strategy for a calculation and explain how they worked out the answer.
5Ps9	Explain methods and justify reasoning orally and in writing; make hypotheses and test them out.
5Ps10	Solve a larger problem by breaking it down into sub-problems or represent it using diagrams.

Prerequisites for learning

Learners need to:

• multiply and divide a number by multiples of ten and 100.

Success criteria

Learners can:

• make a reasonable estimate for the answer
• divide ten times a multiple of a number by a one-digit number.

Vocabulary

key fact, divide, divisible by, estimate, grouping, sharing

Resources

mini whiteboard and pen or paper (per pair)

Refresh

• Choose an activity from Number – Calculation: Mental strategies, Multiplication and division in the Refresh activities.

Discover 🆂🅱

• Read the text in the Student's Book. Ask: **How would you divide $396 among nine people?** Discuss sharing and grouping.

Teach and Learn 🆂🅱 🖥

• Discuss the text in the Student's Book. Write 360 ÷ 2, 376 ÷ 8. Say: **To solve a division problem, you should try to use an efficient strategy. For small numbers, sharing is better.** Demonstrate for 360 ÷ 2: 360 divided equally between two is 180 (halving).

• Say: **For 376 ÷ 8, grouping is better**. Refer to the Example section. Grouping involves repeated subtraction of multiples of the number they are dividing by, until they find the total number of groups. Say: **Always estimate first. Can ten groups of eight be made from 376?** (yes) **Can 20 groups be made?** (yes) **Can 30, 40, 50, 60 groups be made?** (40 groups, 50 cannot) Ask: **Is 376 closer to 320 (40 groups) or 400 (50 groups)?** (400) Say: **So the answer will be approximately 50.**

• Display **Slide 1** and talk through the example. Provide further guided examples, such as 182 ÷ 7, 477 ÷ 9. Then set more problems for learners to solve (228 ÷ 8; 264 ÷ 9, 196 ÷ 8). Ask: **What's different about the last problem?** Establish that there is a remainder. Show how the answer is written with a remainder: 24 r 4.

Practise 🆆🅱

• Workbook: Dividing a 3-digit number by a single-digit number (2) page 112

• Refer to Activity 4 from the Additional practice activities.

Apply 👥 🖥

• Display **Slide 2**. Learners share solutions. (658 ml ÷ 7 = 94 ml)

Review

• Say: **A shop needs to pack 348 doughnuts in boxes. If each box holds 6 doughnuts, how many boxes does the shop need?** (58)

Assessment for learning

• I divide $581 into seven equal shares by grouping. Explain how I do this.

Additional practice activities

Activity 1 ▪▪ [Challenge 2]

Learning objective

- Use knowledge of multiplication and division facts to 10 × 10 to solve number problems.

Resources

paper and pens (per pair)

What to do

- Each learner chooses a digit from 3 to 9 and writes fact triangles for five multiples of that number.
- For 7, for example, a learner may write five fact triangles: 35, 7, 5; 42, 7, 6; 49, 7, 7; 56, 7, 8; 63, 7, 9.
- For each triangle, they write two multiplication facts and two division facts on a separate piece of paper.
- They erase one number from each fact and swap with their partner to identify the missing numbers.

Variation

[Challenge 3] Learners create fact triangles with two-digit numbers.

Activity 2 ▪▪ [Challenge 2]

Learning objective

- Use factors to simplify a multiplication so that it can be solved by repeated doubling.

Resources

mini whiteboard and pen or paper (per pair)

What to do

- On the board, write: 327 × 8, 732 × 4, 629 × 8, 126 × 16.
- Say: **Work out the answer to each calculation by splitting the smaller number into its factors.**
- Demonstrate: 327 × 8. Show that 8 can be split into 2 × 4 and then again into 2 × 2 × 2. Write: 327 × 8 = 327 × 2 × 2 × 2.
- Point out that 2 × 2 × 2 is equivalent to doubling a number three times (327 × 2 × 2 × 2 = 2616).
- Ask learners to use a similar method to work out the remaining calculations (732 × 4 = 2928; 629 × 8 = 5032; 126 × 16 = 2016).

Variation

[Challenge 1] Ask learners to solve a set of simpler multiplications: 125 × 4, 215 × 4, 355 × 8 (500, 860, 2840)

Additional practice activities

Activity 3

Learning objectives

- Set up a 'calculation factory' in which learners work together to use a range of strategies to answer two-digit by two-digit multiplication questions.
- Choose an appropriate strategy for a calculation and explain how to work out the answer.

Resources

mini whiteboard and pen or paper (per pair)

What to do 🖥

- Display **Slide 1**.
- Say: **You are going to work together as a class to find the answers to the problems. You will combine your multiplication skills, working in five separate teams to become part of a 'calculation factory'.**
- Organise the learners into five groups: Partitioning, Grid method, Expanded written method, Halving and doubling and Multiplying by factors.
- Say: **The name of your group is the strategy you will use to solve each problem.**
- Ask two volunteers to be the managers of the 'calculation factory'. Their role is to assess each calculation and decide which team should work on the problem.

Variation

Challenge **1** Provide a set of two-digit by one-digit and three-digit by one-digit multiplications for learners to solve.

Activity 4

Learning objective

- Choose an appropriate strategy for a division calculation and explain how to work out the answer.

Resources

mini whiteboard and pen or paper (per pair)

What to do 🖥

- Display **Slide 2**.
- Pairs of learners decide whether each question is best answered as a sharing problem or a grouping problem.
- They decide who will be the 'sharer' and who will be the 'grouper'.
- They each solve the problems, then compare their answers with those of other learners to confirm that they are correct.
- The 'sharer' and 'grouper' then swap roles and answer the remaining problems.

Variation

Challenge **3** Include divisions where the answer has a remainder.

Unit 12: Multiplication and division 2

Learning objectives

Code	Learning objective
5Nc3	Know multiplication and division facts for the 2× to 10× tables.
5Nc6	Know squares of all numbers to 10 × 10.
5Nc12	Multiply multiples of 10 to 90, and multiples of 100 to 900, by a single-digit number.
5Nc13	Multiply by 19 or 21 by multiplying by 20 and adjusting.
5Nc14	Multiply by 25 by multiplying by 100 and dividing by 4.
5Nc15	Use factors to multiply, e.g. multiply by 3, then double to multiply by 6.
5Nc16	Double any number up to 100 and halve even numbers to 200 [and use this to double and halve numbers with one or two decimal places, e.g. double 3·4 and half of 8·6].
5Nc17	Double multiples of 10 to 1000 and multiples of 100 to 10 000, e.g. double 360 or double 3600, and derive the corresponding halves.
5Nc20*	Multiply or divide three-digit numbers by single-digit numbers.
5Nc21*	Multiply two-digit numbers by two-digit numbers.
5Nc23*	Divide three-digit numbers by single-digit numbers, including those with a remainder (answers no greater than 30).
5Nc26	Decide whether to round an answer up or down after division, depending on the context.

*Learning objective revised and consolidated

Strand 5: Problem solving

5Pt2	Solve single and multi-step word problems (all four operations); represent them, e.g. with diagrams or a number line.
5Pt4	Use multiplication to check the result of a division [, e.g. multiply 3·7 × 8 to check 29.6 ÷ 8].
5Pt6	Estimate and approximate when calculating, e.g. using rounding, and check working.
5Pt7	Consider whether an answer is reasonable in the context of a problem.
5Ps2	Choose an appropriate strategy for a calculation and explain how they worked out the answer.
5Ps3	Explore and solve number problems and puzzles, e.g. logic problems.
5Ps4	Deduce new information from existing information to solve problems.
5Ps5	Use ordered lists and tables to help to solve problems systematically.
5Ps8	Investigate a simple general statement by finding examples which do or do not satisfy it [, e.g. the sum of three consecutive whole numbers is always a multiple of three].
5Ps9	Explain methods and justify reasoning orally and in writing; make hypotheses and test them out.
5Ps10	Solve a larger problem by breaking it down into sub-problems or represent it using diagrams.

Unit overview

In this unit, learners review knowledge of multiplication and division facts for the two to ten times tables and factors of two-digit numbers. They use understanding of number facts and place value to multiply multiples of ten and 100 by a one-digit number. Learners are introduced to compensation strategies for multiplication by 19 or 21, and a multiplying and dividing strategy for 25. Strategies for multiplication are extended to include splitting numbers into factors to multiply. Learners practise doubling and halving numbers, including doubles of multiples of ten to 1000, and multiples of 100 to 10 000, using this to derive corresponding halves. They practise mental and informal written strategies for multiplication, and consider problems where they must decide whether to round an answer up or down after division.

Common difficulties and remediation

Extended practical experience of drawing or arranging objects in arrays will help learners understand how a number breaks down into its factors. At this stage learners are increasingly aware of division as an operation that, unlike multiplication, is not commutative and often involves a more complex procedure. It is important that learners have opportunities to solve problems in which they need to make decisions about what to do with the remainder.

Promoting and supporting language

Learners sometimes confuse the terms 'factors' and 'multiples'. Regularly highlighting the differences between the two should help to avoid confusion. Also adapt language using simpler definitions such as 'Factors are what we multiply to get the number, whereas multiples are what we get after multiplying the number.'

Lesson 1: **Multiplication and division facts**

Learning objectives

Code	Learning objective
5Nc3*	Know multiplication and division facts for the 2× to 10× tables.
5Nc6	Know squares of all numbers to 10 × 10.

Strand 5: Problem solving

5Pt4	Use multiplication to check the result of a division [, e.g. multiply 3.7 × 8 to check 29.6 ÷ 8].
5Ps2	Choose an appropriate strategy for a calculation and explain how they worked out the answer.
5Ps3	Explore and solve number problems and puzzles, e.g. logic problems.
5Ps4	Deduce new information from existing information to solve problems.

Prerequisites for learning

Learners need to:

- count on and back in steps of constant size
- understand the concepts of multiplication and division.

Success criteria

Learners can:

- recall all the multiplication and division facts for the 2× to 10× multiplication tables

- recall squares of all numbers up to 10 × 10.

Vocabulary

multiplication table, multiplication, division, fact family, fact triangle

Resources

paper, rulers and pencils (per pair)

Refresh

- Choose an activity from Number – Calculation: Mental strategies, Multiplication and division in the Refresh activities.

Discover [SB]

- Refer learners to the text and diagram in the Students' Book. Ask: **If the top number changed to 56, how would the bottom numbers change?** Discuss how numbers are related by multiplication and division.

Teach and Learn [SB]

- Discuss the text and Example in the Student's Book. Draw a fact triangle with 28 at the top and 4 and 7 at the bottom. **Demonstrate how the three numbers are related by multiplication and division.**
- Write 4 × ☐ and ☐ × 7. Ask: **How did you find the missing numbers for these?** Elicit that they are division problems 'in disguise'. Remind learners that multiplication and division are **inverse operations.** Refer to the fact triangle and show how the number sentence 4 × 7 = 28 can be rearranged: 28 ÷ 7 = 4, 28 ÷ 4 = 7.
- Draw fact triangles for 8 × 5 = 40 and 6 × 9 = 54. Learners write two multiplication facts and two division facts for each triangle.
- Pairs of learners draw fact triangles up to 10 × 10 with a number missing and asking their partner to identify the missing number.

- Write on the board the sequence of squares (1, 4, 9 …). Ask: **What do these numbers have in common?** Agree they are square numbers. Say: **Square numbers are numbers that create squares. They are formed by multiplying a number by itself. In other words 1, 4, 9, 16, 25, …** List the square numbers as multiplication number sentences in the correct order: 2 × 2 = 4, 3 × 3 = 9, …

Practise [WB]

- Workbook: Multiplication and division facts page 114

Apply 👥

- Distribute paper for learners to draw triangles. Say: **Four children share 24 cupcakes equally. Draw a fact triangle to show how many cupcakes each child will get.**

Review 👥

- Learners take turns to say a number and link it to multiplications and divisions.

Assessment for learning

- Knowing 4 × 8 makes 32 means that I also know 32 ÷ 4 is 8. Explain.

Unit 12 Multiplication and division 2

Lesson 2: Multiplying multiples of 10 and 100 (1)

Learning objectives

Code	Learning objective
5Nc12	Multiply multiples of 10 to 90, and multiples of 100 to 900, by a single-digit number.

Strand 5: Problem solving

5Pt4	Use multiplication to check the result of a division [, e.g. multiply 3·7 × 8 to check 29.6 ÷ 8].
5Pt6	Estimate and approximate when calculating, e.g. using rounding, and check working.
5Ps2	Choose an appropriate strategy for a calculation and explain how they worked out the answer.
5Ps3	Explore and solve number problems and puzzles, e.g. logic problems.
5Ps5	Use ordered lists and tables to help to solve problems systematically.
5Ps9	Explain methods and justify reasoning orally and in writing; make hypotheses and test them out.
5Ps10	Solve a larger problem by breaking it down into sub-problems or represent it using diagrams.

Prerequisites for learning

Learners need to:
- recall multiplication and division facts to 10 × 10

Success criteria

Learners can:
- multiply multiples of 10 to 90, and multiples of 100 to 900, by a one-digit number.

Vocabulary

number fact, multiple, partition, place value

Refresh

- Choose an activity from Number – Calculation: Mental strategies, Multiplication and division in the Refresh activities.

Discover [SB]

- Refer learners to the Student's Book and read the text with the class. Ask: **How would you solve this problem?** Discuss strategies.

Teach and Learn [SB] [🖥]

- Discuss the text and examples in the Student's Book.
- Display **Slide 1.** Point to 30. Ask: **What are the first two multiples of 30?** (30, 60) Record the answers in the 1×, 2× columns. Repeat until all multiples of 30 up to 300 are recorded. Ask: **What patterns can you see?** Establish the numbers increase by 30. Ask: **What is four times 30?** (120) Discuss strategies, but emphasise that learners can solve the problem mentally.
- Write 30 × 4. Rewrite the a problem using partitioning: 10 × 3 × 4. Ask: **How can you solve this now?** Elicit: multiply three by four, making 12, then multiply by ten. Revise using place value knowledge to multiply by ten. Solve 30 × 7 as a class. Write 30 × 5, 30 × 8 on the board for learners to solve.
- Return to **Slide 1** and point to 70. Ask: **What is three times 70? Six times 70?** Prompt learners to find the answers mentally, using partitioning and place value. Expect answers 210, 420.

- Display **Slide 2.** Learners form a line and come to the board, adding answers to the table then joining the line again. Point to the second table. Ask: **How does this change the answers?** Praise learners who identify that the answers increase by the same power of ten. Demonstrate partitioned multiplication: 800 × 7 = 100 × 8 × 7 = 100 × 56 = 5600.

Practise [WB]

- Workbook: Multiplying multiples of 10 and 100 (1) page 116
- Refer to Activity 1 from the Additional practice activities.

Apply [👥] [SB]

- Ask pairs of learners to solve the problem in the Discover section of the Student's Book. Invite learners to share and discuss their solutions. (30 people: (30 × 3, 30 × 6, 30 × 8) 90, 180, 240; 60 people: 180, 360, 480; 90 people: 270, 540, 720)

Review

- Draw three boxes labelled A: $40, B: $300, C: $700. Ask: **What is the total price for five of box A, six of box B and eight of box C?** ($7600)

Assessment for learning

- I have nine buckets of sand, each with mass 600 g. What is the combined mass? (5400 g)

Lesson 3: **Multiplying by 19, 21 or 25 (1)**

Learning objectives

Code	Learning objective
5Nc13	Multiply by 19 or 21 by multiplying by 20 and adjusting.
5Nc14	Multiply by 25 by multiplying by 100 and dividing by 4.

Strand 5: Problem solving

5Pt2	Solve single and multi-step word problems (all four operations); represent them, e.g. with diagrams or a number line.
5Pt6	Estimate and approximate when calculating, e.g. using rounding, and check working.
5Ps2	Choose an appropriate strategy for a calculation and explain how they worked out the answer.
5Ps3	Explore and solve number problems and puzzles, e.g. logic problems.
5Ps8	Investigate a simple general statement by finding examples which do or do not satisfy it [, e.g. the sum of three consecutive whole numbers is always a multiple of three].
5Ps9	Explain methods and justify reasoning orally and in writing; make hypotheses and test them out.
5Ps10	Solve a larger problem by breaking it down into sub-problems or represent it using diagrams.

Prerequisites for learning

Learners need to:

- multiply multiples of 10 to 90.

Success criteria

Learners can:

- multiply by 19 or 21 by multiplying by 20 and adjusting

- multiply by 25 by multiplying by 100 and dividing by four.

Vocabulary

multiple, round, adjust

Refresh

- Choose an activity from Number – Calculation: Mental strategies, Multiplication and division in the Refresh activities.

Discover [SB]

- Read the text and look at the picture in the Student's Book. Ask learners to complete the calculation. (152) Say: **Tell your partner a problem that could be solved in a similar way.**

Teach and Learn [SB]

- Discuss the text and examples in the Student's Book. Write 6×19, 9×19, 7×19 on the left of the board and 6×20, 9×20, 8×20 on the right. Ask: **How would you solve the problems on the left?** Discuss strategies. Ask learners if they think the problems can be solved mentally. Say: **They may look tricky, but you can use a strategy you have used before.**

- Prompt learners to look to the right of the board for a clue. Praise any learner who suggests using the strategy of rounding and adjusting. Write: $6 \times 19 = (6 \times 20) - (6 \times 1)$. Say: **You round 19 to 20 and multiply by six. To adjust, subtract six for multiplying 1×6 too many.** Solve, with learners assisting. (114)

- Point to 9×19. Ask: **How will you solve this problem?** Elicit: multiply by 20 and adjust by subtracting nine. Invite a volunteer to demonstrate.

They should record: $9 \times 19 = (9 \times 20) - 9 = 180 - 9 = 171$. Ask learners to solve the remaining problem, 7×19. (133)

- Write 8×21, 6×21. Ask: **Is there a similar strategy you can use to solve these problems?** Elicit that the answer is to round and adjust, but adding this time. Demonstrate: $8 \times 21 = (8 \times 20) + (8 \times 1) = 168$. Guide learners to solve the remaining problem, 6×21. (126)

- On the board, write: 18×25. Discuss the strategy: multiply by 100, divide by four. Demonstrate: 18×100, then $1800 \div 4 = 1800$ halved, then halved again (450).

Practise [WB]

- Workbook: Multiplying by 19, 21 or 25 page 118

Apply 👥 🖥

- Display **Slide 1**. Agree that the problem is $4 \times 19 \times 21 \times 25$. (39 900)

Review

- On the board, draw three boxes labelled A: $19, B: $21, C: $25. Ask: **What is the total price for seven of box A, nine of box B and eight of box C?** ($522)

Assessment for learning

- I work out 7×19 by multiplying by 20 and adjusting. How do I adjust?

Number – Calculation: Mental strategies, Multiplication and division

145

Lesson 4: **Multiplication by factors (1)**

Learning objectives

Code	Learning objective
5Nc15	Use factors to multiply, e.g. multiply by 3, then double to multiply by 6.

Strand 5: Problem solving

5Pt2	Solve single and multi-step word problems (all four operations); represent them, e.g. with diagrams or a number line.
5Pt6	Estimate and approximate when calculating, e.g. using rounding, and check working.
5Ps2	Choose an appropriate strategy for a calculation and explain how they worked out the answer.
5Ps3	Explore and solve number problems and puzzles, e.g. logic problems.
5Ps8	Investigate a simple general statement by finding examples which do or do not satisfy it [, e.g. the sum of three consecutive whole numbers is always a multiple of three].
5Ps9	Explain methods and justify reasoning orally and in writing; make hypotheses and test them out.
5Ps10	Solve a larger problem by breaking it down into sub-problems or represent it using diagrams.

Prerequisites for learning

Learners need to:
- understand the terms 'factor' and 'multiple'
- double and halve whole numbers.

Success criteria

Learners can:
- multiply by rewriting a calculation using the factors of one number.

Vocabulary

multiple, factor

Resources

1–6 dice, alternatively use Resource sheet 4: 1–6 spinner (per pair)

Refresh

- Choose an activity from Number – Calculation: Mental strategies, Multiplication and division in the Refresh activities.

Discover [SB] 🖥

- Refer learners to the text and arrays in the Student's Book and display **Slide 1.** Complete the solution to the problem. (288) Ask: **How would you use the factor pairs to solve 32 × 12?**

Teach and Learn [SB]

- Discuss the text and examples in the Student's Book. Remind learners of earlier work with factors. Say: **You are going to use factors to work out the answer to a multiplication.** Write: 15, 18. Ask learners for factor pairs and record correct responses under each number: 1 × 15, 3 × 5; 1 × 18, 2 × 9, 3 × 6.
- Write: 15 × 18. Ask: **How could you use factors to rewrite the calculation?** Elicit breaking one of the numbers in the calculation into its factors and then multiplying. Demonstrate by writing 15 × 18 = 15 × 3 × 6.
- Ask: **Can you go any further with factors, for example splitting six?** Agree it can be split further as 15 × 3 × 3 × 2.
- Explore rewriting the calculation in different ways, rearranging terms and discussing the 'easiest'

multiplication to solve. Take suggestions, for example: 15 × 3 × 3 × 2 = 15 × 2 × 3 × 3 = 30 × 9 = 270. Confirm that each expression gives the same answer, 270. Write: 24 × 16, 18 × 36, 19 × 17.
- Ask learners to solve (384, 648, 323) and invite them to discuss strategies. Ask: **What did you notice about the last problem?** Elicit that both numbers are prime, having only one and itself as factors, and cannot be split, so the calculation is unchanged.

Practise [WB]

- Workbook: Multiplication by factors (1) page 120
- Refer to Activity 2 from the Additional practice activities.

Apply 👥🖥

- Display **Slide 2**. Agree the problem is (16 × 32) + (12 × 24). (800 toys)

Review

- On the board, write: 36 × 37. Ask: **Which number should I split into its factors?** Agree 36. Ask: Why not 37? (it is a prime number) Learners solve. (1332)

Assessment for learning

- I can simplify 27 × 36 by splitting 36 into its factors. Which factors should I use?

Lesson 5: **Doubles and halves (1)**

Learning objectives

Code	Learning objective
5Nc16	Double any number up to 100 and halve even numbers to 200 [and use this to double and halve numbers with one or two decimal places, e.g. double 3·4 and half of 8·6].
5Nc17	Double multiples of 10 to 1000 and multiples of 100 to 10 000, e.g. double 360 or double 3600, and derive the corresponding halves.

Strand 5: Problem solving

5Pt2	Solve single and multi-step word problems (all four operations); represent them, e.g. with diagrams or a number line.
5Pt6	Estimate and approximate when calculating, e.g. using rounding, and check working.
5Ps2	Choose an appropriate strategy for a calculation and explain how they worked out the answer.
5Ps3	Explore and solve number problems and puzzles, e.g. logic problems.
5Ps8	Investigate a simple general statement by finding examples which do or do not satisfy it [, e.g. the sum of three consecutive whole numbers is always a multiple of three].
5Ps9	Explain methods and justify reasoning orally and in writing; make hypotheses and test them out.
5Ps10	Solve a larger problem by breaking it down into sub-problems or represent it using diagrams.

Prerequisites for learning

Learners need to:
• double and halve whole numbers.

Success criteria

Learners can:
• double and halve numbers by using known number facts and partitioning.

Vocabulary

doubling, halving, inverse

Refresh

• Choose an activity from Number – Calculation: Mental strategies, Multiplication and division in the Refresh activities.

Discover [SB]

• Refer learners to the Student's Book and read the text together. Ask: **How would you work out the number the boy is thinking of?** Discuss strategies.

Teach and Learn [SB] [bar chart icon]

• Discuss the text and examples in the Student's Book. Revise the terms 'doubling' and 'halving'. Say: **Doubling means the same as multiplying by two. Halving means the same as dividing by two.** Display the **Function machine tool** with machines labelled '× 2' and '÷ 2' to demonstrate.

• Ask: **What is the relationship between doubling and halving?** Establish that halving is the inverse of doubling. Demonstrate by writing outputs and asking learners to determine inputs. Practise doubling any number up to 100 and halving even numbers to 200.

• Ask: **What strategies do you use to double?** Confirm the method, for example: double tens, double units, and combine. Begin with easy numbers, such as 22, 44, 33, moving on to numbers such as 37, 48, 56, then even harder numbers, 86, 67, 98.

• Discuss other strategies, including using knowledge of doubles to halve numbers, for example, 23 + 23 = 46,

so half of 46 is 23. Do the same for multiples of ten: double 20 to double 100; half of 20 to half of 200.

• Demonstrate how to use knowledge of doubling two-digit numbers to double multiples of ten to 1000. Write: 830. Say: **Doubling 830 is easy if you think of it as 83 tens, double (166) then adjust by multiplying by ten (1660).** Repeat for halving 2840 (think 284 tens, halve (142) then adjust (1420).

• Do the same for doubling and halving multiples of 100 to 1000.

Practise [WB]

• Workbook: Doubles and halves (1) page 122

Apply [icons]

• Display **Slide 1**. Invite learners to discuss whether the set of operations always gives the answer three. Ask them to explain why this happens.

Review

• On the board, write 34, 79, 490, 3700; in a second row, write 78, 168, 574. Say: **Double the numbers in the first row and halve the numbers in the second.** (68, 158, 980, 7400; 39, 84, 287)

Assessment for learning

• If I double 36, then halve the answer, I get the number I started with. Why is that?

Number – Calculation: Mental strategies, Multiplication and division

Lesson 6: **Multiplying a 3-digit number by a single-digit number (2)**

Learning objectives

Code	Learning objective
5Nc20*	Multiply [or divide] three-digit numbers by single-digit numbers.

Strand 5: Problem solving

5Pt2	Solve single and multi-step word problems (all four operations); represent them, e.g. with diagrams or a number line.
5Pt6	Estimate and approximate when calculating, e.g. using rounding, and check working.
5Ps2	Choose an appropriate strategy for a calculation and explain how they worked out the answer.
5Ps3	Explore and solve number problems and puzzles, e.g. logic problems.
5Ps9	Explain methods and justify reasoning orally and in writing; make hypotheses and test them out.
5Ps10	Solve a larger problem by breaking it down into sub-problems or represent it using diagrams.

Prerequisites for learning

Learners need to:
• partition a three-digit number.

Success criteria

Learners can:
• make a reasonable estimate for the answer

• multiply a three-digit number by a single-digit number, using a range of strategies.

Vocabulary

multiple, key fact, estimate, partition

Resources

mini whiteboard and pen or paper (per pair)

Refresh

• Choose an activity from Number – Calculation: Mental strategies, Multiplication and division in the Refresh activities.

Discover [SB]

• Ask learners to read the text and look at the picture in the Student's Book. Ask: **How would you solve this problem?** Accept and discuss suggestions. Explain to learners that they will return to the question later.

Teach and Learn [SB]

• Discuss the text and examples in the Student's Book. Say: **You are going to review estimation and calculation of three-digit by one-digit multiplications.**
• On the board, write: 489 × 4. Ask: **How would you estimate the answer?** Discuss strategy of rounding and write 500 × 4 = 2000. Ask: **Who can remember a strategy for calculating the answer?** Elicit partitioning or the grid method. Demonstrate partitioning, with learners assisting. Ask: **How will you partition 489?** (400, 80 and 9) Demonstrate multiplication using partitioning: 489 × 4 = (400 × 4) + (80 × 4) + (9 × 4) = 1600 + 320 + 36 = 1956. Say: **The estimate, 2000, shows that the answer is correct.**

• Invite a volunteer to demonstrate the grid method. Expect them to draw a grid, split the partitions across the boxes, then multiply and add. Write 667 × 9. Say: **Who can remember a written method of calculation?** Praise any learner who suggests the expanded method. On the board, draw a template layout. Demonstrate the method.

Practise [WB]

• Workbook: Multiplying a 3-digit number by a single-digit number (2) page 124
• Refer to Activity 3 from the Additional practice activities.

Apply [SB]

• Learners use the expanded written method to find the total mass pictured in the Discover section. (987 × 8 = 7896 kg)

Review

• Reinforce partitioning, the grid method and also the expanded method taught in the lesson. On the board, write: 389 × 3, 586 × 7. Ask learners to find the solutions. (1167, 4102)

Assessment for learning

• Show me two different methods for calculating 939 × 9.

Lesson 7: **Multiplying a 2-digit number by a 2-digit number (3)**

Learning objectives

Code	Learning objective
5Nc21*	Multiply two-digit numbers by two-digit numbers.

Strand 5: Problem solving

Code	Learning objective
5Pt2	Solve single and multi-step word problems (all four operations); represent them, e.g. with diagrams or a number line.
5Pt6	Estimate and approximate when calculating, e.g. using rounding, and check working.
5Ps2	Choose an appropriate strategy for a calculation and explain how they worked out the answer.
5Ps3	Explore and solve number problems and puzzles, e.g. logic problems.
5Ps9	Explain methods and justify reasoning orally and in writing; make hypotheses and test them out.
5Ps10	Solve a larger problem by breaking it down into sub-problems or represent it using diagrams.

Prerequisites for learning

Learners need to:
* partition numbers into tens and units.

Success criteria

Learners can:
* make a reasonable estimate for the answer
* use various strategies to multiply a two-digit number by a two-digit number.

Vocabulary

multiple, key fact, estimate, partition

Resources

mini whiteboard and pen or paper (per pair)

Refresh

* Choose an activity from Number – Calculation: Mental strategies, Multiplication and division in the Refresh activities.

Discover [SB]

* Direct learners to the Student's Book. Ask: **How would you calculate the number of boxes in the stack?** Discuss strategies.

Teach and Learn [SB]

* Discuss the text and Example in the Student's Book.
* Write 56 × 47. Ask: **How would you estimate the answer?** Discuss strategy of rounding and write 60 × 50 = 3000. Say: **The answer should be close to 3000.**
* Ask: **Who can remember a written strategy for calculation?** Elicit partitioning or the grid method. Demonstrate partitioning, with learners assisting. Record on the board: 56 × 47 = (56 × 40) + (56 × 7). Partition further and complete the calculation: = (50 × 40) + (6 × 40) + (50 × 7) + (6 × 7) = 2000 + 240 + 350 + 42 = 2632. Say: **The estimate, 3000, shows the answer is correct.**
* Ask a volunteer to come and demonstrate the grid method. Expect the volunteer to draw a grid, split partitions across boxes, then multiply and add.

* Write 89 × 66 and draw a template for the expanded written method. Ask: **What is 89 multiplied by 6?** (534) Write 534 in the correct place value columns. Continue for the tens multiplication, finally adding to get the total (5874).

Practise [WB]

* Workbook: Multiplying a 2-digit number by a 2-digit number (3) page 126

Apply 👥 🖥

* Display **Slide 1**. Ask learners to share their solutions to the problem. Stack 1: 47 × 6 × 7 = 1974; Stack 2: 39 × 8 × 6 = 1872. Stack 1 has more boxes.

Review

* Reinforce partitioning, the grid method and also the expanded method taught in the lesson. On the board, write: 59 × 73; 86 × 89. Ask learners to work out the solutions. (4307, 7654)

Assessment for learning

* Show me two different methods for calculating 64 × 88.

Lesson 8: **Dividing a 3-digit number by a single-digit number (3)**

Number – Calculation: Mental strategies, Multiplication and division

Learning objectives

Code	Learning objective
5Nc20*	[Multiply or] divide three-digit numbers by single-digit numbers.
5Nc23*	Divide three-digit numbers by single-digit numbers, including those with a remainder (answers no greater than 30).
5Nc26*	Decide whether to round an answer up or down after division, depending on the context.

Strand 5: Problem solving	
5Pt2	Solve single and multi-step word problems (all four operations); represent them, e.g. with diagrams or a number line.
5Pt4	Use multiplication to check the result of a division [, e.g. multiply $3·7 \times 8$ to check $29.6 \div 8$].
5Pt6	Estimate and approximate when calculating, e.g. using rounding, and check working.
5Pt7	Consider whether an answer is reasonable in the context of a problem.
5Ps2	Choose an appropriate strategy for a calculation and explain how they worked out the answer.
5Ps3	Explore and solve number problems and puzzles, e.g. logic problems.
5Ps4	Deduce new information from existing information to solve problems.
5Ps10	Solve a larger problem by breaking it down into sub-problems or represent it using diagrams.

Prerequisites for learning

Learners need to:

- multiply and divide multiples of 10 and 100
- divide ten times a multiple of a number by a one-digit number.

Success criteria

Learners can:

- divide a three-digit number by a one-digit number using a range of strategies

- solve a division problem that has a remainder
- recognise when to round up or down after division, depending on the problem.

Vocabulary

key fact, estimate, partition, grouping, sharing, rounding

Resources

mini whiteboard and pen or paper (per pair)

Refresh

- Choose an activity from Number – Calculation: Mental strategies, Multiplication and division in the Refresh activities.

Discover SB

- Read the text in the Student's Book as a class. Say: **Describe to your partner two division problems: one where the answer requires rounding up and one that requires rounding down.**

Teach and Learn SB 🖥

- Write: $270 \div 3$, $384 \div 6$. Ask: **Which problem is easier to solve? Why?** Expect learners to say the first problem, as 270 is easily divisible by 3. Ask: **How would you solve the problem mentally?** Expect learners to use the known fact $27 \div 3 = 9$ then scale the answer ten times (90).
- Invite solutions to the second problem. Elicit partitioning as a strategy. Solve, with learners assisting. Ask: **How would you split 384?** Say: **Look for a known fact.** $384 \div 6 = (360 \div 6) + (24 \div 6) = 60 + 4 = 64$. Discuss the expanded written method by referring to the example in the Student Book.
- Say: **Sometimes we need to round the answer to a division up or down to give a sensible answer.**

Display **Slide 1**. Work through the division problem together. (21 boxes will be needed for all of the eggs; there are 20 full boxes with a remainder of 2 eggs)

Practise WB

- Workbook: Dividing a 3-digit number by a single-digit number (3) page 128.
- Refer to Activity 4 from the Additional practice activities.

Apply 👥 🖥

- Display **Slide 2**. Ask learners to share solutions and say whether the answer required rounding up or down. (41, down)

Review

- Reinforce partitioning and the expanded method taught in the lesson. On the board, write: $536 \div 8$; $887 \div 9$. Learners solve in pairs. (67, 98 r 5)

Assessment for learning

- Tell me a division problem where the answer requires rounding up.

Additional practice activities

Activity 1 :: Challenge 2

Learning objective
• Identify factor pairs of a multiple of ten or 100 by using knowledge of number facts.

Resources
squared paper (per pair)

What to do
• On the board, write: 240 chairs.
• Say: **A concert organiser wants to organise seating for halls of different shapes. How many different arrangements of rows and columns are there for 240 chairs?**
• Establish that we can use knowledge of number facts and arrays to find factor pairs for 24 then adjust for ten times bigger, for example: $3 \times 8 = 24$, so $3 \times 80 = 240$ (3 rows of 80).

• On the board, write the following numbers of chairs: 450, 720, 810, 3600, 5600, 6300.
• Ask: **How many different seating arrangements can you find for each number of chairs?**
• Ask learners to share the seating arrangements that they find.

Variation
Challenge 1 Learners work with multiplication and division facts up to 10×10 to find arrangements for 45, 72, 81, 56, 63 seats.

Activity 2 :: Challenge 2

Learning objectives
• Multiply two-digit numbers by 19, 21, 25 using strategies of compensation and division, and multiply two-digit numbers where the multiplication can be simplified using factors.
• Solve a larger problem by breaking it down into sub-problems.

Resources
mini whiteboard and pen or paper (per pair)

What to do 🖥
• Display **Slide 1.**
• Say: **Each box is a packaged gift on sale in a shop. Look at the three lists of gifts to purchase. Use your knowledge of multiplying strategies to work out the total cost for each set of gifts.**
• Learners calculate the total cost of the gifts on each list. Ask: **Which is the most expensive set of gifts?** (Set 1: $1618; Set 2: $1793; Set 3: $1808. Set 3 is the most expensive)

Variation
Challenge 3 Add two additional items: E: $120 and F: $160 to each set with multiple purchases of 19, 21, 25, 16, 18 and 24.

Additional practice activities

Activity 3

Learning objective

- Use understanding of mental and written strategies for doubling, halving and multiplying three-digit numbers by a one-digit number to calculate the output for a number passed through a set of function machines.

Resources

mini whiteboard and pen or paper (per pair)

What to do 🖵

- Display **Slide 2.**
- Ask learners to input the following numbers: 240, 390, 470.
- Learners write the outputs and say which machine always produces the larger number.
- Remind learners to use strategies for calculating halves and doubles, and to use the grid method or expanded written method to multiply. (First machine: 3339, 5439, 6559; Second machine: 1791, 2466, 2826; First machine produces bigger numbers)

Variation

Challenge 3 Add an extra doubling function at the end of each machine.

Activity 4

Learning objective

- Use understanding of written strategies for multiplying a two-digit number by a two-digit number and dividing a three-digit number by a one-digit number to calculate the output for a number passed through a set of function machines.

Resources

mini whiteboard and pen or paper (per pair)

What to do 🖵

- Display **Slide 3**. Ask learners to input the following numbers: 47, 63, 86.
- Learners write the outputs with remainders and say which machine always produces the larger number.
- Remind learners to use the grid method or expanded written method to multiply and divide. (First machine: 29, 38 r 5, 53; Second machine: 37, 49 r 5, 67 r 6; Second machine produces bigger numbers)

Variation

Challenge 3 Add an extra multiplication function at the end of each machine.

Unit 13: Multiplication and division 3

Learning objectives

Code	Learning objective
5Nc12*	Multiply multiples of 10 to 90, and multiples of 100 to 900, by a single-digit number.
5Nc13*	Multiply by 19 or 21 by multiplying by 20 and adjusting.
5Nc14*	Multiply by 25 by multiplying by 100 and dividing by 4.
5Nc15*	Use factors to multiply, e.g. multiply by 3, then double to multiply by 6.
5Nc16*	Double any number up to 100 and halve even numbers to 200 and use this to double and halve numbers with one or two decimal places, e.g. double 3·4 and half of 8·6.
5Nc17*	Double multiples of 10 to 1000 and multiples of 100 to 10 000, e.g. double 360 or double 3600, and derive the corresponding halves.
5Nc22	Multiply two-digit numbers with one decimal place by single-digit numbers, e.g. 3·6 x 7.
5Nc24	Start expressing remainders as a fraction of the divisor when dividing two-digit numbers by single-digit numbers.
5Nc26*	Decide whether to round an answer up or down after division, depending on the context.
5Nc27	Begin to use brackets to order operations and understand the relationship between the four operations and how the laws of arithmetic apply to multiplication.

Strand 5: Problem solving	*Learning objective revised and consolidated
5Pt2	Solve single and multi-step word problems (all four operations); represent them, e.g. with diagrams or a number line.
5Pt4	Use multiplication to check the result of a division, e.g. multiply 3·7 × 8 to check 29·6 ÷ 8.
5Pt6	Estimate and approximate when calculating, e.g. using rounding, and check working.
5Pt7	Consider whether an answer is reasonable in the context of a problem.
5Ps2	Choose an appropriate strategy for a calculation and explain how they worked out the answer.
5Ps3	Explore and solve number problems and puzzles, e.g. logic problems.
5Ps4	Deduce new information from existing information to solve problems.
5Ps5	Use ordered lists and tables to help to solve problems systematically.
5Ps8	Investigate a simple general statement by finding examples which do or do not satisfy it [, e.g. the sum of three consecutive whole numbers is always a multiple of three].
5Ps9	Explain methods and justify reasoning orally and in writing; make hypotheses and test them out.
5Ps10	Solve a larger problem by breaking it down into sub-problems or represent it using diagrams.

Unit overview

In this unit, learners review the multiplication of multiples of ten and 100 by a single-digit number and strategies for multiplying by 19, 21 or 25. They use factors to multiply one or both numbers, focusing on splitting and reordering. Strategies for doubling and halving are consolidated and applied to numbers with one or two decimal places. Learners apply whole number strategies, and use partitioning or grid methods, to multiply a two-digit number with one decimal place by a one-digit number. They extend previous work to expressing the quotient of a division as a fraction. Learners make generalisations about the order of operations and apply these rules to expressions that include more than one operation.

Common difficulties and remediation

Emphasise the relationship between the number of zeros in the power of ten and the number of zeros in the answer. This is important in laying the foundation for multiplying by decimals.

By the end of the unit, learners should be able to model, write and explain division by a one-digit divisor. They should be confident in handling quotients with a remainder and extend this to expressing remainders as fractions. For division of two- and three-digit numbers, learners build confidence through modelling, but this needs to be extended to using a more conceptual understanding of division, where they use multiples of ten to narrow in on the quotient.

Prior to work that introduces order of operations, learners may have the misconception that all multiplications are calculated before divisions and additions are calculated before subtractions. Take them through guided examples where learners refer to a mnemonic, such as BODMAS/BIDMAS, to help them solve operations in the correct order.

Promoting and supporting language

At this stage, learners need to integrate a broad range of mathematical vocabulary into their explanations, such as 'brackets', 'operation', 'quotient', 'factor' and 'multiple'. When a new key word is introduced, ask learners to write a definition in their own words. Sketching a diagram alongside will help embed the definition.

Lesson 1: **Multiplying multiples of 10 and 100 (2)**

Learning objectives

Code	Learning objective
5Nc12*	Multiply multiples of 10 to 90, and multiples of 100 to 900, by a single-digit number.

Strand 5: Problem solving

5Pt2	Solve single and multi-step word problems (all four operations); represent them, e.g. with diagrams or a number line.
5Pt6	Estimate and approximate when calculating, e.g. using rounding, and check working.
5Pt7	Consider whether an answer is reasonable in the context of a problem.
5Ps2	Choose an appropriate strategy for a calculation and explain how they worked out the answer.
5Ps3	Explore and solve number problems and puzzles, e.g. logic problems.
5Ps9	Explain methods and justify reasoning orally and in writing; make hypotheses and test them out.

Prerequisites for learning

Learners need to:
- count on and back in steps of 10 and 100
- recall multiplication and division facts to 10 x 10

Success criteria

Learners can:
- multiply multiples of 10 to 90, and multiples of 100 to 900, by a single-digit number.

Vocabulary

number fact, multiple, partition, place value

Refresh

- Choose an activity from Number – Calculation: Mental strategies, Multiplication and division in the Refresh activities.

Discover [SB]

- Direct learners to the Student's Book. Ask: **Following the example of estimation given for 796 × 4, what is a good estimate for 67 × 6?** Prompt learners to use their knowledge of rounding and multiplying multiples of ten.

Teach and Learn [SB]

- Discuss the text and examples in the Student's Book. Ask: **What do you recall about multiplication of multiples of ten and 100 by a one-digit number?**
- Write: 40 × 9. Ask: **What is the answer to this problem?** If necessary, say: **Use what you know about four times nine.** Elicit the correct answer. (360) Expect learners to explain that, as one factor in the problem (40) is ten times four, then the product of nine and 40 must be ten times larger than the product of nine and four. Encourage learners first to identify and model the one-digit problem (for example: 4 × 9 for 40 × 9) then use this simpler problem to find the product for larger multiples.
- Change the question to 400 × 9. Ask: **How can a one-digit problem help you answer this question?** Expect the response that as 400 is 100 times four, then the product of nine and 400 must be

100 times larger than the product of nine and four. Write 400 × 9 = 100 × 4 × 9 = 100 × 36 = 3600.
- Write: 4 × 5, 7 × 8, 3 × 9. Say: **How many different number sentences can you make from each one-digit multiplication?** Provide two examples: 4 × 50 = 200, 400 × 5 = 2000.

Practise [WB]

- Workbook: Multiplying multiples of 10 and 100 (2) page 130

Apply 👥

- Say: **Water bottles are placed in 6 stacks of 313. Estimate the total number of bottles.** Invite learners to discuss their estimates. (estimate: 1800 (300 × 6))

Review

- On the board, write: Team A: 6 runners, 700m; Team B: 8 runners, 600m; Team C: 9 runners, 500m. Explain that the distance given is how far each competitor runs. Say: **Calculate the total distance run by each team. Which team ran the furthest distance?** (B)

Assessment for learning

- I pour nine jugs of water into a tray. If each jug holds 800ml, how much water is that?

Lesson 2: **Multiplying by 19, 21 or 25 (2)**

Learning objectives

Code	Learning objective
5Nc13*	Multiply by 19 or 21 by multiplying by 20 and adjusting.
5Nc14*	Multiply by 25 by multiplying by 100 and dividing by 4.

Strand 5: Problem solving

5Pt2	Solve single and multi-step word problems (all four operations); represent them, e.g. with diagrams or a number line.
5Pt7	Consider whether an answer is reasonable in the context of a problem.
5Ps2	Choose an appropriate strategy for a calculation and explain how they worked out the answer.
5Ps3	Explore and solve number problems and puzzles, e.g. logic problems.

Prerequisites for learning

Learners need to:

- multiply multiples of 10 to 90.

Success criteria

Learners can:

- multiply by 19 or 21 by multiplying by 20 and adjusting
- multiply by 25 by multiplying by 100 and dividing by 4.

Vocabulary

multiple, round, adjust

Resources

mini whiteboard and pen or paper (per pair)

Refresh

- Choose an activity from Number – Calculation: Mental strategies, Multiplication and division in the Refresh activities.

Discover [SB]

- Direct learners to the Student's Book. Ask: **What is the total cost of the beans?** Discuss strategies used.

Teach and Learn [SB]

- Discuss the text in the Student's Book. Ask: **Who can remember the different strategies you used to multiply by 19, 21 or 25?**
- On the board, write: 8 × 19. Ask: **What is the strategy for multiplying by 19?** Invite any learner who suggests rounding and adjusting to explain their answer. Elicit the explanation 'multiply by 20 and subtract the number'. On the board, write: 8 × 19 = (8 × 20) – (8 × 1). Say: **Round 19 to 20 and multiply by eight. To adjust, subtract eight for multiplying one eight too many.**
- Write 40 × 19. Ask: **How will you solve this problem?** (adjusting and rounding) On the board, write: 40 × 19 = (40 × 20) – 40. Ask: **How will you multiply 40 × 20?** Praise learners who suggest solving a one-digit problem and scaling by powers of ten. Say: **You know four times two is eight; so you also know 40 times 20 is 800.** Continue by writing (40 × 20) – 40 = 800 – 40 = 760.

- On the board, write: 70 × 19, 600 × 19. Ask learners for the solutions. (1330, 11 400)
- On the board, write: 40 × 21. Ask: **Who can remember how to multiply by 21?** Elicit 'round and adjust', but adding this time. Demonstrate: 40 × 21 = (40 × 20) + (40 × 1) = 800 + 40 = 840.
- Write on the board: 70 × 21, 500 × 21. Ask learners for the solutions. (1470, 10 500)
- Review how to multiply by 25, using the 'multiply by 100 and divide by 4' method. Ask learners to multiply 30 × 25, 400 × 25.

Practise [WB]

- Workbook: Multiplying by 19, 21 or 25 (2) page 132
- Refer to Activity 1 from the Additional practice activities.

Apply 🔠 🖥

- Display **Slide 1**. Establish that the calculation to solve is: ($20 × 19) + ($30 × 21) + ($40 × $25) = $2010.

Review

- Draw three boxes labelled A: $19, B: $21, C: $25. Ask: **What is the total price for 30 of box A, 50 of box B and 40 of box C?** ($2620)

Assessment for learning

- I work out 40 × 21 by multiplying by 20 and adjusting. How do I adjust?

Lesson 3: **Multiplication by factors (2)**

Learning objectives

Code	Learning objective
5Nc15*	Use factors to multiply, e.g. multiply by 3, then double to multiply by 6.

Strand 5: Problem solving

Code	Learning objective
5Pt2	Solve single and multi-step word problems (all four operations); represent them, e.g. with diagrams or a number line.
5Pt6	Estimate and approximate when calculating, e.g. using rounding, and check working.
5Ps2	Choose an appropriate strategy for a calculation and explain how they worked out the answer.
5Ps3	Explore and solve number problems and puzzles, e.g. logic problems.
5Ps8	Investigate a simple general statement by finding examples which do or do not satisfy it [, e.g. the sum of three consecutive whole numbers is always a multiple of three].
5Ps9	Explain methods and justify reasoning orally and in writing; make hypotheses and test them out.
5Ps10	Solve a larger problem by breaking it down into sub-problems or represent it using diagrams.

Prerequisites for learning

Learners need to:

- recall the multiplication tables up to 12 × 12 and associated facts involving multiples of ten
- understand the terms 'factor' and 'multiple'
- double and halve whole numbers
- multiply by rewriting a calculation using the factors of one number.

Success criteria

Learners can:

- multiply by rewriting a calculation, using the factors of both numbers.

Vocabulary

multiple, factor, prime number

Resources

mini whiteboard and pen or paper (per pair)

Refresh

- Choose an activity from Number – Calculation: Mental strategies, Multiplication and division in the Refresh activities.

Discover [SB]

- Direct learners to the Student's Book. Ask: **How would you use array diagrams to find factor pairs?**

Teach and Learn [SB]

- Discuss the text and examples in the Student's Book.
- Give learners a whiteboard and pen and ask them to work out 23 × 24. They can draw arrays to help them. Ask: **How did you split the factors?** Expect 23 × 3 × 8 = 23 × 3 × 2 × 2 × 2 = 23 doubled three times multiplied by three. (552) Ask: **Why did you not split 23?** Establish that 23 is 'prime' so can only be split into one and 23, which would not aid the calculation. Say: **If it is possible, you could choose to split both numbers into their factors.** Write on the board: 35 × 8. Ask: **How would you split the numbers?** Invite learners to demonstrate. Expect 35 × 8 = 5 × 7 × 4 × 2. Remind them that multiplication can be done in any order. Ask: **Can you rearrange the numbers to make the calculation easier?** Elicit 5 × 2 × 7 × 4. Establish

that 2 and 5 are important factors, as multiplying by 2 and then 5 is equivalent to multiplying by 10. Say: **You can make the calculation easier if you can multiply factors to make ten.** Demonstrate: 5 × 2 × 7 × 4 = 10 × 28 = 280.

Practise [WB]

- Workbook: Multiplication by factors (2) page 134

Apply 👥

- Say: **11 children buy 18 stickers each. How many stickers is that altogether?** Learners solve the problem using factors, then share their solutions and methods. (198)

Review

- Ask learners to use factors to multiply 64 × 25. (1600)

Assessment for learning

- Show me two ways of simplifying and solving 42 × 35.

Lesson 4: **Doubles and halves (2)**

Learning objectives

Code	Learning objective
5Nc17*	Double multiples of 10 to 1000 and multiples of 100 to 10 000, e.g. double 360 or double 3600, and derive the corresponding halves.

Strand 5: Problem solving	
5Pt2	Solve single and multi-step word problems (all four operations); represent them, e.g. with diagrams or a number line.
5Pt6	Estimate and approximate when calculating, e.g. using rounding, and check working.
5Ps2	Choose an appropriate strategy for a calculation and explain how they worked out the answer.
5Ps3	Explore and solve number problems and puzzles, e.g. logic problems.
5Ps8	Investigate a simple general statement by finding examples which do or do not satisfy it [, e.g. the sum of three consecutive whole numbers is always a multiple of three].
5Ps9	Explain methods and justify reasoning orally and in writing; make hypotheses and test them out.
5Ps10	Solve a larger problem by breaking it down into sub-problems or represent it using diagrams.

Prerequisites for learning

Learners need to:

- recall the multiplication tables up to 12 × 12 and associated facts involving multiples of ten
- double numbers to 100 and halve even numbers to 200.

Success criteria

Learners can:

- double multiples of 10 to 1000 and multiples of 100 to 10 000, and derive corresponding halves by using known facts and partitioning.

Vocabulary

doubling, halving, inverse

Resources

mini whiteboard and pen or paper (per pair) 1–6 dice, alternatively use Resource sheet 4: 1–6 spinner (per pair)

Refresh

- Choose an activity from Number – Calculation: Mental strategies, Multiplication and division in the Refresh activities.

Discover [SB]

- Ask learners to read the text and look at the picture in the Student's Book. Ask: **How would you work out double 240? Half of 240?** Discuss strategies.

Teach and Learn [SB]

- Discuss the Example in the Student's Book. Write Double 6 is 12. Ask: **How can you use this fact to create other double facts?** Elicit that if you know double six is 12, you also know double 60 is 120, double 600 is 1200, etc. Repeat with: Double 4 is 8.

- Write: Half 14 is 7. Ask: **How can you create more half facts?** Elicit if you know half 14 is seven, then half 140 is 70, half 1400 is 700 etc. Write: Half 16 is 8. Ask learners to write other 'half' facts.

- Say: **I want you to double 20 and keep doubling as far as you can go.** Remind learners to use their knowledge of number facts and place value to help. Learners present their lists of doubles. Ask: **How could you work out halves of larger numbers?**

Remind learners of the inverse relationship between halving and doubling. Ask them to write a list of halves, for example: Half of 2560 is 1280.

Practise [WB]

- Workbook: Doubles and halves (2) page 136
- Refer to Activity 2 from the Additional practice activities.

Apply 👥 🖥

- Display **Slide 1**. Invite learners to discuss their conclusions. Ask: **Which number produces the smallest half?** (for 1000 it is 125; for 4800 it is 75; 4800 produces the smallest half.)

Review

- On the board, write: 240, 820, 3600, 5200. Say: **Double and halve each number.** (480, 120; 1640, 410; 7200, 1800; 10 400, 2600)

Assessment for learning

- As I know double 36 is 72, I also know double 3600 is 7200. Explain how I know this.

Lesson 5: **Doubles and halves (3)**

Number – Calculation: Mental strategies, Multiplication and division

Learning objectives

Code	Learning objective
5Nc16*	Double any number up to 100 and halve even numbers to 200 and use this to double and halve numbers with one or two decimal places, e.g. double 3·4 and half of 8·6.

Strand 5: Problem solving

5Pt2	Solve single and multi-step word problems (all four operations); represent them, e.g. with diagrams or a number line.
5Pt4	Use multiplication to check the result of a division [, e.g. multiply 3·7 × 8 to check 29·6 ÷ 8].
5Ps2	Choose an appropriate strategy for a calculation and explain how they worked out the answer.
5Ps3	Explore and solve number problems and puzzles, e.g. logic problems.
5Ps10	Solve a larger problem by breaking it down into sub-problems or represent it using diagrams.

Prerequisites for learning

Learners need to:

- recall the multiplication tables up to 12 × 12 and associated facts involving multiples of ten
- double numbers to 100 and halve even numbers to 200
- double multiples of 10 to 1000 and multiples of 100 to 10 000, and derive the corresponding halves.

Success criteria

Learners can:

- find the double or half of a decimal by doubling or halving the related whole number
- double and halve two-digit numbers and explain how to use this to double and halve related decimals.

Vocabulary

decimal, decimal place, doubling, halving, inverse

Resources

mini whiteboard and pen or paper (per pair)

Refresh

- Choose an activity from Number – Calculation: Mental strategies, Multiplication and division in the Refresh activities.

Discover [SB]

- Direct learners to the Student's Book. Ask: **What is double 4·6 litres? Half 4·6 litres?** Discuss strategies.

Teach and Learn [SB] 🖵

- Discuss the text and example in the Student's Book. Display **Slide 1.** Ask: **Who can come to the board and write the double of one of the numbers?** Invite learners to complete the 'double' column. Expect: 44, 72, 144. Do the same for the 'half' column. (11, 18, 36)
- Display **Slide 2.** Ask: **How would you find double 2·2? Half 2·2?** Prompt by saying: **Look at the number facts table you just completed.** Accept suggestions. Elicit that as they know double and half of 22 they also know double and half of 2·2. By moving the digits one place to right, demonstrate that 2·2 is ten times smaller than 22. Say: **As 2·2 is ten times smaller than 22 then its double and its half will also be ten times smaller.** Demonstrate 44 and 11 becoming ten times smaller on the place value grid.

- Write 1·1, 2·2, 4·4 in the 'half/double' table. Repeat for 3·6 (1·8, 7·2) and 7·2 (3·6, 14·4) by showing the relationship to double and half of 36 and 72.
- Write 0·76 in the table. Ask: **How would you double and halve this number?** Accept suggestions. Praise any learner that suggests doubling and halving 76 and adjusting powers of ten. Ask: **What is double/half 76?** (38, 152). Record the numbers in the table. Ask: **How would you adjust these answers for 0·76?** Agree that 0·76 is 100 times smaller than 76, so adjust by dividing by 100. (0·38, 1·52)

Practise [WB]

- Workbook: Doubles and halves (3) page 138

Apply 👥 🖵

- Display **Slide 3.** Establish that the calculation is: double $3·70 + double $1·85 + half $1·90 + half $5·4. ($14.75)

Review

- On the board, write: 3·8, 2·9, 1·74. Say: **Double and halve each number.** (7·6, 1·9; 5·8, 1·45; 3·48, 0·87)

Assessment for learning

- As I know double 87 is 178, I also know double 8·7 is 17·8. Explain how I know this.

Lesson 6: **Multiplying a decimal by a single-digit number**

Learning objectives

Code	Learning objective
5Nc22	Multiply two-digit numbers with one decimal place by single-digit numbers, e.g. 3·6 × 7.

Strand 5: Problem solving

5Pt2	Solve single and multi-step word problems (all four operations); represent them, e.g. with diagrams or a number line.
5Pt4	Use multiplication to check the result of a division [, e.g. multiply 3·7 × 8 to check 29·6 ÷ 8].
5Ps2	Choose an appropriate strategy for a calculation and explain how they worked out the answer.
5Ps3	Explore and solve number problems and puzzles, e.g. logic problems.
5Ps5	Use ordered lists and tables to help to solve problems systematically.
5Ps9	Explain methods and justify reasoning orally and in writing; make hypotheses and test them out.
5Ps10	Solve a larger problem by breaking it down into sub-problems or represent it using diagrams.

Prerequisites for learning

Learners need to:

- recall the multiplication tables up to 10 × 10
- multiply and divide a number by ten and 100, including decimals
- multiply two-digit number by a single digit number.

Success criteria

Learners can:

- mentally multiply a two-digit number with one decimal place by a one-digit number, using whole-number thinking or partitioning or grid methods.

Vocabulary

decimal, decimal place, partition, grid method

Refresh

- Choose an activity from Number – Calculation: Mental strategies, Multiplication and division in the Refresh activities.

Discover [SB]

- Direct learners to the Student's Book. Ask: **How would you solve this problem?** Accept and discuss strategies.

Teach and Learn [SB]

- Discuss the text and example in the Student's Book. Write 23 × 7 =. Ask: **How would you solve this problem?** Expect learners to recall partitioning and grid methods. Revise these methods on the board.
- Partition and multiply: 23 × 7 = (20 × 7) + (3 × 7) = 140 + 21 = 161.
- Write: 2·3 × 7 =. Ask: **How would you use 23 × 7 to solve 2·3 × 7?** Elicit how to work it out – **I know 23 × 7 = 161, and 2·3 × 7 is ten times smaller, so the answer will be ten times smaller than 161, which is 16·1.**
- Repeat with other examples of whole number multiplications related to two-digit numbers with one decimal place multiplications, for example: 79 × 6 (7·9 × 6), 68 × 4 (6·8 × 4).
- Write: 4·6 × 9. Say: **Another method for multiplying a decimal by a single-digit number**

is to split the number into its whole number and decimal parts, then multiply each partition and add. Write: 4·6 × 9 = (4 × 9) + (0·6 × 9). Ask: **What is 0·6 × 9?** Elicit finding the answer to the related multiplication 6 × 9 = 54 and making it ten times smaller, 5·4. Volunteers demonstrate multiplication of partitions: = 36 + 5·4 = 41·4. Demonstrate the grid method.

Practise [WB]

- Workbook: Multiplying a decimal by a single-digit number page 140
- Refer to Activity 3 from the Additional practice activities.

Apply 👥 🖥

- Display **Slide 1**. Invite learners to explain the methods they used to solve the problem. (eight crates)

Review

- Reinforce the methods taught in the lesson. On the board, write: 7·8 × 3; 5·9 × 7. Ask learners to find the solutions, in pairs. (23·4, 41·3)

Assessment for learning

- Show me two different methods for calculating 6·6 × 8.

Lesson 7: Writing a remainder as a fraction

Number – Calculation: Mental strategies, Multiplication and division

Learning objectives

Code	Learning objective
5Nc24	Start expressing remainders as a fraction of the divisor when dividing two-digit numbers by single-digit numbers.
5Nc26*	Decide whether to round an answer up or down after division, depending on the context.

Strand 5: Problem solving

Code	Learning objective
5Pt2	Solve single and multi-step word problems (all four operations); represent them, e.g. with diagrams or a number line.
5Pt4	Use multiplication to check the result of a division [, e.g. multiply 3·7 × 8 to check 29·6 ÷ 8].
5Pt6	Estimate and approximate when calculating, e.g. using rounding, and check working.
5Pt7	Consider whether an answer is reasonable in the context of a problem.
5Ps2	Choose an appropriate strategy for a calculation and explain how they worked out the answer.
5Ps3	Explore and solve number problems and puzzles, e.g. logic problems.
5Ps5	Use ordered lists and tables to help to solve problems systematically.
5Ps10	Solve a larger problem by breaking it down into sub-problems or represent it using diagrams.

Prerequisites for learning

Learners need to:

- multiply and divide a number by multiples of ten and 100.

Success criteria

Learners can:

- make a reasonable estimate for the answer
- calculate the fraction remainder in a division calculation.

Vocabulary

remainder, fraction

Resources

mini whiteboard and pen or paper (per pair)

Refresh

- Choose an activity from Number – Calculation: Mental strategies, Multiplication and division in the Refresh activities.

Discover [SB]

- Read the text in the Student's Book and look at the diagram. Ask: **How would you divide 15 chocolates between 2 people?**

Teach and Learn [SB] [chart]

- Discuss the text and Example in the Student's Book. Display the **Number cards tool** set to random cards 10–99. Choose a random card number and record the number on the board. Set the tool to random cards (0–9), choose a card and record it on the board as a division problem with the first number, for example: 45 ÷ 6. Ask: **What is the multiple of six that is closest to, but not more than, 45?** (42) **How many sixes are there in 42?** (7) **What number is left?** (3) Record on the board: 45 ÷ 6 = 7 r 3.

- Ask: **How would you write the remainder as a fraction?** Demonstrate that a remainder of three, when dividing by six, can be written as $\frac{3}{6}$. and simplified to $\frac{1}{2}$. Say: **This is no longer a remainder but part of the answer.** So 45 divided by six is $7\frac{1}{2}$.

- Repeat for two more randomly generated division problems, for example 99 ÷ 8.

Practise [WB]

- Workbook: Writing a remainder as a fraction page 142

Apply [icons]

- Display **Slide 1**. Invite learners to explain their solution. ($76 \div 8 = 9\frac{1}{2}$; $79 \div 4 = 19\frac{3}{4}$; children at Pavel's party have smaller fractional pieces as $\frac{1}{2} < \frac{3}{4}$)

Review

- Reinforce methods taught in the lesson. On the board, write: 83 ÷ 8; 61 ÷ 7. Let the learners work out the solutions, writing remainders as fractions. ($10\frac{3}{8}$; $8\frac{5}{7}$)

Assessment for learning

- Show me why the answer to the division $75 \div 6$ is $12\frac{1}{2}$.

Lesson 8: **Order of operations**

Learning objectives

Code	Learning objective
5Nc27	Begin to use brackets to order operations and understand the relationship between the four operations and how the laws of arithmetic apply to multiplication.

Strand 5: Problem solving

5Pt2	Solve single and multi-step word problems (all four operations); represent them, e.g. with diagrams or a number line.
5Pt7	Consider whether an answer is reasonable in the context of a problem.
5Ps2	Choose an appropriate strategy for a calculation and explain how they worked out the answer.
5Ps3	Explore and solve number problems and puzzles, e.g. logic problems.
5Ps4	Deduce new information from existing information to solve problems.
5Ps8	Investigate a simple general statement by finding examples which do or do not satisfy it [, e.g. the sum of three consecutive whole numbers is always a multiple of three].
5Ps9	Explain methods and justify reasoning orally and in writing; make hypotheses and test them out.
5Ps10	Solve a larger problem by breaking it down into sub-problems or represent it using diagrams.

Prerequisites for learning

Learners need to:

- calculate mentally with all operations.

Success criteria

Learners can:

- apply the BODMAS ruler to calculate a set of operations in the correct order.

Vocabulary

brackets, operation, BODMAS

Resources

mini whiteboard and pen or paper (per pair)

Refresh

- Choose an activity from the Refresh activities.

Discover [SB]

- Direct learners to the Student's Book and read the text and look at the pictures. Say: **Tell your partner why you think there are two different answers.** Discuss ideas. Say: **We will discuss the confusion over which answer is correct later in the lesson.**

Teach and Learn [SB]

- Discuss the text and example in the Student's Book. Write $6 + 3 \times 5 =$. Say: **Work out the answer.** Establish that there may be two different answers. Ask: **Why do you think there are two different answers to the same calculation?** Elicit that, without guidance, people may interpret the order of operations differently.
- Say: **When a calculation involves a mixture of operations, you need a set of rules to tell you the order to complete them. Division is completed first, then multiplication, followed by addition and subtraction in that order.** Record this on the board, using initial letters of operations: D, M, A, S. Return to $6 + 3 \times 5$. Show that the rules give the answer 21.
- Write: $8 + 24 \div 6 - 2 =$. Say: **Tell your partner what you think the order of operations should be.** Allow discussion, then say: **The order is division first.** Complete $24 \div 6 = 4$ and record: $8 + 4 - 2$. Say: **Next comes addition.** Complete $8 + 4 = 12$. Say: **Finally, complete the subtraction.** $12 - 2 = 10$.

- Write $160 - 16 \times 9 + 8 =$. Ask learners to solve it. (8)
- Write: $(72 - 47) \times 2 + 9 =$. Say: **Some calculations include brackets that indicate an operation that must be completed first.** On the board, demonstrate how brackets override other operations: $(72 - 47) \times 2 + 9 = (25) \times 2 + 9 = 59$.
- Write BODMAS. Explain how this mnemonic states the order: brackets (B), order (O), division (D), multiplication (M), addition (A), subtraction (S). Say: **O is for 'orders', squaring, for example.**
- An alternative to BODMAS is BIDMAS, in which the I stands for index/indices.

Practise [WB]

- Workbook: Order of operations page 144
- Refer to Activity 4 from the Additional practice activities.

Apply 👥 🖥

- Display **Slide 1**. Invite learners to discuss the solution. $(25 + 5 \times 5 - 12 = 38)$

Review

- Reinforce the BODMAS rule. Ask learners to solve: $16 - 5 \times 2 + 3$. (3)

Assessment for learning

- I solve $4 \times 6 + 18 \div 2$ and get 21. Is that right? Why/Why not?

Additional practice activities

Activity 1 ●● [Challenge 2]

Learning objective

• Use knowledge of multiplying by 19, 21 or 25 and multiplying multiples of 10 and 100 to find a set of numbers that, when multiplied and added, make a given total.

Resources

1–6 dice, alternatively use Resource sheet 4: 1–6 spinners (per pair)

What to do 🖥

• Display **Slide 1**.
• Learners take turns to create and record a four-digit 'target' number by rolling a dice four times.
• From the board, they select and record four numbers from the top row and four numbers from the bottom row.
• They multiply and add the numbers, in any order, to achieve a total as close as possible to the target number.

Variation

[Challenge 3] Learners create a five-digit target number.

Activity 2 ●● [Challenge 2]

Learning objective

• Use knowledge of factors and halves of multiples of 10 and 100 to simplify multiplication problems.

Resources

mini whiteboard and pen or paper (per pair)

What to do 🖥

• Display **Slide 2**.
• Learners work in pairs to find solutions.

Variation

[Challenge 1] Learners work with two-digit numbers: 24, 70, 66, 360 and 740. The instruction 'divided by 10' is removed.

Additional practice activities

Activity 3 Challenge 2

Learning objectives

- Use knowledge of halving and doubling two-digit numbers with one decimal place and multiplying decimals with one decimal place by one-digit numbers to calculate the price of a set of supermarket items.
- Solve a larger problem by breaking it down into sub-problems, or represent it using diagrams.

Resources

mini whiteboard and pen or paper (per pair)

What to do 🖥

- Display **Slide 3**.

	Cheap Supermarket	**Expensive Supermarket**
Tin of beans	$2.60	$8.40
Box of eggs	$3.80	$13.80
Bag of oranges	$4.40	$17.60

- Say: **These were the prices at Cheap Supermarket and Expensive Supermarket last week. Since then prices at Cheap Supermarket have doubled and prices at Expensive Supermarket have halved.**
- Say: **Sarah wants to buy some items at the lowest price possible: three tins of beans, four boxes of eggs and seven bags of oranges.**

- Ask: **Calculate how much the set of items would cost if they were bought in each supermarket. Remember to adjust for the rise and fall of prices. Which supermarket should Sarah shop at?** (total costs: Cheap Supermarket: $107.60; Expensive Supermarket: $101.80)

Variation

Challenge **3** Introduce some odd numbers as prices for example a bag of oranges at Expensive Supermarket is $17.70.

Activity 4 •• Challenge 2

Learning objectives

- Express remainders as a fraction of the divisor when dividing two-digit numbers by single-digit numbers.
- Begin to use brackets to order operations

What to do

- On the board, write:
  ◊ $83 \div 7 = 11\frac{5}{7}$ (Answer should be: $11\frac{6}{7}$)
  ◊ $95 \div 8 = 11\frac{3}{4}$ (Answer should be: $11\frac{7}{8}$)
  ◊ $96 \div 9 = 10\frac{1}{3}$ (Answer should be: $10\frac{2}{3}$)
- Say: **A calculator that expresses the remainders of division problems as a fraction appears to be malfunctioning. Whenever a set of numbers is input, the calculator outputs the wrong remainder.**
- Point to the examples on the board. Say: **Check each calculation and write the correct remainder.**

Variation

Challenge **1** Use the following calculations, writing incorrect calculator answers for each: $35 \div 3$, $37 \div 4$, $53 \div 5$.

Unit 14: 2D shape, including symmetry

Learning objectives

Code	Learning objective
5Gs1	Identify and describe properties of triangles and classify as isosceles, equilateral or scalene.
5Gs2	Recognise reflective and rotational symmetry in regular polygons.
5Gs3	Create patterns with two lines of symmetry, e.g. on a pegboard or squared paper.
5Gs5	Recognise perpendicular and parallel lines in 2D shapes, drawings and the environment.

Strand 5: Problem solving

5Pt5	Recognise the relationships between different 2D [and 3D] shapes, e.g. a face of a cube is a square.
5Ps4	Deduce new information from existing information to solve problems.
5Ps7	Identify simple relationships between shapes, e.g. these triangles are all isosceles because ...
5Ps9	Explain methods and justify reasoning orally and in writing; make hypotheses and test them out.

Unit overview

This unit introduces the properties of parallel and perpendicular lines and develops a broader understanding of symmetry and geometric shape. Triangles are investigated, classified and compared. Both reflective and rotational symmetry are examined, using mirrors and by turning shapes. Patterns are created and completed using two lines of symmetry. Finally, the properties of parallel and perpendicular lines are investigated.

Common difficulties and remediation

Common misconceptions of triangles include assuming that triangles have one point at the top and two points at the bottom, or that the bottom of a triangle is flat. It is important that teachers expose learners to visual examples of triangles in different orientations and introduce properties that define a shape. Learners need to see non-triangles, such as a three-sided object with a wavy line, or a three-sided object with an opening.

The definition of an isosceles triangle used in the lesson is a triangle that has exactly two equal sides. An equilateral triangle is a special case of an isosceles triangle, having all three sides equal. To avoid confusion, it is worth stating that when we call a triangle isosceles we are referring to one that has a third side unequal to the other two.

Vertical and horizontal lines are generally recognised as perpendicular to one another without problem, but when oriented differently are not as easily perceived as forming right angles by learners. To perceive them more clearly, learners would benefit from extended practice in visualising the rotation of perpendicular lines.

Promoting and supporting language

At every stage, learners require a mathematical vocabulary to access questions and problem-solving exercises. If appropriate, when a new key word is introduced, ask learners to write a definition in their books, drawing a box around it for emphasis. Encourage learners to write the definition in their own words, for example: 'An isosceles triangle is one with only two sides equal.' In addition, be aware of terms that have more than one definition, particularly meanings outside mathematics, for example: 'reflect' and 'reflection'.

Encourage frequent usage of mathematical language to help embed vocabulary, for example, modelling use of the word 'perpendicular' when learners refer to 'two lines arranged at right angles' as they make the transition to mathematically precise language. Also, help learners with pronunciation of tricky words, such as, 'isosceles' pronounced *ahy-sos-uh-leez* and 'scalene' pronounced *skey-leen*.

Throughout the lesson it is important to encourage learners to seek clarification and confirmation of the mathematical language. This may involve prompting learners to call out and complete definitions, for example, following a demonstration of parallel lines by saying: **We know the lines are parallel because they are the same distance apart from each other all the way along their length. Even if you make them longer they will never meet.** Always praise learners when they ask for terms to be defined. A few times during each lesson, ask for a volunteer to summarise what has been learned so far, so that they get experience of explaining concepts in their own words. You may find it useful to refer to the audio glossary on Collins Connect.

Lesson 1: **Triangles**

Learning objectives

Code	Learning objective
5Gs1	Identify and describe properties of triangles and classify as isosceles, equilateral or scalene.

Strand 5: Problem solving

5Ps4	Deduce new information from existing information to solve problems.
5Ps7	Identify simple relationships between shapes, e.g. these triangles are all isosceles because ...

Prerequisites for learning

Learners need to:

- classify triangles using criteria such as the number of right angles and whether or not they are regular
- compare and order angles less than 180°
- identify simple relationships between shapes.

Success criteria

Learners can:

- use a right angle to compare angles in a triangle
- use a ruler to measure the length of the sides of a triangle
- identify equilateral, isosceles and scalene triangles.

Vocabulary

equilateral, isosceles, scalene, regular, irregular, right angle

Resources

ruler (per learner); right-angle tester, e.g. book or sheet of paper (per pair); coloured pencils (per learner); Resource sheet 18: Sorting triangles (per learner); 2D triangle shapes: equilateral, isosceles, scalene (per learner)

Refresh

- Choose an activity from Geometry – Shapes and geometric reasoning in the Refresh activities.

Discover ᔕᏰ

- Discuss the text and images in the Student's Book. Ask learners to use a right-angle tester, such as the corner of a book or sheet of paper, to compare the length and angles of the triangular roofs in the pictures. Ask: **How are the triangles different?** Prompt use of correct mathematical vocabulary in responses, for example: 'All three sides are equal; one of the angles is a right angle; the triangle is irregular.'

Teach and Learn ᔕᏰ ᴰᴸ 🖥

- Display **Slide 1.** Discuss the text and images in the Student's Book and go through the characteristics of the three types of triangle.
- Ask learners to complete Resource sheet 18: Sorting triangles to find examples of each type of triangle.
- Display the **Geoboard tool** showing a triangle. Ask volunteers to use the tool to draw and demonstrate isosceles and scalene triangles. Ask: **Is it possible to draw an equilateral triangle with the tool? Why not?** Point out the difference between the horizontal/vertical distance and oblique distance.

Practise 🆆🅱

- Workbook: Triangles page 146

Apply 👥 or 👨‍👨‍👦

- Remind learners that a shape tessellates if multiple copies can fit together to cover a flat surface without leaving any gaps. Ask learners to find out which of the three types of triangle will tessellate, using their 2D scalene, equilateral and isosceles triangles.

Review

- Hold up each type of triangle in turn: equilateral, isosceles and scalene. Ask learners to take turns to describe one property of the triangle to their partner. Expect them to find at least two properties each. Invite learners to share their descriptions with the class.

Assessment for learning

- How are triangles different?
- How do you know which is an irregular triangle?
- What makes an equilateral triangle?
- What makes an isosceles triangle?
- What is a scalene triangle?
- Which types of triangle can have a right angle? (isosceles and scalene) An angle greater than 90°? (isosceles and scalene)

Geometry – Shapes and geometric reasoning

Lesson 2: **Symmetry in regular polygons**

Geometry – Shapes and geometric reasoning

Learning objectives

Code	Learning objective
5Gs2	Recognise reflective and rotational symmetry in regular polygons.

Strand 5: Problem solving	
5Pt5	Recognise the relationships between different 2D [and 3D] shapes, e.g. a face of a cube is a square.
5Ps4	Deduce new information from existing information to solve problems.

Prerequisites for learning

Learners need to:
- identify and sketch lines of symmetry in 2D shapes and patterns
- find examples of shapes and symmetry in the environment and in art.

Success criteria

Learners can:
- use a mirror to identify reflective symmetry
- identify a shape with rotational symmetry

- rotate a regular polygon to find how many times it can be turned to fit on itself.

Vocabulary

reflective symmetry, rotational symmetry, line of symmetry, mirror line, regular polygon

Resources

ruler (per learner); 2D shape set (per pair or group); mirror (per pair or group); coloured pencils (per learner)

Refresh

- Choose an activity from Geometry – Shapes and geometric reasoning in the Refresh activities.

Discover [SB]

- Discuss the text and images in the Student's Book. Ask: **How does each shape have reflective symmetry?** Encourage use of a mirror to determine symmetry. Ask: **Where can we find similar symmetrical shapes?** Prompt use of correct vocabulary, for example: 'This shape has reflective symmetry; it has two lines of symmetry; it has a vertical line of symmetry.'

Teach and Learn [SB] [barchart]

- Discuss the text and examples in the Student's Book, then display the **Symmetry tool**, showing a pentagon. Ask: **How many lines of symmetry does a pentagon have?** (5) Demonstrate with the tool.

- Display the **Rotate and reflect tool**, showing a pentagon. Ask: **How many times can a pentagon be rotated and still look the same?** (four times, five including the full turn.) Demonstrate with the tool.

- Repeat for the hexagon/octagon. Discuss the diagrams and examples. Ask: **Why do these shapes have rotational symmetry?**

Practise [WB]

- Workbook: Symmetry in regular polygons page 148
- Refer to Activity 1 from the Additional Practice Activities.

Apply [pair icon] or [group icon]

- Pairs find five shapes/objects in the classroom with reflective symmetry. They arrange them in ascending order of number of lines of symmetry. They then find four shapes/objects with rotational symmetry and arrange them in order of the number of times the shape can be rotated and still remain the same. Learners share their solutions with the class.

Review [barchart]

- Display the **Symmetry tool**. Say: **Identify a shape by the number of times it can be rotated and still look the same. The number will include a full turn.** Begin with number three and ask learners to identify the shape. Invite volunteers to demonstrate that the shape they suggested can be rotated three times and still look the same. Repeat for a different number of turns.

Assessment for learning

- Which polygons have four/six/eight lines of symmetry? How do you know? Find something that has a similar number of lines.
- What is rotational symmetry? Explain using a shape. How many times can the shape be rotated and still remain the same?

Lesson 3: **Symmetrical patterns**

Learning objectives

Code	Learning objective
5Gs3	Create patterns with two lines of symmetry, e.g. on a pegboard or squared paper.

Strand 5: Problem solving

5Ps4	Deduce new information from existing information to solve problems.
5Ps7	Identify simple relationships between shapes, e.g. these triangles are all isosceles because ...

Prerequisites for learning

Learners need to:

- identify and sketch lines of symmetry in 2D shapes and patterns
- find examples of shapes and symmetry in the environment and in art.

Success criteria

Learners can:

- 'flip' a shape over a mirror line to find the position of the image

- build a symmetrical pattern by visualising a shape 'flipped' over two lines of symmetry at right angles.

Vocabulary

reflective symmetry, line of symmetry, mirror line, image

Resources

mirror (per pair); squared paper (per pair); coloured pencils (per pair); different coloured whiteboard pens (per class)

Refresh

- Choose an activity from Geometry – Shapes and geometric reasoning in the Refresh activities.

Discover [SB]

- Discuss the images in the Student's Book. Ask: **How has each pattern been constructed?** Elicit that each pattern has two lines of symmetry – horizontal and vertical. Ask learners to find other examples around them – in the classroom, in books and so on.

Teach and Learn [SB] [graph]

- Discuss the text and images in the Student's Book. Let the learners use a mirror to find the lines of symmetry in the pictures.

- Display the **Rotate and reflect tool** showing the vertical mirror line and a small square on one side of the line. Demonstrate the reflection of a square in the mirror line. Say: **The reflected figure is called the 'image'.**

- Project the blank square grid from the **Rotate and reflect** tool on to the whiteboard. Use a whiteboard pen to draw perpendicular mirror lines. Use a coloured pen to colour in a square. Ask: **What will be the position of the image?** Build up a pattern of squares reflected around quadrants of perpendicular mirror lines.

Practise [WB]

- Workbook: Symmetrical patterns page 150

Apply [icon]

- Provide pairs with squared paper and coloured pencils. Learners colour a three by three square grid to make a symmetrical pattern. They copy the pattern three times and place the four grids together to make a bigger square. They investigate whether the pattern is still symmetrical. Discuss solutions as a class.

Review [graph]

- Display the **Rotate and reflect tool**. Say: **I will position two shapes on either side of a mirror line. Tell me if the images and objects will coincide after reflection.** Set a different shape on either side of a vertical mirror line. Say: **Raise your hand if you think the images and objects will coincide.** Invite a volunteer to point out where they think this will occur and to click the 'reflect' button. Repeat for different shapes, positions and mirror lines of different orientation.

Assessment for learning

- I have shaded part of a pattern that has two lines of symmetry. Show me how you complete the pattern. How did you find the image of each shaded cell?
- Explain to me how this pattern is symmetrical.

Geometry – Shapes and geometric reasoning

Lesson 4: **Perpendicular and parallel lines**

Geometry – Shapes and geometric reasoning

Learning objectives

Code	Learning objective
5Gs5	Recognise perpendicular and parallel lines in 2D shapes, drawings and the environment.

Strand 5: Problem solving

5Ps4	Deduce new information from existing information to solve problems.
5Ps9	Explain methods and justify reasoning orally and in writing; make hypotheses and test them out.

Prerequisites for learning

Learners need to:

- read and record the vocabulary of position and direction
- follow and give instructions involving position and direction
- use a ruler to draw a straight line.

Success criteria

Learners can:

- identify lines that are equidistant and do not meet

- identify a right angle
- recognise angles that are less than or greater than a right angle.

Vocabulary

parallel lines, perpendicular lines, right angle

Resources

ruler (per learner); book or website that shows flags of the world (per pair); squared paper (per learner); right angle tester, e.g. book or sheet of paper (per learner); red pencil (per learner)

Refresh

- Use the activity *Do the lines cross?* from the Refresh activities.

Discover [SB]

- Discuss the text and images in the Student's Book. Say: **Describe the red lines in the photograph on the left**. Elicit that the lines never meet. Ask: **What do you notice about the distance between the lines?** Elicit that the lines remain the same distance apart at every point. Ask: **What do you notice about the red lines in the photograph on the right?** Elicit that the lines meet at right angles.

Teach and Learn [SB] [graph icon]

- Discuss the text and images in the Student's Book. Display the **Co-ordinates tool**. Demonstrate how parallel lines can be plotted and show lines in different orientations. Ask volunteers to draw parallel lines on the tool.
- On the tool, draw pairs of perpendicular lines in different orientations.
- Ask: **Can you see any examples of parallel/ perpendicular lines around the classroom?** Ask: **How do you know the lines are parallel/ perpendicular?** Discuss answers and encourage correct mathematical language, for example: 'I know these lines are parallel because they point in the same direction and never meet.'

Practise [WB]

- Workbook: Perpendicular and parallel lines page 152
- Refer to Activity 2 from the Additional Practice Activities.

Apply [pair icon]

- Ask: **Which flags of the world feature parallel or perpendicular lines?** Learners work in pairs to find examples in books or websites. They should use right angle testers to check for perpendicular lines. Ask: **Are there any flags that have both types of lines?** Learners make copies of flags and label the lines.

Review

- Draw a two column table on the whiteboard. Add the title: 'Parallel and perpendicular lines' and label the columns: 'Parallel lines', 'Perpendicular lines'. Ask volunteers to use a ruler to draw examples of parallel and perpendicular lines in the correct part of the table.

Assessment for learning

- How would you check if two lines are parallel/ perpendicular?
- Give me an example of something that has parallel/perpendicular lines.

Additional practice activities

Activity 1

Learning objective
- Describe a shape in terms of its properties and identify a shape from its description.

Resources
2D shapes including regular polygons, isosceles/scalene triangles (per pair); books (per pair); word cards: side, angle, reflective symmetry, rotational symmetry, regular/irregular (per pair)

What to do
- Players sit opposite each other with a book barrier between them.
- They take turns to choose a shape and a word card. They use the word(s) to describe the shape, for example: 'The shape has two sides the same length.'

- If their partner identifies the shape, the player scores three points. If not, they choose a second card and describe the shape using the new word. If they are correct they score two points. If not, a third attempt is made, scoring one point.
- Once the shape has been guessed, turn passes to the other player. The winner is the player with more points after five rounds.

Variations
Challenge 1 Play with a reduced number of shapes/word cards.

Challenge 3 Remove the 'angle' card, and replace with 'acute angle', 'obtuse angle' and 'right angle'.

Activity 2

Learning objective
- Investigate a pattern reflected over parallel and perpendicular mirror lines.

Resources
squared paper (per pair); ruler (per pair); coloured pencils (per pair)

What to do
- Learners use a ruler to mark five parallel 'mirror' lines, eight cells in length, four cells apart. They shade cells to the left of the first mirror line to form a pattern no more than four cells wide.
- Learners predict what the pattern will look like after five reflections and how many lines of symmetry there will be. They write their predictions, then reflect the pattern five times.
- Learners draw a pair of perpendicular lines in the centre of the paper. They shade cells in the top left quadrant to form a pattern no bigger than a 4 × 4 grid then investigate its reflection.

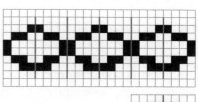

Variations
Challenge 1 Reduce to three mirror lines, shortened to four cells in length and three cells apart.

Challenge 3 Reflection: five vertical lines and one horizontal line. Reflect in any order.

Unit 15: 3D shape

Learning objectives

Code	Learning objective
5Gs4	Visualise 3D shapes from 2D drawings and nets, e.g. different nets of an open or closed cube.

Strand 5: Problem solving

5Pt5	Recognise the relationships between different [2D and] 3D shapes, e.g. a face of a cube is a square.
5Ps4	Deduce new information from existing information to solve problems.
5Ps7	Identify simple relationships between shapes, [e.g. these triangles are all isosceles because ...]
5Ps9	Explain methods and justify reasoning orally and in writing; make hypotheses and test them out.

Unit overview

This unit promotes deeper understanding of the properties of 3D shapes. Learners develop skills in visualising 3D shapes from 2D drawings. They recognise nets of an open or closed cube and explore visualisation of other 3D shapes from nets.

Common difficulties and remediation

Some learners confuse the names of 2D shapes and 3D shapes, typically naming the 3D shape by its two-dimensional face. Provide activities encouraging learners to name 3D shapes and learn their attributes, for example, sorting games that include questions such as: **What are the characteristics of this shape? Why did you sort the shapes into these groups?**

Look out for learners who find it difficult to identify nets that will form a cube. This is typical of learners who have had insufficient experience visualising 3D shapes, particularly how they can be opened up and folded closed again. Consequently, they find it difficult to 'see' how a net can be folded to make a 3D shape. These learners require opportunities to draw and make nets of shapes. They should experience unfolding a range of packaging, such as tubes and prism-shaped packages.

Promoting and supporting language

At every stage, learners require a mathematical vocabulary to access questions and problem-solving exercises. You may find it useful to refer to the audio glossary on Collins Connect. If appropriate, when a new key word is introduced, ask learners to write a definition in their books, drawing a box around it for emphasis. They should use their own words, for example: 'A cube is a three-dimensional shape with six identical square faces joined at their edges.' In addition, be aware of terms that have more than one definition, particularly meanings outside mathematics, for example, 'face'. You should help learners distinguish the term from other meanings by introducing 2D shapes as faces/surfaces of 3D shapes. Learners may identify squares as surfaces on cubes, or triangles as faces on a triangular-based pyramid.

Encourage frequent usage of mathematical language to help embed vocabulary. For example, model the use of the word 'vertex' (singular)/'vertices' (plural) when learners refer to 'a corner/corners of a shape' as they make the transition to mathematically precise language. Also, help learners with pronunciation of tricky words, such as 'prism' pronounced *priz-uhm*.

Throughout the lesson it is important to encourage learners to seek clarification and confirmation of the mathematical language. This may involve prompting learners to call out and complete definitions, for example, following a demonstration of nets, by saying: **A net is ... what a 3D shape would look like if it were opened out flat.**

Lesson 1: **Visualising 3D shapes**

Learning objectives

Code	Learning objective
5Gs4	Visualise 3D shapes from 2D drawings and nets, e.g. different nets of an open or closed cube.

Strand 5: Problem solving

5Ps4	Deduce new information from existing information to solve problems.

Prerequisites for learning

Learners need to:

• identify and describe a range of 3D shapes

• find examples of 3D shapes in the environment and in art.

Success criteria

Learners can:

• visualise a 3D shape from a 2D drawing.

Vocabulary

face, edge, vertex, vertices

Resources

paper and pencil (per learner); set of 3D shapes (per pair or group); drawings of a range of 2D shapes (per pair)

Refresh

• Choose an activity from Geometry – Shapes and geometric reasoning in the Refresh activities.

Discover SB

• Discuss the images in the Student's Book. Ask: **Which 3D shape does each drawing represent?** Lead a discussion, encouraging learners to comment on the features shown in the drawing they use to identify a shape. Prompt use of correct geometrical vocabulary.

Teach and Learn SB 👥

• Discuss the diagrams in the Student's Book. Demonstrate how to draw a cube and a cuboid on the board. Indicate edges and faces that are hidden from view by using dotted lines and explain what these lines indicate.

• Draw a hexagonal prism. Ask: **What 3D shape does this drawing represent?** Invite learners to discuss the properties of the shape they identify from the drawing and how they use these to identify the shape. Repeat for a pentagonal prism.

• Ask learners to work in pairs. Each learner draws a 2D shape. They swap papers with their partner and identify the 3D shape represented by the drawing.

Practise WB

• Workbook: Visualising 3D shapes page 154

Apply 👥

• Provide drawings of 2D shapes and ask learners to work in pairs to identify the 3D shapes represented.

Review 📊

• Display the **Geoboard tool**. Show a 2D representation of a 3D shape and ask learners to identify the shape. Hold up the correct 3D shape to confirm the answers. Repeat for three more shapes.

Assessment for learning

• Point to the correct 3D shape that this 2D drawing represents.

Geometry – Shapes and geometric reasoning

Lesson 2: **Nets**

Geometry – Shapes and geometric reasoning

Learning objectives

Code	Learning objective
5Gs4	Visualise 3D shapes from 2D drawings and nets, e.g. different nets of an open or closed cube.

Strand 5: Problem solving	
5Ps4	Deduce new information from existing information to solve problems.

Prerequisites for learning

Learners need to:

- visualise 3D objects from 2D drawings
- visualise, from the front, top and from the side, 2D drawings of 3D shapes made with interlocking cubes.

Success criteria

Learners can:

- mentally fold a cube from a 2D drawing of its net
- decide whether a net will fold into a closed or open cube by visualising whether it has five or six faces.

Vocabulary

face, net, open cube, closed cube, vertex

Resources

squared paper (per learner); ruler (per learner); scissors (per learner); coloured pencils (per learner)

Refresh

- Choose an activity from Geometry – Shapes and geometric reasoning in the Refresh activities.

Discover [SB]

- Discuss the text and images in the Student's Book. Ask: **What does each object look like when its net is folded up?** Prompt use of correct vocabulary, for example: 'The object will have six square faces. The faces come together at right angles to form a cube.'

Teach and Learn [SB] [bar chart icon]

- Discuss the text and images in the Student's Book. Display the **Nets tool**, showing a cube. Demonstrate a net for a closed cube. Ask learners to draw nets for a closed cube and an open cube on squared paper, then cut them out and fold them into cubes. Ask: **How would you change the net for an open cube into one for a closed cube?**
- If time allows, learners can make two more examples of open and closed cubes. Note: there are eight different nets for an open cube and eleven nets for a closed cube, not including rotations and reflections.

Practise [WB]

- Workbook: Nets page 156
- Refer to Activity 1 from the Additional Practice Activities.

Apply [icons]

- Display **Slide 1**, showing three views of a puzzle cube. Discuss the various views. Learners copy the net of the cube and colour the faces.

Review [icon]

Display **Slide 2**. Ask learners to visualise the folded cube, then answer these questions.

- **Which shape is on the face opposite the star?** (triangle)
- **Which shapes are next to the moon?** (star, triangle, circle, arrow)
- **Which shape is the arrow pointing to?** (star)
- **Which shapes would be at the opening if the arrow tab was removed?** (star, arrow, triangle, circle)

Assessment for learning

- Which are nets of open cubes? Closed cubes? How would you turn the closed cubes into open cubes?
- Here are three views of a closed cube. Draw the net for this cube. How did you work it out?

Lesson 3: **Constructing 3D shapes**

Learning objectives

Code	Learning objective
5Gs4	Visualise 3D shapes from 2D drawings and nets, e.g. different nets of an open or closed cube.

Strand 5: Problem solving

5Pt5	Recognise the relationships between different [2D and] 3D shapes, e.g. a face of a cube is a square.
5Ps4	Deduce new information from existing information to solve problems.

Prerequisites for learning

Learners need to:

- identify, describe, visualise, draw and make a range of 3D shapes
- visualise 3D objects from 2D nets and drawings.

Success criteria

Learners can:

- estimate the number of cubes it takes to build or complete a cuboid
- use a 2D drawing to visualise the edges and vertices of a 3D shape

- position and fix modelling materials to form the edges and vertices of a range of 3D shapes.

Vocabulary

pyramid, prism, octahedron, face, vertex, edge, skeleton

Resources

interlocking cubes (per group); modelling materials, e.g. straws, sticks, modelling clay (per group)

Refresh

- Choose an activity from Geometry – Shapes and geometric reasoning in the Refresh activities.

Discover SB

- Discuss the text and images in the Student's Book. Elicit that the 'skeleton' shows the edges and vertices clearly. Ask: **What 3D shape is shown in each photograph?** Learners should state the name and some facts about each shape, using correct mathematical vocabulary to describe the shape, including faces, vertices and edges. Ask: **Can you think of other examples of these shapes outside the classroom?**

Teach and Learn SB 📊

- Display the **Shape set tool**, showing a cuboid. Discuss the number and shape of faces of a cuboid. Repeat for pyramids and a range of prisms, discussing the number of vertices and edges.
- Draw a table on the board with column headings for faces, edges and vertices and rows for each shape. Ask volunteers to complete the table. Ask: **What materials would you use to construct a skeleton of a cube?** Discuss examples.

Practise WB

- Workbook: Constructing 3D shapes page 158

Apply 👥 SB

- Ask learners to work in pairs. Say: **Look at the objects in the Discover section. What are they used for?**
- Provide modelling materials, such as cubes and straws, and ask pairs of learners to build one or more of the shapes.

Review 👥

- Learners work in pairs. They each use interlocking cubes to create a cuboid. They remove several cubes from the shape.
- Learners exchange cuboids and estimate the number of cubes that need to be added to complete the shape. They complete it and compare the actual number to the estimate.

Assessment for learning

- Build a shape with nine edges from modelling clay and straws. What shape is it?
- Build a shape with eight vertices. What is it?
- Here is part of a cuboid. How many more cubes will complete it?

Geometry – Shapes and geometric reasoning

Lesson 4: **Relationships between 3D shapes**

Learning objectives

Code	Learning objective
5Gs4	Visualise 3D shapes from 2D drawings and nets, e.g. different nets of an open or closed cube.

Strand 5: Problem solving

5Pt5	Recognise the relationships between different [2D and] 3D shapes, e.g. a face of a cube is a square.
5Ps4	Deduce new information from existing information to solve problems.
5Ps7	Identify simple relationships between shapes, [e.g. these triangles are all isosceles because ...]
5Ps9	Explain methods and justify reasoning orally and in writing; make hypotheses and test them out.

Prerequisites for learning

Learners need to:
- identify, describe, visualise, draw and make a range of 3D shapes
- know some of the properties of cubes, cuboids, pyramids and prisms.

Success criteria

Learners can:
- use knowledge of faces, vertices and edges to identify similarities in 3D shapes

- deduce the name of a pyramid from the shape of its base
- deduce the name of a prism from the shape of its two identical ends.

Vocabulary

net, face, vertex, edge, point (apex)

Resources

set of 3D shapes: cube, cuboid, pyramids, prisms, tetrahedron, octahedron (per group); paper; coloured pencils (per learner)

Refresh

- Choose an activity from Geometry – Shapes and geometric reasoning in the Refresh activities.

Discover 〔SB〕

- Remind learners that in Stage 4 they have already learned to classify a range of 2D and 3D shapes based on their properties. They can also recognise the similarities and differences between 2D and 3D shapes.
- Discuss the text and images in the Student's Book. Ask: **What properties are shared by prisms? By pyramids?** Encourage use of correct mathematical vocabulary to describe each shape, including faces (flat, identical, base), vertices and edges, point or apex. Ask: **Where do you find these shapes in the world around you?**

Teach and Learn 〔SB〕 〔▢〕 〔▦〕

- Discuss the text and images in the Student's Book. Display the **Nets tool**, showing square-and triangular-based pyramids, then a pentagonal pyramid. Use the tool to highlight the properties the shapes have in common and how they are named after the shape of their base. Repeat for triangular and hexagonal prisms.
- Display **Slide 1** showing an octagonal prism and a hexagonal pyramid. Say: **Use your knowledge of prism and pyramid properties to name these shapes.**

Practise 〔WB〕

- Workbook: Relationships between 3D shapes page 160
- Refer to Activity 2 from the Additional practice activities.

Apply 〔▣▣〕

- Provide a set of 3D shapes including a range of prisms and pyramids. Learners draw Venn diagrams and sort the shapes according to their properties. Ask learners to explain how they sorted the shapes, for example: prisms, pyramids, neither.

Review 〔▣▣▣▣〕

- Divide the class into two groups. Present a 3D shape and ask learners in each group to describe one of its properties. Each group takes turns to describe a different property, scoring a point for every correct one described. Keep a tally of points on the board. When both groups are unable to state any new properties, introduce a new shape. The team with the higher score at the end of the session is the winner.

Assessment for learning

- Sort these 3D shapes into two or three groups based on their properties.

Additional practice activities

Activity 1

Learning objective

- Use knowledge of nets to put together the net of a closed cube from a choice of pieces.

Resources

squared paper (per pair); ruler (per pair); scissors (per pair)

What to do

- Learners work in pairs to play a game, 'Complete the nets'.
- They each draw three different nets for a closed cube on squared paper.
- They cut out the nets and then cut each one into two pieces.
- Partners swap all their net pieces. They work out the different ways to stick the two parts together so that the two pieces make a cube when folded.

Variations

Challenge 1 The game is played with nets for open cubes.

Challenge 3 Players use four nets each and try to find more than one way of pairing pieces together to make nets for closed cubes.

Activity 2

Learning objective

- Describe the properties of prisms and pyramids and recognise the relationship between the shapes in each group.

Resources

Number cards 3, 4, 5, 6 and 8 from Resource sheet 2: 0–100 number cards (per pair)

What to do

- Pairs play the game 'Prisms and pyramids' and choose to be 'prisms' or 'pyramids'.
- Learners each take a number card. The card number represents either the number of sides of a pyramid base, or the number of sides of a prism end.
- Players say the total number of edges or vertices for their shape. The higher number scores a point. An extra point is scored for naming the shape.

Variations

Challenge 1 Players have access to 3D solids to help them count vertices and edges.

Challenge 3 Play with number cards 3 to 10 and state the number of edges/vertices for unfamiliar shapes.

175

Unit 16: Angles

Learning objectives

Code	Learning objective
5Gs6	Understand and use angle measure in degrees; measure angles to the nearest 5°; identify, describe and estimate the size of angles and classify them as acute, right or obtuse.
5Gs7	Calculate angles in a straight line.
Strand 5: Problem solving	
5Pt7	Consider whether an answer is reasonable in the context of a problem.
5Ps4	Deduce new information from existing information to solve problems.
5Ps7	Identify simple relationships between shapes [, e.g. these triangles are all isosceles because ...]
5Ps8	Investigate a simple general statement by finding examples which do or do not satisfy it [, e.g. the sum of three consecutive whole numbers is always a multiple of three].
5Ps9	Explain methods and justify reasoning orally and in writing; make hypotheses and test them out.

Unit overview

This unit reviews the simple properties of angles and introduces the measurement and classification of angles as acute, right or obtuse. Estimation of angles is encouraged and checked by correct use of a protractor. Examples of angles in the real world are examined, such as the angle made by the two hands of an analogue clock. Finally, the properties of angles on a straight line are investigated.

Common difficulties and remediation

Some learners may find a traditional semi-circular protractor difficult to use. Give learners individual support and practice, confirming that they are following the key points in using a protractor as demonstrated. Circular protractors or measured angle templates may be preferable, if available. Use of a circular protractor, in particular, reinforces the relationship between the dynamic and static aspect of angle. Also, the positioning of the centre of the protractor placed on the point of the angle is more obvious. Look out for learners who confuse the two scales of the protractor, inner and outer, for example: drawing a 135° angle instead of a 45° angle or a 65° angle in place of a 115° angle. To remedy this, discuss acute and obtuse angles with learners, asking them to consider angle types to determine if the angle drawn matches that required.

Some learners mistakenly assume that the size of an angle is proportional to the length of the directional lines at either limit of the rotation. To remedy this, have learners measure multiple versions of the same angle but with a range of directional line lengths.

Promoting and supporting language

At every stage, learners require a mathematical vocabulary to access questions and problem-solving exercises. You may find it useful to refer to the audio glossary on Collins Connect. If appropriate, when a new key word is introduced, ask learners to write a definition in their books, drawing a box around it for emphasis. Encourage learners to write the definition in their own words, for example: 'A straight angle is absolutely straight; it measures 180°.' In addition, be aware of terms that have more than one definition, particularly meanings outside mathematics, for example: degree, angle, acute.

Encourage frequent usage of mathematical language to help embed vocabulary. For example, use 'acute' and 'obtuse' alongside phrases 'less than a right angle/90 degrees' and 'greater than 90 degrees but less than 180 degrees/straight angle' as learners make the transition to mathematically precise language.

Throughout the lesson it is important to encourage learners to seek clarification and confirmation of the mathematical language. This may involve prompting students to call out and complete definitions, for example saying 'adds to 180 degrees' every time 'angles on a straight line' is mentioned, or praising learners when they ask for terms to be defined. A few times each lesson, ask for a volunteer to summarise what has been learned so far so they get experience of explaining concepts in their own words.

How learners interpret mathematical text should also be addressed. Provide frequent opportunities for learners to read terms and phrases out loud and to explain, in their own words, what the text means.

Lesson 1: **Measuring angles**

Learning objectives

Code	Learning objective
5Gs6	Understand and use angle measure in degrees; measure angles to the nearest 5°; identify, describe and estimate the size of angles [and classify them as acute, right or obtuse].

Strand 5: Problem solving

5Pt7	Consider whether an answer is reasonable in the context of a problem.
5Ps4	Deduce new information from existing information to solve problems.

Prerequisites for learning

Learners need to:

- know that angles are measured in degrees and that one whole turn is 360° or four right angles
- compare and order angles less than 180°.

Success criteria

Learners can:

- align a protractor correctly and understand that you always count up from 0°
- put the baseline of the protractor on the arm of the angle

- ensure the vertex of the angle is on the centre point of the protractor
- ensure they measure the angle between the arms and not outside the arms of the angle.

Vocabulary

angle, protractor

Resources

protractor (per learner); books, ramp, toy car (per group)

Refresh

- Choose an activity from Geometry – Shapes and geometric reasoning in the Refresh activities.

Discover [SB]

- Discuss the image in the Student's Book. Ask: **How would you decide which pizza slice is the largest? The smallest?** Take answers. Agree they could compare slices by placing them one on top of another. However, a more accurate method is to treat each slice as a sector of a circle and measure the angle at the centre point. The greater the angle, the larger the slice.

Teach and Learn [SB] 🖥 📊

- Direct learners to the text and images in the Student's Book and display **Slide 1.** Look at the pictures in the example and discuss what it is they are measuring.
- Ask: **What is an angle? What are the different types?** Agree that when any two straight lines meet, an angle is formed, which is a measure of an amount of turn: 360° – full turn, 180° – half turn and 90° – quarter turn.
- Display the **Geometry set tool** with three triangles. Ask learners to work with a partner to estimate the sizes of angles. Demonstrate how to use a protractor to measure the angles to the nearest 5°.
- Draw identical angles with arms of different lengths. Use the protractor to demonstrate that the angle size is not affected by arm length.

Practise [WB]

- Workbook: Measuring angles page 162

Apply 👥

- Groups of learners build a ramp with books and send a toy car down it, timing its descent. Ask: **How does the angle of a ramp affect how long it takes a toy car to travel its length?** They increase the angle three times and record the time taken. Lead a discussion of results and conclusions.

Review 📊

- Ask volunteers to draw four random angles less than 180° on the board. Display the **Geometry set tool**, reproducing the angles. Learners write down estimates of the size of each angle. Use the tool to measure the angles. Ask: **How accurate were your estimates?**

Assessment for learning

- Here are some angles. Which angles are close to 160°? 170°? 50°? 40°? How do you know?
- How do you use a protractor to measure an angle to the nearest 5°?

Geometry – Shapes and geometric reasoning

Lesson 2: **Angle size**

Geometry – Shapes and geometric reasoning

Learning objectives

Code	Learning objective
5Gs7	Understand and use angle measure in degrees; measure angles to the nearest 5°; identify, describe and estimate the size of angles [and classify them as acute, right or obtuse].

Strand 5: Problem solving

5Pt7	Consider whether an answer is reasonable in the context of a problem.
5Ps4	Deduce new information from existing information to solve problems.
5Ps7	Identify simple relationships between shapes [, e.g. these triangles are all isosceles because ...]

Prerequisites for learning

Learners need to:

- know that an acute angle is less than a right angle and is between 0° and 90°
- know that an obtuse angle is less than the angle on a straight line, is greater than a right angle and is between 90° and 180°.

Success criteria

Learners can:

- use a right-angle measurer, such as a set square, to show which angles are less than, or greater than, a right angle

- compare the size of angles, arranging them in ascending and descending order
- estimate the size of an angle.

Vocabulary

right angle, straight angle, degrees

Resources

rulers (per learner); scissors (five per pair)

Refresh

- Choose an activity from Geometry – Shapes and geometric reasoning in the Refresh activities.

Discover 🆂🅱 📊

- Display the **Clock tool** set to one o'clock. Point out the angle made by the minute and hour hands. Ask: **What is the size of this angle?** Move the hour hand to two o'clock, three o'clock and so on, up to six o'clock. Ask: **How does moving the hands change the angle?** Ask volunteers to describe each angle.
- Look at the images of the clocks in the Student's Book. For each image, ask: **Is this angle greater than/the same as/more than a right angle?**

Teach and Learn 🆂🅱 💻 📊

- Direct learners to the text and diagrams in the Student's Book and display **Slide 1.** Read the text together and discuss the diagrams. Display the **Geometry set tool** with the trapezium and the pink triangle. Use a set square to show which angles are less than, or greater than, a right angle. Display the pink parallelogram. Ask: **Which angles are greater than a right angle? Less than a right angle? Can you estimate the size of the angles?** Use a protractor to measure the angles and check estimates.
- In pairs, learners each draw two different angles on paper, collect them together and put them in ascending and descending order.

Practise 🆆🅱

- Workbook: Angle size page 164

Apply 👥

- Give pairs of learners five pairs of scissors. They open each pair so the blades form angles: two angles less than a right angle, one at a right angle and two angles greater than a right angle, but less than a straight angle. They arrange the angles in ascending order.

Review 📊

- Ask learners to tell a partner how to use a right angle to decide whether an angle is equal to, less than or greater than a right angle.
- Display the **Clock tool** showing two o'clock. Ask: **What is the size of the angle formed between the hands at two o'clock?** (less than a right angle) **At three o'clock?** (right angle) **At four and five o'clock?** (greater than a right angle, less than a straight angle) **At six o'clock?** (straight angle)

Assessment for learning

- How do you know if an angle is less than or greater than a right angle?
- How would you arrange a set of angles in descending order?

Lesson 3: **Classifying angles**

Learning objectives

Code	Learning objective
5Gs6	Understand and use angle measure in degrees; measure angles to the nearest 5°; identify, describe and estimate the size of angles and classify them as acute, right or obtuse.

Strand 5: Problem solving

5Pt7	Consider whether an answer is reasonable in the context of a problem.
5Ps4	Deduce new information from existing information to solve problems.
5Ps8	Investigate a simple general statement by finding examples which do or do not satisfy it [, e.g. the sum of three consecutive whole numbers is always a multiple of three].

Prerequisites for learning

Learners need to:

- know that angles are measured in degrees and that one whole turn is 360°, or four right angles
- compare and order angles less than 180°.

Success criteria

Learners can:

- use a right angle to identify an acute angle and know that it is an angle between 0° and 90°

- use a right angle to identify an obtuse angle and know that it is less than the angle on a straight line, is greater than a right angle and is between 90° and 180°.

Vocabulary

right angle, acute angle, obtuse angle

Resources

paper (per learner), protractor (per learner), ruler (per learner); coloured pencils (per learner)

Refresh

- Choose an activity from Geometry – Shapes and geometric reasoning in the Refresh activities.

Discover 📱

- Direct learners to the Students' Book and look at the image. Ask: **Which painting is correctly hung? Why is this the best angle?** Elicit that the middle painting is hung correctly – horizontally with the top (or bottom) at right angles to the vertical. Hanging at angles less than or greater than 90° would make the painting look crooked.

Teach and Learn 📱 🖥 📊

- Direct learners to the text and diagrams in the Student's Book and display **Slide 1.** Discuss the diagrams. Each learner joins their wrists together and splits their palms to form an angle like the ones in the diagrams. Say: **Show me an acute angle/right angle/obtuse angle. Show me the largest obtuse angle/smallest acute angle you can make.**

- Display the **Geometry set tool,** showing the green triangle and the purple rhombus. Use a protractor to show the right angle and two acute angles of the triangle, and the two obtuse and two acute angles of the rhombus.

Practise 📒

- Workbook: Classifying angles page 166
- Refer to Activity 1 from the Additional Practice Activities.

Apply 👥

- Say: **Write all the capital letters of the alphabet that are written with straight lines.** Learners draw the letters on paper with a ruler. They identify and label the angles of each letter as right, acute or obtuse and then use a protractor to check their classification. Ask: **Is there a capital letter that has all three types of angle?** (no) **Do any have both acute and obtuse angles?** (A, K, Y)

Review

- On the board draw a circular pizza and ask learners to identify the type of angle the following fractional sectors make with the centre of the circle: $\frac{1}{5}$, $\frac{1}{4}$, $\frac{2}{5}$, $\frac{3}{8}$ (acute, right, obtuse, obtuse).

Assessment for learning

- Here is a triangle and a quadrilateral. Identify all the angles. How did you do this?
- Draw an irregular pentagon with four obtuse and one acute angle. How did you draw this shape?

Geometry – Shapes and geometric reasoning

Lesson 4: **Angles on a straight line**

Geometry – Shapes and geometric reasoning

Learning objectives

Code	Learning objective
5Gs7	Calculate angles in a straight line.

Strand 5: Problem solving

5Ps4	Deduce new information from existing information to solve problems.
5Ps8	Investigate a simple general statement by finding examples which do or do not satisfy it [, e.g. the sum of three consecutive whole numbers is always a multiple of three].
5Ps9	Explain methods and justify reasoning orally and in writing; make hypotheses and test them out.

Prerequisites for learning

Learners need to:

- say whether an angle is more or less than a right angle
- understand that two right angles equal a straight line.

Success criteria

Learners can:

- identify two angles at a point on a straight line

- recall that angles on a straight line total 180°
- deduce the size of one angle on a straight line, given the size of the other angle.

Vocabulary

right angle, straight angle, angles at a point

Resources

paper strips, ruler, scissors (per pair); protractor (per learner)

Refresh

- Choose an activity from Geometry – Shapes and geometric reasoning in the Refresh activities.

Discover [SB]

- Discuss the text and image in the Student's Book. Say: **Angles that meet on a straight line always add up to 180°. Elicit that 145° + 35° = 180°.**

Teach and Learn [SB] [screen] [chart]

- Direct learners to the text and diagram in the Student's Book and display **Slide 1**. Explain that the angle of a straight line is always 180°. So when there are a pair of angles on a straight line they should both add up to 180°. Display the **Geometry set tool** showing the green triangle. Demonstrate that each angle is acute and meets at a point on a straight line with an obtuse angle where both angles add to 180°.

- In pairs, learners draw angles on a straight line then measure to confirm the two angles add to 180°. Do the same on the board. Label one angle 67° and the other x. Ask: **What is the value of x?** (113°)

Practise [WB]

- Workbook: Angles on a straight line page 168
- Refer to Activity 2 from the Additional Practice Activities.

Apply [screen] [people]

- Display **Slide 2**. Learners draw a line at the same angle across two strips of paper. They measure angles e, f, g and h and cut the lines, making pieces A, B, C and D. Learners predict the size of combined angle e + h. They join A and D and measure (e + h total 180°; A + D form a strip identical to the original A + B). Repeat for B and C and angles f and g. Learners turn over D and join with A, forming angle x. They predict, then measure, angle x (double angle e).

Review [chart]

- Display the **Geoboard tool**. Draw three examples of angles at a point on a straight line. Demonstrate that the sum of each pair is 180°. Draw three more examples. Provide one angle and ask learners to find the other.

Assessment for learning

- A pair of angles meet on a straight line. What is the sum of both angles?
- Here is the size of one of the angles. What is the size of the other?

Additional practice activities

Activity 1 👥 Challenge 2

Learning objective

• Construct a shape with intersecting lines and use a protractor to measure the angles made.

Resources

protractor (per learner); ruler (per learner); paper (per learner); book (per learner)

What to do

• Learners use a ruler to draw a square and then several intersecting lines inside the square (see diagram for example).

• They label each of the angles with a letter.

• They construct a table with rows for each letter and columns headed 'estimated size' and 'actual size'.

• They estimate and record the size of each angle. Then they measure the angles with a protractor and add the measurements to the table.

• Learners comment on how close their estimates are.

Variation

Challenge 1 Learners comment on whether each angle is less than, or greater than, a right angle. They use a right angle tester, such as the corner of a book, to test their predictions.

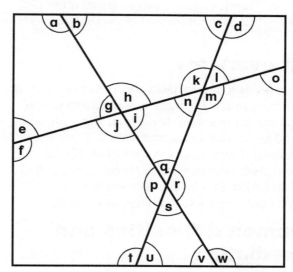

Activity 2 👥 Challenge 2

Learning objective

• Calculate the angles in a straight line.

Resources

Resource sheet 19: Angles on a straight line (per learner); protractor (per learner)

What to do

• Give each learner a copy of Resource sheet 19: Angles on a straight line.

• Learners measure the angles with a protractor and add the measurements to the table.

• They check their answers by adding the two numbers together.

Variation

Challenge 1 Learners comment on whether each angle is less than or greater than a right angle. They use a right angle tester, such as the corner of a book, to test their predictions.

Geometry – Shapes and geometric reasoning

181

Unit 17: Position and movement

Learning objectives

Code	Learning objective
5Gp1	Read and plot co-ordinates in the first quadrant.
5Gp2	Predict where a polygon will be after reflection where the mirror line is parallel to one of the sides, including where the line is oblique.
5Gp3	Understand translation as movement along a straight line, identify where polygons will be after a translation and give instructions for translating shapes.
Strand 5: Problem solving	
5Ps4	Deduce new information from existing information to solve problems.
5Ps7	Identify simple relationships between shapes [, e.g. these triangles are all isosceles because ..].
5Ps8	Investigate a simple general statement by finding examples which do or do not satisfy it [, e.g. the sum of three consecutive whole numbers is always a multiple of three].
5Ps9	Explain methods and justify reasoning orally and in writing; make hypotheses and test them out.

Unit overview

This unit reviews co-ordinates and introduces learners to reflection, where the mirror line is parallel to one of the sides of the shape, or oblique. Translations are explored as straight line movements in either a vertical or horizontal direction, or a combination of both. Practical tasks establish that a shape is unchanged after translation. Learners investigate the tiling patterns made by single or compound shapes.

Common difficulties and remediation

Learners may identify four lines of symmetry in a rectangle. When teachers discuss symmetry with learners they often use the term 'half', for example: 'Can this shape be folded so that the two halves fit exactly on top of each other?' The ambiguity of the word 'half' often leads learners to assume that half of a shape is the reflection of it, causing confusion when a rectangle is investigated. The diagonals can be confused with lines of symmetry because the resulting shapes, two right-angled triangles formed by the fold, are halves of the rectangle. It is important that learners have practical experience of folding shapes and using mirrors to assess symmetry.

Some learners find it difficult to use a mirror to test for symmetry, as they forget to project the image beyond the mirror. The correct way of using mirrors needs to be modelled by teachers. Learners may benefit from use of semi-transparent plastic mirrors that provide a clearer test of symmetry by allowing learners to continue to see their original image at all times.

Promoting and supporting language

At every stage, learners require a mathematical vocabulary to access questions and problem-solving exercises. You may find it useful to refer to the audio glossary on Collins Connect. When a new key word is introduced, you could encourage learners to write a definition in their books, in their own words, for example: 'A shape is symmetrical when you can fold it in half so that one half exactly covers the other half.' Learners could draw a box around the definition for emphasis. Be aware of terms that have more than one definition, particularly meanings outside mathematics, for example: co-ordinate, plot, reflect, translate.

Encourage frequent usage of mathematical language to help embed vocabulary, for example modelling use of the word 'vertex' when learners refer to 'corner', or 'mirror images' when they refer to 'two equal halves', to assist their transition to mathematically precise language.

Throughout the lesson encourage learners to seek clarification and confirmation of the mathematical language. Prompt learners to call out and complete definitions, for example, following a demonstration of a translation by saying: **with the size, shape and orientation of the shape unchanged,** or praising learners when they ask for terms to be defined.

The way learners interpret mathematical text should also be addressed. Provide frequent opportunities for learners to read terms and phrases out loud and to explain, in their own words, what the text means.

Lesson 1: **Reading and plotting co-ordinates**

Learning objectives

Code	Learning objective
5Gp1	Read and plot co-ordinates in the first quadrant.

Strand 5: Problem solving	
5Ps4	Deduce new information from existing information to solve problems.

Prerequisites for learning

Learners need to:

- understand that the term co-ordinates applies to a pair of numbers that denote the exact position of the intersection of two lines in a grid of squares
- describe and identify the position of a square on a grid of squares where rows and columns are numbered and/or lettered.

Success criteria

Learners can:

- plot points using the convention (x, y) where x shows the distance from the origin along the horizontal x-axis and y shows the distance from the origin up or down the vertical y-axis.

Vocabulary

x-axis, y-axis, co-ordinates, origin

Refresh

- Choose an activity from Geometry – Position and movement in the Refresh activities.

Discover 📖 🖥️

- Direct learners to the Student's Book and display **Slide 1**. Ask: **How would you describe the position of a village on the map? A lake?** Ask learners to describe the location of other features on the map. Ask: **What would happen if everyone described grid positions in different ways, for example changing the order in which the row and column numbers are stated?** Elicit the need for a convention that establishes the order.

Teach and Learn 📖 📊

- Read and discuss the diagram in the Student's Book. Ask learners to point to (2, 3) on the grid. Repeat for different co-ordinates.
- Display the **Co-ordinates tool**. Discuss how horizontal and vertical lines of the grid are constructed from the origin (0, 0) and that the point at which two lines cross is given two numbers, called co-ordinates.
- Explain that the horizontal co-ordinate is written first, then the vertical co-ordinate.
- Write several co-ordinates in the form (x, y) and ask volunteers to locate and mark them on the grid. Circle several points on the grid and ask volunteers to write matching co-ordinates. Ask volunteers to comment on co-ordinates for points along horizontal and vertical lines.

Practise 📙

- Workbook: Reading and plotting co-ordinates page 170

Apply 👥 📖 🖥️

- Display **Slide 1** again and ask learners to look at the map in the Student's Book Discover section. Learners write questions about the map for their partner to answer, for example: 'What are the co-ordinates of the town?'

Review 🖥️

- Display **Slide 2**. Ask learners to write the co-ordinates of the landmarks. Then as a class create four sets of co-ordinates that correspond to different landmarks and to plot these objects on the grid.

Assessment for learning

- How would you describe the position of an object on a numbered grid?
- What do you call the axes of a numbered grid?
- Explain what a pair of co-ordinates tells you. Give me an example.

Lesson 2: **Shapes from co-ordinates**

Geometry – Position and movement

Learning objectives

Code	Learning objective
5Gp1	Read and plot co-ordinates in the first quadrant.

Strand 5: Problem solving

5Ps4	Deduce new information from existing information to solve problems.
5Ps7	Identify simple relationships between shapes [, e.g. these triangles are all isosceles because ...]

Prerequisites for learning

Learners need to:

- know that co-ordinates are used to plot the position of a point on a grid
- read and plot co-ordinates in the first quadrant.

Success criteria

Learners can:

- plot specific points on a co-ordinate grid in the first quadrant

- apply their knowledge of polygons to locate the position of a missing vertex and complete the shape.

Vocabulary

x-axis, *y*-axis, co-ordinates, vertex (vertices)

Resources

squared paper (per pair), ruler (per learner),

Refresh

- Choose an activity from Geometry – Position and movement in the Refresh activities.

Discover 🆂🅱 🖥

- Direct learners to the Student's Book and display **Slide 1**. Ask: **How would you describe the position of the pink shape shown on the map?** Elicit that they would use a system of co-ordinates. Point to each vertex in turn and ask the learners to copy you by putting their finger on the picture in the Student's Book. Read the co-ordinates of each vertex together.

Teach and Learn 🆂🅱 📊

- Ask learners to look at the image in the Student's Book. Ask: **If you were given the co-ordinates of three vertices of a square area on a map, how would you work out the fourth vertex?**
- Display the **Co-ordinates tool**. Plot points (3, 2), (3, 6), (7, 6), (7, 2) to form a square. Say: **Write the co-ordinates of four vertices of a square.** Repeat for a triangle. Ask: **How many co-ordinates are needed to draw a square? A triangle?** Plot points (2, 5), (7, 5), (7, 7) for a rectangle. Explain that three vertices have been plotted. Ask: **What are the co-ordinates of missing vertex D?** Say: **You know AC = BD as opposite sides are equal. As AC is 3, BD is 3. Add 3 to the *y*-co-ordinate of vertex B to find the co-ordinates for D, so (8, 6).**

Practise 🆆🅱

- Workbook: Shapes from co-ordinates page 172
- Refer to Activity 1 from the Additional practice activities.

Apply 👥 🖥

- Display **Slide 2**. Learners write the coordinates of the vertices of a square and a rectangle.

Review 🖥

- Display **Slide 3**. Write on the board the coordinates for three of four vertices of a square (2, 3), (4, 3), (4, 1). Ask: **What are the missing coordinates?** (2, 1) Draw the shape on the grid. Write the coordinates for two of three vertices of an isosceles triangle that has a horizontal base: (4, 0), (5, 3). Ask: **What are the missing coordinates?** (6, 0) Draw the shape on the grid.

Assessment for learning

- Draw a right-angled triangle on a grid (and then a square and a rectangular). What are the co-ordinates of the vertices?
- Here are three sets of co-ordinates for a rectangle. What are the co-ordinates for the missing vertex? How do you know?

Lesson 3: **Testing for symmetry**

Learning objectives

Code	Learning objective
5Gp2	Predict where a polygon will be after reflection where the mirror line is parallel to one of the sides, including where the line is oblique.

Strand 5: Problem solving

Code	Learning objective
5Ps7	Identify simple relationships between shapes [, e.g. these triangles are all isosceles because ...]
5Ps8	Investigate a simple general statement by finding examples which do or do not satisfy it [, e.g. the sum of three consecutive whole numbers is always a multiple of three].

Prerequisites for learning

Learners need to:

- identify a vertical line of symmetry in polygons.

Success criteria

Learners can:

- use folding or mirrors to find horizontal, vertical and diagonal lines of symmetry in polygons.

Vocabulary

line of symmetry, mirror line, reflection, symmetrical, horizontal, vertical

Resources

squared paper (per pair); selection of 2D shapes: quadrilaterals (square, rectangle, kite, parallelogram, two irregular quadrilaterals), triangles (equilateral, isosceles, scalene right-angled, isosceles right-angled), regular and irregular pentagons and hexagons (per group); mirror (per learner); ruler (per learner)

Refresh

- Choose an activity from Geometry – Position and movement in the Refresh activities.

Discover [SB]

- Ask learners to look at the pictures in the Student's Book. Ask: **Which of the pictures are symmetrical? How do you know?** Ask learners to describe methods they would use to test for symmetry. Prompt them to comment on the characteristics of a symmetrical figure. Ask: **Which polygons have this property?**

Teach and Learn [SB] [▨]

- Read and discuss the diagrams in the Student's Book. Provide a selection of 2D shapes. Say: **Draw around some of these shapes on paper and cut them out.** Ask: **Does your paper shape fold so that one half fits exactly onto the other half?** Discuss results and ask learners to check the shapes for symmetry with a mirror.

- Display the **Symmetry tool**. Demonstrate lines of symmetry of a square, a rectangle and an isosceles triangle. Ask: **Can you locate the lines of symmetry on a pentagon and hexagon?** Discuss the asymmetry of a scalene triangle and a parallelogram.

Practise [WB]

- Workbook: Testing for symmetry page 174

Apply [▨▨]

- Ask learners to draw a vertical mirror line on squared paper, then place shapes they think are symmetrical with its 'line of symmetry' on the mirror line. They draw around the shape and confirm symmetry by finding the distance of each vertex from the line of symmetry. They count the number of squares and check the vertices of the reflected shape are at equal distances from the line of symmetry.

Review

- Provide a set of 2D shapes and ask learners to sort them into three sets by the number of lines of symmetry: none, one, more than one. Discuss the results and ask learners to comment on the properties of shapes that have no lines of symmetry.

- Hold up a regular octagon. Ask: **How many lines of symmetry does it have?** (8) Write the word 'octagon' on the shape. Draw an irregular octagon. Ask: **Is the shape symmetrical? Why?**

Assessment for learning

- Demonstrate how you would show that a shape is symmetrical
- Here is a hexagon. How many lines of symmetry does it have? How do you know?

Lesson 4: **Reflection (1)**

Learning objectives

Code	Learning objective
5Gp2	Predict where a polygon will be after reflection where the mirror line is parallel to one of the sides [, including where the line is oblique].

Strand 5: Problem solving

5Ps8	Investigate a simple general statement by finding examples which do or do not satisfy it [, e.g. the sum of three consecutive whole numbers is always a multiple of three].
5Ps9	Explain methods and justify reasoning orally and in writing; make hypotheses and test them out.

Prerequisites for learning

Learners need to:

• reflect a 2D shape in a line of symmetry.

Success criteria

Learners can:

• reflect a shape by finding the distance of each vertex from the mirror line

• check the vertices of the reflected shape and its image are at corresponding distances from the mirror line to confirm reflection.

Vocabulary

reflection, image, line of symmetry, parallel, vertex (vertices)

Resources

set of 2D shapes (per pair); squared paper (per pair); ruler (per learner); colouring pencils (per pair); mirror (per learner)

Refresh

• Choose an activity from Geometry – Position and movement in the Refresh activities.

Discover [SB] 💻

• Display **Slide 1** and discuss the text and images in the Student's Book. Ask: **What do you notice about each reflection?** Ask learners to come up to the board and draw a central line across all three images. Elicit that every point is the same distance from the central line and the reflection is the same size as the original image.

Teach and Learn [SB] 📊

• Discuss the text and diagrams in the Student's Book. Display the **Rotate and reflect tool**. Click the vertical mirror line and position a square three grid squares to the left. Ask a volunteer to show where the shape is after reflection in the mirror line. Expect them to count squares, to check that the vertices and their images are the same perpendicular distance from the mirror line.

• Position an isosceles triangle to the left of the line. Ask a volunteer to show where the shape is after reflection. Agree that the shape and size of the triangle remains unchanged by reflection.

• If time allows, select the horizontal line of symmetry on the **Rotate and reflect tool** and reflect another polygon.

Practise [WB]

• Workbook: Reflection (1) page 176

• Refer to Activity 2 from the Additional practice activities.

Apply

• Ask learners to draw a vertical mirror line on squared paper, then a face constructed from polygons on the left of the line. They colour the face and reflect it. Ask: **What will happen if you reflect the face across multiple mirror lines?** Learners predict, draw several mirror lines, then reflect the shape across.

Review 📊

• Display the **Rotate and reflect tool**. Click the vertical mirror line and position a T-shape to the right of the mirror line. Ask a volunteer to show where the shape is after reflection in the mirror line. Select the horizontal mirror line and reflect a trapezium.

Assessment for learning

• What word do we use to describe the reflection of a point or of a shape? (image)

• Demonstrate reflection of a triangle in a mirror line. How did you do this?

• How would you confirm the shape and size of an image are identical to the object?

Lesson 5: **Reflection (2)**

Learning objectives

Code	Learning objective
5Gp2	Predict where a polygon will be after reflection where the mirror line is parallel to one of the sides, including where the line is oblique.

Strand 5: Problem solving

5Ps8	Investigate a simple general statement by finding examples which do or do not satisfy it [, e.g. the sum of three consecutive whole numbers is always a multiple of three].
5Ps9	Explain methods and justify reasoning orally and in writing; make hypotheses and test them out.

Prerequisites for learning

Learners need to:

- predict and draw where a polygon will be after reflection where the mirror line is parallel to one of the sides.

Success criteria

Learners can:

- reflect a shape across a mirror line that is not horizontal or vertical by:
  ◊ measuring an imaginary line from the vertex of the shape to the mirror line, touching the mirror line at a right angle

  ◊ measuring the same distance again on the opposite side of the mirror and plotting the vertex of the image.

Vocabulary

reflection, image, mirror line, oblique, right angle

Resources

set of 2D shapes (per pair); squared paper (per pair); ruler (per learner); mirror (per learner)

Refresh

- Choose an activity from Geometry – Position and movement in the Refresh activities.

Discover [SB]

- Direct learners to the text and diagram in the Student's Book. Provide mirrors and 2D shapes and ask learners to investigate what happens to shapes reflected in an angled mirror, not vertical or horizontal. Ask: **Is the image the same size? Is the image the same distance from the mirror as the object?**

Teach and Learn [SB] [bar chart icon]

- Discuss the text and diagrams in the Student's Book. Display the **Rotate and reflect tool**. Select an oblique line of symmetry and position a square at a distance of two grid squares perpendicular to the line. Ask a volunteer to show where the square is after reflection. Expect them to count squares to check vertices and that their images are the same perpendicular distance from the mirror line.
- Position an isosceles triangle to the left of the line. Ask a volunteer to show where the shape is after reflection. Ask: **What has happened to the position, size and shape of the image?** Agree that the shape and size of the triangle remains unchanged by the reflection.
- If time allows, select the other oblique line of symmetry on the **Rotate and reflect tool** and reflect another polygon.

Practise [WB]

- Workbook: Reflection (2) page 178

Apply [group icon]

- Ask learners to draw a vertical, a horizontal and two oblique lines on the squared paper. They investigate patterns formed by repeated reflection of a shape in a clockwise direction across multiple mirror lines. Learners investigate patterns formed by both regular and irregular polygons, including compound shapes.
- The best patterns are formed from shapes touching both mirror lines. Ask: **Where have you seen these types of patterns?**

Review [bar chart icon]

- Display the **Rotate and reflect tool**. Click an oblique mirror line and position a right-angled triangle to the right of the mirror line. Ask a volunteer to show where the shape is after reflection in the mirror line. Select the other oblique mirror line and reflect a trapezium.

Assessment for learning

- Demonstrate the reflection of a point in an oblique mirror line. How did you do this?
- Show me how to reflect a rectangle in an oblique mirror line

Geometry – Position and movement

187

Lesson 6: **Understanding translation**

Learning objectives

Code	Learning objective
5Gp3	Understand translation as movement along a straight line, identify where polygons will be after a translation and give instructions for translating shapes.

Strand 5: Problem solving	
5Ps8	Investigate a simple general statement by finding examples which do or do not satisfy it [, e.g. the sum of three consecutive whole numbers is always a multiple of three].

Prerequisites for learning

Learners need to:

* identify right angles in 2D shapes
* move in directions to the left or right and up or down on a grid of squares
* know and be able to apply the terms 'horizontal' and 'vertical'.

Success criteria

Learners can:

* say that a translation is the movement of an object or image in a straight line without changing its size, shape or orientation
* translate a 2D shape a given number of units up, down, left or right.

Vocabulary

translate, translation, orientation, vertex (vertices), tiling pattern

Resources

construction blocks (per group); squared paper (per pair); ruler (per learner)

Refresh

* Choose an activity from Geometry – Position and movement in the Refresh activities.

Discover [SB]

* Discuss the text and images in the Student's Book. Ask: **Where have you seen patterns like these before?** (fish scales, repeated wallpaper pattern, repeated architecture, child in sled being pulled from one place to another) Say: **A translation is movement along a straight line. Look at the sled being dragged – have you experienced a translation? Where did you move from/to?** Ask learners to identify patterns in the classroom where an object has been moved in a straight line, creating one or more identical images.

Teach and Learn [SB] [bar]

* Read and discuss the diagrams in the Student's Book. Briefly explain to the learners the convention for labelling the object (e.g. A) and the image (e.g. A'). Display the **Rotate and reflect tool** with an empty grid and a rectangle displayed on the far left. Demonstrate a translation five squares in a straight line to the right. Agree the rectangle remains unchanged in shape, size and orientation.
* Ask volunteers to demonstrate a translation of different shapes up/down/left/right.
* Demonstrate creating a tiling pattern with multiple translations of a triangle.

* If time allows, ask learners to create their own pattern on squared paper. Ask: **How do you know that you have made a correct translation of the shape?**

Practise [WB]

* Workbook: Understanding translation page 180

Apply [group] [screen]

* Display **Slide 1**. Learners create a straight path of grid squares/tiles between construction blocks, with some marked with an 'X'. One learner stands to the left of the path; the other to the right. The learner on the right guides their partner, avoiding crosses representing 'danger' by calling out instructions, e.g: **translation, three squares to the right.**

Review [bar]

* Display the **Rotate and reflect tool** with a blank grid and a triangle displayed. Translate five squares to the right. Then translate the image five squares down and rotate it 90°. Ask: **Which of the two images is NOT a translation of original image?** (the rotated triangle). Elicit the shape has a different orientation and has, therefore, been changed.

Assessment for learning

* Demonstrate the translation of a rectangle in all four directions. How do you know each translation is correct?
* Explain the difference between a reflection and a translation.

Lesson 7: **Shape translation (1)**

Learning objectives

Code	Learning objective
5Gp3	Understand translation as movement along a straight line, identify where polygons will be after a translation and give instructions for translating shapes.

Strand 5: Problem solving

5Ps4	Deduce new information from existing information to solve problems.
5Ps8	Investigate a simple general statement by finding examples which do or do not satisfy it [, e.g. the sum of three consecutive whole numbers is always a multiple of three].
5Ps9	Explain methods and justify reasoning orally and in writing; make hypotheses and test them out.

Prerequisites for learning

Learners need to:

• move a shape in directions to the left or right and up or down on a grid of squares.

Success criteria

Learners can:

• translate a 2D shape using a combination of movements

• identify where a shape will be after a translation

• create 2D shapes that, following translations, form a tiling pattern.

Vocabulary

translate, translation, orientation, vertex (vertices), tiling pattern

Resources

simple 2D shapes (per pair); ruler (per learner); squared paper (per learner)

Refresh

• Choose an activity from Geometry – Position and movement in the Refresh activities.

Discover SB

• Discuss the pictures in the Student's Book. Ask: **How do you think the tiling patterns are created? What translation instructions could be used to make these patterns?** Provide 2D shapes and ask learners to work in pairs to construct their own tiling patterns. They share their patterns and translation instructions.

Teach and Learn SB 📊

• Read and discuss the diagrams. Discuss the text and diagrams in the Student's Book. Remind learners of the convention for labelling the object (for example, A) and the image (for example, A'). Display the **Rotate and reflect tool** with an empty grid and a triangle displayed on the far left. Demonstrate a translation of five squares in a straight line to the right and four squares down. Agree that the triangle remains unchanged in shape, size and orientation. Ask volunteers to demonstrate the translation of different shapes using a combination of two directions.

• Demonstrate how to create a tiling pattern using a shape translated in two directions across a 2 × 2 grid. Provide squared paper and ask learners to create their own pattern.

Practise WB

• Workbook: Shape translation (1) page 182

• Refer to Activity 3 from the Additional practice activities.

Apply 👥

• Each learner in a pair has some squared paper. They both draw a square in the same position on their sheets of paper. Learners must then sit so that they cannot see each other's paper. They take turns to describe a translation of the original square and draw it on their sheets. They should attempt to make a repeating pattern.

• After an agreed number of turns each they compare their drawings. Are their patterns the same? Have they created a repeating pattern?

Review

• Divide the class in half. Ask half of the learners to draw and shade four squares on squared paper to form an L-shape. They investigate how the shape can be translated to form a tiling pattern. Ask the other half of the learners to draw a T-shape made from four squares. Lead a discussion where learners comment on the translation instructions they used; prompt them to describe how the vertex of each shape moves each time.

Assessment for learning

• Demonstrate the translation of a triangle, using a combination of two directions. How do you know each translation is correct?

• How do you translate a T-shape to form a tiling pattern? Why does this work?

Geometry – Position and movement

Lesson 8: **Shape translation (2)**

Geometry – Position and movement

Learning objectives

Code	Learning objective
5Gp3	Understand translation as movement along a straight line, identify where polygons will be after a translation and give instructions for translating shapes.

Strand 5: Problem solving

Code	Learning objective
5Ps4	Deduce new information from existing information to solve problems.
5Ps8	Investigate a simple general statement by finding examples which do or do not satisfy it [, e.g. the sum of three consecutive whole numbers is always a multiple of three].
5Ps9	Explain methods and justify reasoning orally and in writing; make hypotheses and test them out.

Prerequisites for learning

Learners need to:

- translate a 2D shape using a combination of movements in two directions: up or down, and left or right.

Success criteria

Learners can:

- identify where a compound shape will be after a translation that involves a combination of movements in two directions: up or down, and left or right.

- create compound shapes that, following translations to the left or right and up or down, form a tiling pattern.

Vocabulary

translate, translation, compound shape, unit shape, tiling pattern

Resources

squared paper (per pair); plain paper (per pair); ruler (per learner); set of 2D shapes (per pair); chessboard and knight (per class)

Refresh

- Choose an activity from Geometry – Position and movement in the Refresh activities.

Discover 🆂🖥

- Direct learners to the text and diagrams in the Student's Book. Ask: **How do you think these tiling patterns were created?** Provide 2D shapes and ask learners to work in pairs to construct their own compound shapes. Ask: **What compound shapes can you make that fit together without any gaps?** Learners discuss and demonstrate shapes that fit together.

Teach and Learn 🆂📊

- Read and discuss the diagram in the Student's Book. Display the **Co-ordinates tool**. Draw an L-shape (outlining three squares) and demonstrate its translation in two ways: one square right, one up and one square left, one down. Point out how the L-shapes interlock. Then translate the original shape two squares right, one down and show how this image can be translated to make further L-shapes that interlock. Ask: **What other compound shapes will do this?**

- Provide 2D shapes and ask learners to investigate making tiling patterns.

Practise 🆆🅱

- Workbook: Shape translation (2) page 184
- Refer to Activity 4 from the Additional practice activities.

Apply 👥

- Demonstrate a chessboard and the movement of the knight. Ask learners to mark an 8 × 8 grid on squared paper to represent a chessboard. On plain paper they draw and cut out a knight piece. They move the knight four squares in any legal move, then shade the squares and translate the L-shape multiple times. Ask: **What is the largest number of squares that can be covered by the repeated translation of a knight move, without crossing the edge of the board?**

Review 📊

- Display the **Rotate and reflect tool**. Ask learners to identify shapes to use for making tiling patterns. Lead a discussion in which learners comment on the shapes that successfully tile.

Assessment for learning

- Show me how to construct a compound shape that will tile.
- How would you translate a compound shape made from a hexagon and a square to form a tiling pattern?

Additional practice activities

Activity 1

> **Learning objectives**
> - Use co-ordinates to draw and complete shapes.
> - Solve problems involving the co-ordinates of missing vertices.

Resources

squared paper (per pair); ruler (per pair)

What to do

- Learners draw and label a 10 × 10 co-ordinate grid on squared paper. They plot points and connect lines to make the following letters. They find the missing co-ordinate:
  ◊ Letter H: (2, 2) (2, 4) (2, 6) (5, 6) (5, 2) *(5, 4)*; Letter E: (5, 4) (3, 4) (3, 6) (3, 8) (5, 8) *(5, 6)*
  ◊ Letter M: (3, 5) (3, 9) (7, 9) (7, 5) *(5, 5)*; Letter W: (1, 5) (2, 1) (4, 1) (5, 5) *(3, 5)*

Variations

Challenge 1 Provide all the required co-ordinates and ask learners to identify the shape.

Challenge 3 Identify the missing co-ordinates from three- and four-letter words made up of straight-sided letters, for example: WIT, LET, FEZ, FIZZ, HILL.

Activity 2

> **Learning objective**
> - Predict where a polygon will be after reflection where: (i) the mirror line is parallel to one of the sides; (ii) one or more vertices touch the mirror lines; (iii) a shape crosses the mirror line.

Resources

squared paper (per pair); ruler (per pair)

What to do

- Provide squared paper and ask learners to investigate reflection of shapes that touch the mirror line. Agree that a vertex touching the mirror line remains unchanged by reflection.
- Ask learners to investigate the reflection of shapes that cross the mirror line. Learners predict the position of the image first, then reflect. They comment on their discoveries.

Variation

Challenge 3 Ask learners to investigate the reflection of irregular polygons that cross the mirror line.

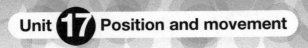
Additional practice activities

Activity 3

Learning objectives

- Recognise where a shape will be after a translation.

Resources

2D shapes: square, rectangle, equilateral triangle, pentagon, hexagon, octagon (per pair); squared paper (per pair)

What to do

- Learners draw around and colour a shape on squared paper and provide their partner with instructions for translating the shape to form a tiling pattern. Ask: **Which shapes do/do not translate to make a tiling pattern? Why?**

Variation

Challenge **3** Ask learners to investigate the translation of compound shapes, for example, a square and a rectangle, or a square and an equilateral triangle.

Activity 4

Learning objective

- Recognise where a shape will be after a translation.

Resources

squared paper (per pair)

What to do

- On squared paper, learners mark an 8 × 8 square, divide it into four smaller 4 × 4 grids, then divide into four smaller 2 × 2 grids.
- In the top left-hand corner of four squares, they make a design with four black triangles, one triangle per square, in any direction. They translate the design to the right, then down, and then down and to the right, to complete a 4 × 4 square. The 4 × 4 square is then translated to complete the 8 × 8 square.

Variation

Challenge **1** Learners work with three squares instead of four triangles.

Unit 18: Length

Learning objectives

Code	Learning objective
5Ml1	Read, choose, use and record standard units to estimate and measure length, [mass and capacity] to a suitable degree of accuracy.
5Ml2	Convert larger to smaller metric units (decimals to one place), [e.g. change 2·6 kg to 2600 g.]
5Ml3	Order measurements in mixed units.
5Ml4	Round measurements to the nearest whole unit.
5Ml7	Draw and measure lines to the nearest centimetre and millimetre.
Strand 5: Problem solving	
5Pt1	Understand everyday systems of measurement in length, [weight, capacity, temperature and time] and use these to perform simple calculations.
5Pt2	Solve single and multi-step word problems (all four operations); represent them, e.g. with diagrams or a number line.
5Pt7	Consider whether an answer is reasonable in the context of a problem.
5Ps1	Understand everyday systems of measurement in length, [weight, capacity, temperature and time] and use these to perform simple calculations.
5Ps2	Choose an appropriate strategy for a calculation and explain how they worked out the answer.
5Ps4	Deduce new information from existing information to solve problems.
5Ps5	Use ordered lists and tables to help to solve problems systematically.
5Ps9	Explain methods and justify reasoning orally and in writing; make hypotheses and test them out.
5Ps10	Solve a larger problem by breaking it down into sub-problems or represent it using diagrams.

Unit overview

This unit reviews standard units of length and introduces conversion of larger to smaller metric units by multiplication. Learners order measurements in mixed units by converting to the same unit. Rules for rounding numbers to the nearest whole unit are recalled and applied to metric units of length. Drawing and measuring lines to the nearest centimetre and millimetre are discussed and examined.

Common difficulties and remediation

Typically, measuring lines incorrectly results from measuring from the edge of the ruler rather than the zero mark, or measuring from 1 cm instead of the zero mark. Such errors indicate a learner does not understand the zero concept and lacks experience of reading measurement scales that begin at 0, not 1. Provide examples of how measurements are incorrectly adjusted when the measurement is not made from the zero mark. Also, look out for learners who measure from the wrong end of the ruler.

It is important to emphasise to learners that, when measuring length, they should look at the end-to-end unit spaces, not count the lines. Explain that the small marks on a ruler represent millimetres. Some learners, when asked to use the ruler to find how many millimetres in one centimetre, mistakenly assume that there are eight millimetres in one centimetre. This is because they ignore the longer mark at 5 millimetres.

Similar errors occur when learners count not only the 5 millimetre mark, but the end marks as well.

Promoting and supporting language

Whenever a key word is introduced, ask learners to write the definition in their own words, for example: 'Rounding means giving a number to the nearest 1, 10, 100 and so on, and not as an exact amount, especially when you are not worried about being totally accurate.' In addition, be aware of terms that have more than one definition, particularly meanings outside mathematics, for example 'unit' and 'round', as well as the words 'large' and 'small' when used in the context of the 'size' of a metric unit.

Encourage frequent use of mathematical language to help embed vocabulary. For example, model the use of the phrase 'ascending order' when learners refer to 'ordering from smallest to largest' as they make the transition to mathematically precise language.

Throughout the unit it is important to encourage learners to seek clarification and confirmation of the mathematical language. This may involve prompting learners to call out and complete definitions, for example saying 'multiply by 100' every time converting metres to centimetres is mentioned, or praising learners when they ask for terms to be defined.

A few times each lesson, ask a volunteer to summarise what has been learned thus far, so they get experience of explaining concepts in their own words.

Unit **18** Length

Lesson 1: **Measuring length**

Learning objectives

Code	Learning objective
5Ml1	Read, choose, use and record standard units to estimate and measure length, [mass and capacity] to a suitable degree of accuracy.

Strand 5: Problem solving

5Pt1 / 5Ps1	Understand everyday systems of measurement in length, [weight, capacity, temperature and time] and use these to perform simple calculations.
5Pt7	Consider whether an answer is reasonable in the context of a problem.

Prerequisites for learning

Learners need to:

- choose and use a suitable unit to estimate and measure a length of an object and the distance between two objects.

Success criteria

Learners can:

- measure length by lining up the object to be measured with the zero mark of a ruler and reading the scale

- read the scale and interpret abbreviations on a ruler calibrated in millimetres and centimetres.

Vocabulary

millimetre (mm), centimetre (cm), metre (m), kilometre (km), scale, zero mark

Resources

ruler or tape measure (per learner); 4 classroom objects (per learner); tray of objects less than 10 cm in length (per pair)

Refresh

- Choose an activity from Measure – Length, mass and capacity in the Refresh activities.

Discover SB

- Discuss the image and text in the Student's Book and say: **Imagine that you need to fit a piece of glass in a small window frame. How would you make sure the glass is the correct size?** Accept comments and discuss the most appropriate unit of measurement for the task. Ask: **What would be the problem measuring the length of the glass in metres?**

Teach and Learn SB

- Discuss the text and illustrations in the Student's Book. Ask: **What other objects would you measure in mm/cm/m/km?** Display the **Geometry set tool** with the ruler and the green triangle. Ask: **How can you prove the triangle is scalene?** Demonstrate how to measure the sides, lining up the vertex on the zero mark each time. Show all the sides are different lengths.

- Provide groups with trays of objects less than 10 cm in length. Learners work in pairs to estimate, measure and record the lengths of the items, in centimetres and millimetres. Discuss an example and explain that the same procedure is used to measure objects against a metre scale.

Practise WB

- Workbook: Measuring length page 186

Apply

- Each learner completes a standing jump from a chalk mark. The group estimates the distance jumped, then measures it. They record the measurements in a table on the board. Ask: **Who jumped the furthest?**

Review

- Ask learners to explain how they use a ruler to measure the length of a pencil. Ask: **Why is a metre rule not appropriate for this measurement?**

- Learners measure the heights of library books of different sizes. They place them in same height groups.

- Ask learners to suggest a length/width/thickness that could be measured in millimetres. They estimate and measure.

Assessment for learning

- How would you measure the length of a table? What unit would you use?

- Tell me an example of something you would measure in kilometres/metres/centimetres/millimetres? What unit of measurement would you use for the length of a garden fence, and so on?

Lesson 2: **Converting units**

Learning objectives

Code	Learning objective
5Ml2	Convert larger to smaller metric units (decimals to one place), [e.g. change 2·6 kg to 2600 g.]

Strand 5: Problem solving

5Pt1 / 5Ps1	Understand everyday systems of measurement in length, [weight, capacity, temperature and time] and use these to perform simple calculations.
5Pt2	Solve single and multi-step word problems (all four operations); represent them, e.g. with diagrams or a number line.
5Pt7	Consider whether an answer is reasonable in the context of a problem.
5Ps2	Choose an appropriate strategy for a calculation and explain how they worked out the answer.
5Ps4	Deduce new information from existing information to solve problems.
5Ps5	Use ordered lists and tables to help to solve problems systematically.
5Ps9	Explain methods and justify reasoning orally and in writing; make hypotheses and test them out.
5Ps10	Solve a larger problem by breaking it down into sub-problems or represent it using diagrams.

Prerequisites for learning

Learners need to:
- measure and estimate length using metric units.

Success criteria

Learners can:
- convert km to m by multiplying by 1000
- convert m to cm by multiplying by 100
- convert cm to mm by multiplying by 10
- use place value to multiply by 1000/100/10.

Vocabulary

millimetre (mm), centimetre (cm), metre (m), kilometre (km), place value, metric unit

Resources

centimetres ruler or tape measure (per pair)

Refresh

- Choose an activity from Measure – Length, mass and capacity in the Refresh activities.

Discover [SB]

- Direct learners to the text and image in the Student's Book. Say: **Imagine you are walking on a path 2·8 kilometres long, from village A in the direction of village B. The distance between the two villages is 2900 metres.** Ask: **Will the path reach the village? How do you know?**

Teach and Learn [SB]

- Discuss the text and Example in the Student's Book. On the board, write: 1 m = 100 cm, 1 cm = 10 mm. Ask: **Which is the larger unit?** (metres) Elicit that converting metres to centimetres, we multiply the larger unit by 100. Say: **Convert 4 m/7 m to cm.**
- Repeat for centimetres to millimetres, conversion factor 10. Say: **Convert 42 cm/7 cm to mm.**
- Ask: **How many metres are there in a kilometre?** Write: 1 km = 1000 m. Ask: **How do we convert kilometres to metres?** (multiply by 1000) Convert 5/8/9·6 km to m (5000/8000/9600 m) For all examples, demonstrate the place value method: moving digits one, two or three places to the left.

Practise [WB]

- Workbook: Converting units page 188
- Refer to Activity 1 from the Additional practice activities.

Apply 👥💻

- Display **Slide 1**. Discuss the diagram, which shows the dimensions and folding instructions for a pop-up card. Ask learners to convert the dimensions to millimetres.

Review

- On the board, write: 1 km = ☐ m; 1 m = ☐ cm; 1 cm = ☐ mm. Ask volunteers to complete the conversions. Ask: **When converting kilometres to metres/metres to centimetres/centimetres to millimetres, what number do you multiply by?**
- Write: A road is 4·3 km long. Ask: **What is the length of the road in metres?** (4300 m) Write: A footpath is 36·7 m long. Ask: **What is the length of the path in centimetres?** (3670 cm) Write: A book is 51·6 cm in height. Ask: **What is the height in millimetres?** (516 mm)

Assessment for learning

- A playground is 37·3 metres long. What is this measurement in centimetres? (3730 cm)
- A pencil is 17·8 cm long. What is this measurement in millimetres? (178 mm)

Lesson 3: **Ordering and rounding length**

Measure – Length, mass and capacity

Learning objectives

Code	Learning objective
5MI3	Order measurements in mixed units.
5MI4	Round measurements to the nearest whole unit.

Strand 5: Problem solving

5Pt1 / 5Ps1	Understand everyday systems of measurement in length, [weight, capacity, temperature and time] and use these to perform simple calculations.
5Pt2	Solve single and multi-step word problems (all four operations); represent them, e.g. with diagrams or a number line.
5Pt7	Consider whether an answer is reasonable in the context of a problem.
5Ps4	Deduce new information from existing information to solve problems.
5Ps5	Use ordered lists and tables to help to solve problems systematically.
5Ps9	Explain methods and justify reasoning orally and in writing; make hypotheses and test them out.
5Ps10	Solve a larger problem by breaking it down into sub-problems or represent it using diagrams.

Prerequisites for learning

Learners need to:
- estimate and measure length in standard units (m, cm, mm)
- round numbers to the nearest 10 or 100
- use knowledge of place value to convert between units of length.

Success criteria

Learners can:
- convert km to m/m to cm/cm to mm by using known conversion factors

- round length measurements to the nearest metre or centimetre
- order measurements in mixed units by converting them to the same unit.

Vocabulary

millimetre (mm), centimetre (cm), metre (m), kilometre (km), round, place value

Resources

ruler (per learner); metre ruler or tape measure (per group); wooden plank, books, three toy cars of different mass (per group)

Refresh

- Choose an activity from Measure – Length, mass and capacity in the Refresh activities.

Discover 📖

- Discuss the text and image in the Student's Book. Remind learners of previous work on place value. Explain that it is important to understand place value when talking about rounding numbers, including measurements.

Teach and Learn 📖 📊

- Write on the board 37·2 cm, 127·8 mm, 0·5 m. Ask: **How can you round each measurement to the nearest unit?**
- Ask learners to look at the text and Example in the Student's Book. Review the rules for rounding and establish that the rounding digit is the units. Learners, in pairs, round the numbers. Invite them to share their answers and solutions. (37 cm, 128 mm, 1 m)
- Write on the board 33·5 cm, 336 mm and 0·3 m. Ask: **How would you place these measurements in ascending order – smallest to largest?** Elicit that they would convert measurements to the same unit. Learners convert and order the lengths (336 mm, 33·5 cm, 0·3 m).
- Display the **Geometry set tool** with the octagon, pentagon and T-shape. Ask volunteers to measure

one side of each shape and round the measurement to the nearest whole unit.

Practise 📓

- Workbook: Ordering and rounding length page 190

Apply 👥

- Help each group to construct a ramp of fixed height from books and a plank. Learners place three cars of different sizes at the top of the ramp. They record the distance travelled by each car and round the figures to the nearest whole unit.

Review

- Ask: **Why does 3·8 metres round up to 4 metres?** Repeat for 2·1 cm (2 cm) and 9·5 mm (10 mm). Say: **Round these numbers to the nearest whole unit: 43·2 cm** (43 cm); **111·5 mm** (112 mm).
- On the board, draw five children labelled with heights: 135·2 cm, 134·9 cm, 1351 mm, 1353 mm, 1·3 m. Learners place the heights in ascending order. (1·3 m, 134·9 cm, 1351 mm, 135·2 cm, 1353 mm)

Assessment for learning

- Do we round this number up or down?
- Place these measurements in ascending order: 2·1 cm, 23 mm, 2·5 cm, 22 mm, 2·4 cm.

Lesson 4: **Measuring lines**

Learning objectives

Code	Learning objective
5MI7	Draw and measure lines to the nearest centimetre and millimetre.

Strand 5: Problem solving

Code	Learning objective
5Pt1 / 5Ps1	Understand everyday systems of measurement in length, [weight, capacity, temperature and time] and use these to perform simple calculations.
5Pt2	Solve single and multi-step word problems (all four operations); represent them, e.g. with diagrams or a number line.
5Pt7	Consider whether an answer is reasonable in the context of a problem.
5Ps4	Deduce new information from existing information to solve problems.
5Ps5	Use ordered lists and tables to help to solve problems systematically.
5Ps10	Solve a larger problem by breaking it down into sub-problems or represent it using diagrams.

Prerequisites for learning

Learners need to:

- draw and measure length using standard units (m, cm, mm).

Success criteria

Learners can:

- measure length by lining up the object to be measured with the zero mark of a ruler and reading the scale

- find the value of each interval on a ruler and use this to give approximate values of readings between divisions to the nearest millimetre or centimetre.

Vocabulary

millimetre (mm), centimetre (cm), zero mark

Resources

ruler (per learner); string (per pair); scissors (per pair); pegs (per group); three scarves or similar rectangular shaped item of clothing (per group)

Refresh

- Choose an activity from Measure – Length, mass and capacity in the Refresh activities.

Discover [SB]

- Discuss the text and illustrations in the Student's Book. Ask: **What does each red line represent? For what situation would it be better to measure an object in millimetres, rather than centimetres? Why?** Learners discuss the degree of accuracy required for different measurements.

Teach and Learn [SB] [chart] [partners]

- Read and discuss the text and examples in the Student's Book. Display the **Geometry set tool** with the purple rhombus and the ruler. Demonstrate the measurement of a side to the nearest millimetre/ centimetre (side length is between 74 and 75 mm, but physically closer to 74 mm; this is 7 cm rounded to the nearest cm).

- Provide string. Learner 1 cuts the string; Learner 2 measures and records it to the nearest millimetre, then the nearest centimetre. Learner 1 uses measurements in mm/cm to draw lines of same length. Ask: **Are they the same length as the string? Why?**

Practise [WB]

- Workbook: Measuring lines page 192
- Refer to Activity 2 from the Additional practice activities.

Apply [group]

- Ask: **What is the shortest length of string that will make a washing line to carry three scarves?** Learners measure the length of one scarf and use this figure to calculate the washing line length required. They measure and cut this length of string, then test it by tying the string between supports and adding pegs and scarves.

Review [chart]

- Display the **Geometry set tool** with the red rectangle and the ruler. Invite volunteers to measure the lengths of the sides to the nearest millimetre. They use the measurements to calculate the perimeter. (33 cm: 10 cm × 2 + 6·5 cm × 2)

- Draw four lines in ascending order of length where successive lines have a difference of 1·6 cm.

Assessment for learning

- Draw a line that is between 5 and 10 cm in length. Swap lines with a partner. Measure the line to the nearest millimetre/nearest centimetre. How did you make sure you measured correctly?

Measure – Length, mass and capacity

197

Additional practice activities

Activity 1 👥 **Challenge 2**

Learning objectives

• Convert larger to smaller metric units.
• Solve multi-step word problems that involve length.

Resources

paper (per pair); rulers (per pair)

What to do

• On the board, write: Rectangle A – 12·2 cm ×
197 mm; Rectangle B – 18·7 cm × 111 mm;
Rectangle C – 15·4 cm × 143 mm. Learners work
out which of these rectangle combinations joined
together at their shorter sides will produce a shape
with the largest perimeter: (1) two rectangle As;
(2) two rectangle Bs; (3) three rectangle Cs.

• Ask learners to share their solutions.

Variation

Challenge 1 Learners compare the perimeters of single
rectangles, i.e. one rectangle A compared to one
rectangle B compared to one rectangle C.

Activity 2 👥 **Challenge 2**

Learning objectives

• Round measurements to the nearest whole units.
• Investigate a simple general statement by finding examples that do or do not satisfy it.

Resources

paper (per pair); rulers (per pair); six pieces of string
(per pair); sticky labels (per pair)

What to do

• On the board, write: The order of length rounded
to the nearest centimetre is identical to the order of
length rounded to the nearest millimetre. Learners
discuss this statement and predict whether it is true
or not, explaining their reasons.

• String pieces are held taut and an oblique cut
is made across to create six pieces of different
lengths. They are labelled A to F and arranged
randomly in a pile.

• Learners take turns to pick a string and measure
its length to the nearest centimetre and nearest
millimetre. Ask: **How do the two orders compare?**
What does this tell you about the statement?

Variation

Challenge 1 Reduce the number of strings to three.

Unit 19: Mass

Learning objectives

Code	Learning objective
5Ml1	Read, choose, use and record standard units to estimate and measure [length,] mass [and capacity] to a suitable degree of accuracy.
5Ml2	Convert larger to smaller metric units (decimals to one place), e.g. change 2·6 kg to 2600 g.
5Ml3	Order measurements in mixed units.
5Ml4	Round measurements to the nearest whole unit.
5Ml5	Interpret a reading that lies between two unnumbered divisions on a scale.
5Ml6	Compare readings on different scales.
Strand 5: Problem solving	
5Pt1	Understand everyday systems of measurement in [length,] weight, [capacity, temperature and time] and use these to perform simple calculations.
5Pt2	Solve single and multi-step word problems (all four operations); represent them, e.g. with diagrams or a number line.
5Pt7	Consider whether an answer is reasonable in the context of a problem.
5Ps1	Understand everyday systems of measurement in [length,] weight, [capacity, temperature and time] and use these to perform simple calculations.
5Ps2	Choose an appropriate strategy for a calculation and explain how they worked out the answer.
5Ps4	Deduce new information from existing information to solve problems.
5Ps5	Use ordered lists and tables to help to solve problems systematically.
5Ps9	Explain methods and justify reasoning orally and in writing; make hypotheses and test them out.
5Ps10	Solve a larger problem by breaking it down into sub-problems or represent it using diagrams.

Unit overview

This unit reviews standard units of mass and introduces conversion of kilograms to grams by multiplication. Learners order measurements in mixed units by converting to the same unit. Rules for rounding numbers to the nearest whole unit are recalled and applied to metric units of mass. Learners are encouraged to remember gram equivalents of one half, one quarter and one tenth of a kilogram. Practical examples of interpreting readings that lie between pairs of unnumbered divisions on a scale are provided, as are readings taken on different scales.

Common difficulties and remediation

Many learners find the distinction between weight and mass confusing. Help them to understand that weight is a force, and that use of the term in sentences such as 'The weight of the ball is 250 g' is scientifically incorrect.

Look out for learners who have difficulty reading weighing scales accurately. This is a good opportunity to review 'place value' with the kilogram as the 'unit' and the grams as 'tenths, hundredths and thousandths'. Learners also require experience of comparing readings on different scales. This helps to avoid errors when comparing and calculating with sets of measurements expressed in mixed units. Using different scales, such as circular scales, can support estimation skills and problem solving, set in the context of mass.

Some learners may convert kilograms to grams incorrectly, for example, treating 7·3 kg as 7003 g not 7300 g. Explaining that 7·3 kg can be written as 7·300 kg can help remedy this.

Promoting and supporting language

Whenever a new key word is introduced, ask learners to write a definition in their own words, for example: 'A scale division is the smallest quantity marked on a scale; on a weighing scale it might be 10 g, 50 g or 100 g.' Sketching a diagram alongside will help embed the definition. In addition, be aware of terms that have more than one definition, particularly meanings outside mathematics, for example: unit, mass and scale.

Encourage frequent use of mathematical language to help embed vocabulary. For example, model the use of the question 'What is the mass of the object?' when learners ask 'how heavy?' or 'how light?' something is, as they make the transition to mathematically precise language.

Throughout, encourage learners to seek clarification and confirmation of the mathematical language. Prompting learners to call out and complete definitions, for example, saying 'multiply by 1000' when 'converting kilograms to grams' is mentioned. Praise learners who ask for terms to be defined.

In each lesson, ask volunteers to summarise what they have learned, to give them experience of explaining concepts.

Lesson 1: **Measuring mass**

Measure – Length, mass and capacity

Learning objectives

Code	Learning objective
5MI1	Read, choose, use and record standard units to estimate and measure [length,] mass [and capacity] to a suitable degree of accuracy.

Strand 5: Problem solving	
5Pt1 / 5Ps1	Understand everyday systems of measurement in [length,] weight, [capacity, temperature and time] and use these to perform simple calculations.
5Pt7	Consider whether an answer is reasonable in the context of a problem.

Prerequisites for learning

Learners need to:
- know that mass is measured in kilograms and grams
- know that 1 kg is equivalent to 1000 g
- read and interpret scales.

Success criteria

Learners can:
- make estimates of mass, using an appropriate unit (kg and g)

- make judgments about the type of equipment to be used and determine the mass of different substances, reading a scale to the nearest division.

Vocabulary

mass, kilogram (kg), gram (g), scale, division

Resources

weighing scales calibrated in 100 g divisions (per group); packets of pasta, pulses or grains (per class); packages labelled X, Y and Z weighing 1·4 kg, 0·6 kg and 1·1 kg respectively (per group)

Refresh

- Choose an activity from Measure – Length, mass and capacity in the Refresh activities.

Discover [SB]

- Discuss the text and images in the Student's Book. Say: **Imagine you need to send a parcel overseas.** Ask: **How would you ensure the correct postage is paid?** Discuss the importance of measuring mass accurately. Ask: **What is the best unit for measuring the mass of small book/ laptop computer? Why is it important to use the correct unit?**

Teach and Learn [SB] [bar chart icon]

- Discuss the text and Example in the Student's Book. Ask: **What other objects would you measure in g/ kg?** Display the **Weighing tool** set at 500 g in 10 g divisions. Place two 100 g and one 10 g weights on the pan. Ask: **What is the total mass?**
- Repeat for a scale set at 200 kg in 100 g divisions. Put on 1 kg, one 500 g and two 100 g weights. Pour 700 g of pasta onto the weighing scales. Ask: **Follow the pointer. What is the mass?** Add more pasta and explain how to read the scale to the nearest 100 g.

Practise [WB]

- Workbook: Measuring mass page 194
- Refer to Activity 1 from the Additional practice activities.

Apply [group/computer icons]

- Display **Slide 1**. Give out weighing scales and packages X, Y and Z. Learners should weigh the objects and find the postage price. Ask: **Is it cheaper to post items separately, or in one combined package?**

Review [bar chart icon]

- Display the **Weighing tool** or actual scales. Point to any scale interval below 1 kg. Ask learners to imagine they are measuring pasta and the pointer stops here. Say: **Explain how to find the mass of pasta in grams.** Repeat for a reading over 1 kg. Ask learners to explain how to find the mass in kilograms.
- Show the class the scales and pasta. Ask volunteers to add pasta to show these masses: 350 g, 0·7 kg, 1·1 kg.

Assessment for learning

- Show me how you would measure the mass of three large books/a small bag of pasta? What unit(s) would you use?

Lesson 2: **Converting units**

Learning objectives

Code	Learning objective
5Ml1	Read, choose, use and record standard units to estimate [and measure length,] mass [and capacity] to a suitable degree of accuracy.
5Ml2	Convert larger to smaller metric units (decimals to one place), e.g. change 2·6 kg to 2600 g.

Strand 5: Problem solving

5Pt1 / 5Ps1	Understand everyday systems of measurement in [length,] weight, [capacity, temperature and time] and use these to perform simple calculations.
5Pt2	Solve single and multi-step word problems (all four operations); represent them, e.g. with diagrams or a number line.
5Pt7	Consider whether an answer is reasonable in the context of a problem.
5Ps2	Choose an appropriate strategy for a calculation and explain how they worked out the answer.
5Ps4	Deduce new information from existing information to solve problems.
5Ps5	Use ordered lists and tables to help to solve problems systematically.
5Ps10	Solve a larger problem by breaking it down into sub-problems or represent it using diagrams.

Prerequisites for learning

Learners need to:

- use metric units (g and kg) to measure and estimate mass
- know that 1 kg is equivalent to 1000 g.

Success criteria

Learners can:

- convert kilograms to grams by multiplying by 1000

- use place value to multiply by 1000 (the digits move three places to the left).

Vocabulary

mass, kilogram (kg), gram (g), place value

Resources

weighing scales calibrated in 100 g divisions (per group); two books of different mass (per group)

Refresh

- Choose an activity from Measure – Length, mass and capacity in the Refresh activities.

Discover 🆂🅱

- Discuss the text and images in the Student's Book. Say: **You are preparing a meal in a kitchen and need 0·4 kg of pasta. You weigh the pasta and the scales read 400 g.** Ask: **Do you have enough? How do you know?**

Teach and Learn 🆂🅱

- Discuss the text and Example in the Student's Book. Write: 1 kg = 1000 g. Ask: **Which is the larger unit?** (kilograms) Say: **To convert kilograms to grams, multiply the number of kilograms by 1000.** Demonstrate the place-value method: moving digits three places to the left. Say: **Convert 0·6 kg/4·1 kg to grams. (600 g/4100 g)** Ask: **What is the total mass in grams of 1·4 kg and 300 g of flour?** Say: **Convert to the same unit, grams, to find the sum.** (1400 g + 300 g = 1700 g) Learners ask their partners similar mixed unit addition questions.

Practise 🆆🅱

- Workbook: Converting units page 196

Apply 👥

- Weigh two books of different mass, A and B, before the lesson. Provide groups with weighing scales and books A and B. On the board, write the mass of book A in kilograms. Say: **If you were told to weigh both books together, not separately, and record the mass in grams, how would you find the mass of book B?** Discuss solutions. (B = (A g + B g) – (A kg × 1000))

Review

- On the board, write: 0·5 kg = ☐ g, 13·4 kg = ☐ g. Ask volunteers to complete the sentences. Ask: **When converting kilograms to grams, what number do you multiply by?** (1000)
- Write: A box has a mass of 0·9 kg. Ask: **What is the mass of the box in grams?** (900 g) Ask: **Which has the greater mass, 7·6 kg or 7500 g?** (7·6 kg) **How do you know?**

Assessment for learning

- I have a weighing scale that only measures in kilograms. I want to measure a box of cereal in grams. What should I do?

Lesson 3: **Ordering and rounding mass**

Measure – Length, mass and capacity

Learning objectives

Code	Learning objective
5MI3	Order measurements in mixed units.
5MI4	Round measurements to the nearest whole unit.

Strand 5: Problem solving

5Pt1 / 5Ps1	Understand everyday systems of measurement in [length,] weight, [capacity, temperature and time] and use these to perform simple calculations.
5Pt2	Solve single and multi-step word problems (all four operations); represent them, e.g. with diagrams or a number line.
5Pt7	Consider whether an answer is reasonable in the context of a problem.
5Ps4	Deduce new information from existing information to solve problems.
5Ps5	Use ordered lists and tables to help to solve problems systematically.
5Ps10	Solve a larger problem by breaking it down into sub-problems or represent it using diagrams.

Prerequisites for learning

Learners need to:

- use standard units (g, kg) to estimate and measure mass
- round any number to the nearest 10 or 100
- use knowledge of place value to convert between units of mass.

Success criteria

Learners can:

- use a known conversion factor to convert kg to g
- round measurements to the nearest kg and 100 g
- order measurements in mixed units by converting them to the same unit.

Vocabulary

mass, kilogram (kg), gram (g), round, decimal place

Resources

weighing scales calibrated in 100 g divisions (per group); parcels labelled A, B, C (A and B between 100 g and 450 g, C between 500 g and 900 g) (per group)

Refresh

- Choose an activity from Measure – Length, mass and capacity in the Refresh activities.

Discover [SB]

- Discuss the text and image in the Student's Book. Ask: **In what other situations might you need to order mass?** Review prior knowledge of kilogram to gram conversion. Ask: **Why is it important to convert units of mass when ordering mixed units?**

Teach and Learn [SB] [bar]

- Discuss the text and the Example in the Student's Book. On the board, write: 2436 g, 2·4 kg, 2634 g and 2·5 kg. Ask: **How can you place these measurements in ascending order, lightest to heaviest?** Elicit that learners must convert measurements to the same unit, then order them. (2·4 kg, 2436 g, 2·5 kg, 2634 g).
- Display the **Weighing tool** with 762 g added. Ask: **What is the mass to the nearest 100 g?** (800 g) Review the rules of rounding and demonstrate 'nearest 100 g' on a number line 700 to 800. Repeat for 715 g, 749 g and 750 g. Set the scale to 5 kg in 0·5 kg divisions. Ask learners to round these masses to the nearest kilogram: 3·9 kg, 3·3 kg, 3·5 kg.

Practise [WB]

- Workbook: Ordering and rounding mass page 198

Apply [group]

- Weigh three parcels before the lesson, round to the nearest 100 g and write the rounded measurements on the board. Provide groups with parcels and weighing scales. Say: **Match each rounded measurement to its parcel, then order the parcels, lightest to heaviest.**

Review [screen]

- Ask: **Why does 5055 g round up to 5100 g when rounding to the nearest 100 g?** Repeat for 5049 g (5000 g) and 5050 g (5100 g). Say: **Round these measurements to the nearest kilogram: 7·6 kg, 19·2 kg, 85·5 kg.** (8 kg/19 kg/86 kg)
- Display **Slide 1**. Learners place the boxes in ascending order of mass. (6026 g, 6·1 kg, 6·2 kg, 6260 g)

Assessment for learning

- Sort the following measurements into those that round down and those that round up to the nearest 100 g: 471 g, 450 g, 437 g.
- Place the following measurements in ascending order: 5·1 kg, 5055 g, 4·9 kg, 5505 g, 5·5 kg. Which measurement is in the middle? (5·1 kg)

Lesson 4: **Reading weighing scales**

Learning objectives

Code	Learning objective
5MI5	Interpret a reading that lies between two unnumbered divisions on a scale.
5MI6	Compare readings on different scales.
Strand 5: Problem solving	
5Pt1 / 5Ps1	Understand everyday systems of measurement in [length,] weight, [capacity, temperature and time] and use these to perform simple calculations.
5Pt2	Solve single and multi-step word problems (all four operations); represent them, e.g. with diagrams or a number line.
5Pt7	Consider whether an answer is reasonable in the context of a problem.
5Ps4	Deduce new information from existing information to solve problems.
5Ps9	Explain methods and justify reasoning orally and in writing; make hypotheses and test them out.
5Ps10	Solve a larger problem by breaking it down into sub-problems or represent it using diagrams.

Prerequisites for learning

Learners need to:

- estimate and measure mass using standard units (g, kg)
- round numbers to the nearest 10 or 100
- read and interpret scales.

Success criteria

Learners can:

- find the value of each interval on a set of weighing scales and use this to give approximate values of readings between divisions

- record estimates and measurements involving halves, quarters or tenths
- compare readings on different scales.

Vocabulary

mass, kilogram (kg), gram (g), scale, division

Resources

elastic band (per group), small plastic bag (per group), several objects of different mass between 100 g and 1 kg (per pair)

Refresh

- Choose an activity from Measure – Length, mass and capacity in the Refresh activities.

Discover 🆂🅱

- Discuss the text and images in the Student's Book. Say: **Think of a task you completed that involved weighing.** Ask: **What would have happened if you had not read the measurements accurately?** Ask learners to describe the different types of scales they have used. Ask about their graduations.

Teach and Learn 🆂🅱 📊

- Direct learners to the text and images in the Student's Book. Display the **Weighing tool** set at 5 kg in 0·25 kg divisions with 2·25 kg mass added. Ask: **What is the mass in grams? How do you know?** Divisions are 0·25 kg. 2·25 kg converts to 2250 grams. Repeat for 3·5 kg (3500 g) and 4·75 kg (4750 g).

- Set the **Weighing tool** at 2 kg in 10 g divisions with 1110 g added. Ask: **What is the mass? How do you know?** Discuss 10 g divisions.

- Set it now at 2 kg in 20 g divisions with 1480 g added. Learners read the scale and respond. Show 10 g. Say: **The pointer is between divisions. How do you know what the reading is?**

Practise 🆆🅱

- Workbook: Reading weighing scales page 200
- Refer to Activity 2 from the Additional practice activities.

Apply 👥💻

- Display **Slide 1** and ask pairs of learners to set up the experiment as shown on the slide. They rest a pencil across the gap between two tables, attach a small plastic bag to the elastic band, then loop the band over the pencil. All pairs weigh, record and add identical light objects to the bag. They weigh and add more objects until the band snaps. Pairs calculate the total mass of objects minus the mass of the first object and compare results.

Review 📊

- Display the **Weighing tool**. On the board, write: 1750 g. Ask learners to use the tool to show this reading on different scales. Show two readings on different scales: 2·5 kg (in 0·25 kg divisions) and 1160 g (in 20 g divisions). Say: **Calculate the difference between masses and share your solutions.**

Assessment for learning

- Show me the same measurement on two different scales. How do you know they show the same reading?
- Show 1·25 kg and 2320 g on the **Weighing tool**. What is the difference between the masses? Explain your solution.

Measure – Length, mass and capacity

Measure – Length, mass and capacity

Additional practice activities

Activity 1

Learning objective

• Estimate mass in kilograms and grams; measure accurately using kilograms and grams.

Resources

four classroom objects, labelled A to D, wrapped as parcels (per group); kitchen scales (per group)

What to do

• Weigh the parcels before the activity and write the measurements on the board, two in grams, two in kilograms. Provide each group with parcels and ask them to estimate each mass in kilograms, matching the measurements on the board with the parcel letter.

• They weigh each parcel and check against their estimate. Ask: **What is the combined weight of all four parcels?**

Variation

Challenge 3 Learners arrange the parcels in order of estimated mass. They measure and check the order.

Activity 2

Learning objectives

• Round measurements made on different scales to the nearest whole unit.
• Solve simple problems involving the rounding and unit conversion of measurements.

Resources

two weighing scales with different scale intervals, for example 10 g, 20 g, 50 g, 100 g, 0·25 kg (per pair); selection of objects between 100 g and 2 kg (per pair)

What to do

• Without their partner seeing, learners place one item on each of the two different scales. They record the masses, round to the nearest whole unit and find the sum. They remove one item and write the sum missing the mass of the removed item, for example, 1·25 kg + = 1720 g.

• Their partner must calculate the mass of the missing item and point to where the reading would be on the scale. They weigh the item to see if they are correct. Partners swap roles.

Variation

Challenge 1 They record the mass of both items in grams.

Unit 20: Capacity

Learning objectives

Code	Learning objective
5MI1	Read, choose, use and record standard units to estimate and measure [length, mass and] capacity to a suitable degree of accuracy.
5MI2	Convert larger to smaller metric units (decimals to one place) [, e.g. change 2 6kg to 2600 g.]
5MI3	Order measurements in mixed units.
5MI4	Round measurements to the nearest whole unit.
5MI5	Interpret a reading that lies between two unnumbered divisions on a scale.
5MI6	Compare readings on different scales.
Strand 5: Problem solving	
5Pt1/5Ps1	Understand everyday systems of measurement in [length, weight,] capacity, [temperature and time] and use these to perform simple calculations.
5Pt2	Solve single and multi-step word problems (all four operations); represent them, e.g. with diagrams or a number line.
5Pt7	Consider whether an answer is reasonable in the context of a problem.
5Ps1	Understand everyday systems of measurement in [length, weight,] capacity, [temperature and time] and use these to perform simple calculations.
5Ps2	Choose an appropriate strategy for a calculation and explain how they worked out the answer.
5Ps4	Deduce new information from existing information to solve problems.
5Ps5	Use ordered lists and tables to help to solve problems systematically.
5Ps9	Explain methods and justify reasoning orally and in writing; make hypotheses and test them out.
5Ps10	Solve a larger problem by breaking it down into sub-problems or represent it using diagrams.

Unit overview

This unit reviews standard units of capacity and introduces conversion of litres to millilitres by multiplication. Learners order measurements in mixed units by converting to the same unit. Rules for rounding numbers to the nearest whole unit are recalled and applied to metric units of capacity. Learners are encouraged to remember millilitre equivalents of one half, one quarter and one tenth of a litre. Examples of interpreting readings that lie between two unnumbered divisions on a scale are provided, as are readings taken on different scales.

Common difficulties and remediation

Many learners find the distinction between capacity and volume confusing. Help them to understand that volume is how much space an object takes up and capacity is how much space an object has inside – or, how much it can hold. For example, a glass may have a capacity to hold 350 ml, but the liquid volume of milk inside may only be 150 ml.

Some learners believe the volume of liquid has changed when a set amount is poured from one container to another of different size. They believe that the container with the higher level holds more. Teachers need to provide learners with extended opportunities to transfer liquids between containers.

Look out for learners who make errors when reading capacity scales. Some learners pick the jar or cylinder up and hold it at an angle when reading the scale; others read the scale from a height, therefore introducing parallax errors. Teachers need to provide learners with opportunities to use and read a range of measuring scales, some at different orientations. Learners need to be instructed to keep a container on a flat surface and look at the base of the liquid meniscus at eye level.

Promoting and supporting language

Whenever a new key word is introduced, ask learners to write a definition in their own words, for example: 'Capacity is the amount a container can hold when filled.' In addition, be aware of terms that have more than one definition, particularly meanings outside mathematics, for example: capacity, level and scale.

Encourage frequent usage of mathematical language to help embed vocabulary. For example, modelling use of the question 'What is the capacity of the container?' when learners ask 'how much liquid will fill it?' as they make the transition to mathematically precise language.

Throughout the unit it is important to encourage learners to seek clarification and confirmation of the mathematical language. This may involve prompting learners to call out and complete definitions, for example, saying 'multiply by 1000' every time converting litres to millilitres is mentioned, or praising learners when they ask for terms to be defined.

A few times each lesson, ask a volunteer to summarise what has been learned thus far so they get experience of explaining concepts in their own words.

Unit **20** Capacity

Lesson 1: **Measuring capacity**

Learning objectives

Code	Learning objective
5Ml1	Read, choose, use and record standard units to estimate and measure [length, mass and] capacity to a suitable degree of accuracy.

Strand 5: Problem solving

5Pt1 / 5Ps1	Understand everyday systems of measurement in [length, weight,] capacity, [temperature and time] and use these to perform simple calculations.
5Pt7	Consider whether an answer is reasonable in the context of a problem.

Prerequisites for learning

Learners need to:

- know that capacity is measured in litres (*l*) and millilitres (ml)
- know that one litre is equivalent to 1000 ml
- read and interpret scales.

Success criteria

Learners can:

- make estimates of capacity using an appropriate unit (*l* and ml)

- make judgments about the type of equipment to be used and determine the capacity of different containers, reading a scale to the nearest division.

Vocabulary

capacity, litre (*l*), millilitre (ml), scale, division

Resources

calibrated measuring jugs: 2 × 500 ml, 2 × 250 ml, 1 × 100 ml (per group); containers of various sizes, funnel, water (per group); paper or tags for labels (per class); blue coloured pencil (per learner)

Refresh

- Choose an activity from Measure – Length, mass and capacity in the Refresh activities.

Discover ⬛

- Discuss the text and images in the Student's Book. Ask: **Why is it important to measure capacity accurately?** Ask learners to suggest other examples where it is important to measure capacity accurately.

Teach and Learn ⬛ ◨

- Ask learners to look at the text and images in the Student's Book. Ask: **What other containers would you measure in ml/*l*?** Review the difference between 'capacity' and 'volume'. Display the Capacity tool with a 500 ml measuring container. Say: **I am pouring out the water that fills a container**. Pour 300 ml. Ask: **What is the capacity of the container?** Repeat for 450 ml.
- Change to a 10 000 ml container. Say: **I am pouring out …** Pour six litres. Ask: **What is the capacity of the container?** Repeat for 8·5 litres.
- Provide groups with containers, measuring jugs funnels and water. Say: **Find the capacity of each container. Remember to read the scale at eye level.**

Practise ◨

- Workbook: Measuring capacity page 202

Apply ◨◨◨

- Provide groups with containers, litre measuring jugs funnels, water and price labels. On the board, write: x per 100 ml. (x is your currency unit). Say: **You are selling bottles of juice. Fill each container with water (juice). Price it per 100 ml. Use the figure on the board. Add a price label**. Ask: **How did you calculate the prices?**

Review ◨

- Display the **Capacity tool** with the 500 ml container. Ask: What is the largest capacity you can measure with this jug? (up to 500 ml) Repeat for other jugs. Fill a 1000 ml jug to 400 ml. Ask: **How much water is there? How much more water will it take to fill the container?** (600 ml) Accept answers and ask a volunteer to demonstrate.

Assessment for learning

- Demonstrate how you would measure the capacity of three different sized containers? What unit(s) would you use?

Lesson 2: **Converting units**

Learning objectives

Code	Learning objective
5Ml2	Convert larger to smaller metric units (decimals to one place) [, e.g. change 26 kg to 2600 g.]

Strand 5: Problem solving

5Pt1 / 5Ps1	Understand everyday systems of measurement in [length, weight,] capacity, [temperature and time] and use these to perform simple calculations.
5Pt2	Solve single and multi-step word problems (all four operations); represent them, e.g. with diagrams or a number line.
5Pt7	Consider whether an answer is reasonable in the context of a problem.
5Ps2	Choose an appropriate strategy for a calculation and explain how they worked out the answer.
5Ps4	Deduce new information from existing information to solve problems.
5Ps5	Use ordered lists and tables to help to solve problems systematically.
5Ps10	Solve a larger problem by breaking it down into sub-problems or represent it using diagrams.

Prerequisites for learning

Learners need to:

- use standard units (ml and *l*) to measure and estimate capacity
- know that one litre is equivalent to 1000 millilitres.

Success criteria

Learners can:

- convert litres to millilitres by multiplying by 1000

- use place value to multiply by 1000 (the digits move three places to the left).

Vocabulary

capacity, litre (*l*), millilitre (ml), place value

Resources

litre measuring jug (per group); funnel, water (per group); containers A and B (capacity difference about one litre); blue coloured pencil (per learner)

Refresh

- Choose an activity from Measure – Length, mass and capacity in the Refresh activities.

Discover 🆂🅱 🖥

- Display **Slide 1**. Discuss the text and images in the Student's Book. Accept comments. Say: **You are preparing a cool drink and need 1·5 litres of orange juice. You pour the juice into a measuring jug and the scale reads 1400 ml.** Ask: **Do you have enough? How do you know?**

Teach and Learn 🆂🅱 📊

- Write: 1 litre (*l*) = 1000 millilitres (ml). Ask: **Which is the larger unit?** (litres) To convert litres to millilitres, multiply the number of litres by 1000. Demonstrate the place value method: moving digits three places to the left.
- Display the **Capacity tool** with 1000 ml container set on litres. Learners should convert 0·7 *l*/1·9 *l*/4·3 *l* to millilitres. Switch the tool to 'show as ml'. Use it to show conversions of 0·1 *l*/0·25 *l*/0·5 *l*/0·75 *l* to millilitres. Ask: **What is the total capacity in millilitres for two containers: 0·75 *l* and 350 ml?** We convert to the same unit, millilitres, to find the sum (750 ml + 350 ml = 1100 ml). Learners ask their partner similar mixed unit addition questions.

Practise 🆆🅱

- Workbook: Converting units page 204
- Refer to Activity 1 from the Additional practice activities.

Apply 👥

- Before the lesson, measure the capacity of two containers, A and B. Provide litre measuring jugs, containers A and B, funnels and water. Write the combined capacity of the containers in litres. Say: **Measure the capacity of container A in millilitres and use this to find the capacity of B without measuring it.** Discuss solutions. ((A + B) × 1000 ml) – A ml)

Review

- Write: 0·25 *l* = ☐ml, 7·6 *l* = ☐ml. Ask volunteers to complete the sentences. Ask: **When converting litres to millilitres, what do you multiply by?** (1000)
- Write: A bottle has a capacity of 0·8 *l*. Ask: **What is the capacity of the bottle in millilitres?** (800 ml) **Which is the greater capacity, 5·25 *l* or 5205 ml?** (5·25 *l*) **How do you know?**

Assessment for learning

- I have a measuring jug with a scale marked in litres. I want to measure the capacity of a cup in millilitres. What should I do?

Unit **20** Capacity

Measure – Length, mass and capacity

Lesson 3: **Ordering and rounding capacity**

Learning objectives

Code	Learning objective
5MI3	Order measurements in mixed units.
5MI4	Round measurements to the nearest whole unit.

Strand 5: Problem solving

5Pt1 / 5Ps1	Understand everyday systems of measurement in [length, weight,] capacity, [temperature and time] and use these to perform simple calculations.
5Pt2	Solve single and multi-step word problems (all four operations); represent them, e.g. with diagrams or a number line.
5Pt7	Consider whether an answer is reasonable in the context of a problem.
5Ps4	Deduce new information from existing information to solve problems.
5Ps5	Use ordered lists and tables to help to solve problems systematically.
5Ps10	Solve a larger problem by breaking it down into sub-problems or represent it using diagrams.

Prerequisites for learning

Learners need to:
- use standard units (*l*, ml) to estimate and measure capacity
- round any number to the nearest 10 or 100
- use knowledge of place value to convert between units of capacity.

Success criteria

Learners can:
- use a known conversion factor to convert litres to ml

- round measurements to the nearest litre and 100 ml.

Vocabulary

capacity, litre (*l*), millilitre (ml), round, decimal place

Resources

plastic bottles of various capacities, labelled A to E (per group); litre measuring jug (per group); funnel, water (per group)

Refresh

- Choose an activity from Measure – Length, mass and capacity in the Refresh activities.

Discover 〔SB〕 💻

- Display **Slide 1**. Discuss the text and image in the Student's Book. Recap litre to millilitre conversion. Ask: **When ordering mixed units, why is it important to convert to the same unit?**

Teach and Learn 〔SB〕 📊

- Discuss the text and Example in the Student's Book. Write: 5·1*l*, 5009 ml, 5*l*, 5011 ml. Ask: **How will you place these measurements in ascending order, smallest to largest?** Learners convert the measurements to the same unit, then order them (5*l*, 5009 ml, 5011 ml, 5.1*l*).
- Display the **Capacity tool**, 1000 ml with 749 ml filled. Ask: **What is the reading to the nearest 100 ml?** (700 ml) Review the rules of rounding and demonstrate 'nearest 100 ml' on a number line 700 to 800 on the board. Repeat for 795 ml, 750 ml and 708 ml.
- Use a 2000 ml jug, scale to litres. Say: **Round 1·1*l*, 1·5*l* and 1·9*l* to the nearest litre.**

Practise 〔WB〕

- Workbook: Ordering and rounding capacity page 206.

Apply 👥

- Provide containers A to E, litre measuring jugs, funnels and water. Say: **Imagine you are setting up a drinks stall at a school fair. You need to arrange bottles in order of capacity and label them with capacity rounded to the nearest 100 ml. How would you do it?**

Review 📊

- Display the **Capacity tool**. Ask learners to imagine that all the measuring jugs displayed, from 100 ml to 1000 ml, are filled with water. Ask: **How many ways can you fill a 1000 ml/750 ml/500 ml container?** (500 ml: 1 × 500 ml; 2 × 150 ml plus 2 × 100 ml and so on).
- Ask: **Round these measurements to the nearest 100 ml: 321 ml, 2450 ml, 4049 ml.** (300 ml / 2500 ml / 4000 ml)

Assessment for learning

- Sort the following measurements into those that round down and those that round up to the nearest 100 ml: 2050 ml, 555 ml, 3303 ml. (down: 3303; up: 2050, 555)
- Explain how you would order a set of mixed units, litres and millilitres.

Lesson 4: **Reading capacity scales**

Learning objectives

Code	Learning objective
5MI5	Interpret a reading that lies between two unnumbered divisions on a scale.
5MI6	Compare readings on different scales.

Strand 5: Problem solving

5Pt1 / 5Ps1	Understand everyday systems of measurement in [length, weight,] capacity, [temperature and time] and use these to perform simple calculations.
5Pt2	Solve single and multi-step word problems (all four operations); represent them, e.g. with diagrams or a number line.
5Pt7	Consider whether an answer is reasonable in the context of a problem.
5Ps4	Deduce new information from existing information to solve problems.
5Ps9	Explain methods and justify reasoning orally and in writing; make hypotheses and test them out.
5Ps10	Solve a larger problem by breaking it down into sub-problems or represent it using diagrams.

Prerequisites for learning

Learners need to:

- estimate and measure capacity using standard units (ml, *l*)
- round numbers to the nearest 10 or 100
- read and interpret scales.

Success criteria

Learners can:

- find the value of each interval on a measuring jar and use this to give approximate values of readings between divisions

- record estimates and measurements involving halves, quarters or tenths
- compare readings on different scales.

Vocabulary

capacity, litre (*l*), millilitre (ml), scale, division

Resources

litre measuring jug (per pair); funnel, water, card strips and tubes (per pair)

Refresh

- Choose an activity from Measure – Length, mass and capacity in the Refresh activities.

Discover [SB]

- Discuss the text and image in the Student's Book. Say: **Think of other examples where it is important to read a capacity scale accurately.** Ask: **What would happen if there was too much or too little of the thing being measured?** Discuss the consequences of inaccurate readings. Learners describe the different types of capacity scales they have used, and the graduations on the scales.

Teach and Learn [SB] [chart]

- Ask learners to read the text and look at the images of the containers in the Student's Book. Display the **Capacity tool**, 1000 ml jug set for litres with 0·5 *l* of water. Ask: **What is the capacity? How do you know?** Each division is 0·1 *l* (one tenth) and the level reading is 0·5 *l*. Switch to millilitres and show this is 500 ml.
- Change the level to 0·25 *l*. Ask: **What is the capacity?** As the level is equidistant between divisions 0·2 and 0·3 this is 0·25 *l*. Switch to millilitres and show this is 250 ml. Repeat for 0·75 *l*.

Practise [WB]

- Workbook: Reading capacity scales page 208

- Refer to Activity 2 from the Additional practice activities.

Apply [pair icon] [screen icon]

- Display **Slide 1** and ask pairs of learners to set up the experiment, making the card bridge as strong as possible. They add water in 25 ml portions (using eye-level reading to check when the required amount has been added). The actual reading is recorded alongside the structural state of the bridge. Learners record the amount of water at the point when the bridge collapses and compare their results with others.

Review [chart]

- Display the **Capacity tool**. On the board, write: 1750 ml. Ask learners to use the tool to show the reading on different scales. Show two readings on different scales: 275 ml (in 25 ml divisions) and 3·4 *l* (in 100 ml divisions). Ask learners to calculate the difference between capacities and share solutions.

Assessment for learning

- Show me the same measurement on two different scales. How do you know both scales show the same reading?
- Show 0·75 *l* and 7250 ml on the **Capacity tool**. What is the difference between the measurements? Explain your solution.

Measure – Length, mass and capacity

Additional practice activities

Activity 1 Challenge 2

Learning objective

• Estimate capacity in litres and millilitres; measure accurately using litres and millilitres.

Resources

calibrated measuring jugs: 2 × 1000 ml, 2 × 500 ml, 2 × 250 ml, 1 × 100 ml (per group); funnel, water, four containers, labelled A to D (per group)

What to do

• Measure the capacity of the containers before the activity and write the measurements on the board, two in millilitres, two in litres. Provide containers and ask groups to estimate each capacity in litres, matching the measurements on the board with the letter on the container.

• Learners measure the capacity of each container and check against their estimate. Ask: **What is the combined capacity of all four containers?**

Variation

Challenge 3 Learners order containers by estimated capacity. They then measure and check the order.

Activity 2 Challenge 2

Learning objectives

• Solve simple problems involving the rounding and ordering of measurements.

• Choose an appropriate strategy for a calculation involving measurements.

Resources

calibrated measuring jugs: 2 × 1000 ml, 2 × 500 ml, 2 × 250 ml, 1 × 100 ml (per group); funnel, water (per group); three containers, labelled X, Y and Z (approx. 1, 2 and 4 litres) (per group); cups A – 50 ml, B – 225 ml, C – 150 ml (per group)

What to do

• Say: **You are preparing cups of water for a party. Cups come in three sizes.** Write: Cup A – 50·2 ml; Cup B – 0·225 l; Cup C – 149·5 ml. Provide containers X, Y and Z each filled with water.

• Learners round sizes to the nearest whole unit and arrange them in order of size.

• They work out the maximum number of cups of each size that can be poured from each container. They record the results and share their solutions.

Variation

Challenge 1 Use only cup sizes A and B, and smaller capacity containers, 0·5 l and 1 l.

Unit 21: Time

Learning objectives

Code	Learning objective
5Mt1	Recognise and use the units for time (seconds, minutes, hours, days, months and years).
5Mt2	Tell and compare the time using digital and analogue clocks using the 24-hour clock.
5Mt3	Read timetables using the 24-hour clock.
5Mt4	Calculate time intervals in seconds, minutes and hours using digital or analogue formats.
5Mt5	Use a calendar to calculate time intervals in days and weeks (using knowledge of days in calendar months).
5Mt6	Calculate time intervals in months or years.
Strand 5: Problem solving	
5Pt1	Understand everyday systems of measurement in [length, weight, capacity, temperature and] time and use these to perform simple calculations.
5Pt2	Solve single and multi-step word problems (all four operations); represent them, e.g. with diagrams or a number line.
5Pt7	Consider whether an answer is reasonable in the context of a problem.
5Ps1	Understand everyday systems of measurement in [length, weight, capacity, temperature and] time and use these to perform simple calculations.
5Ps2	Choose an appropriate strategy for a calculation and explain how they worked out the answer.
5Ps4	Deduce new information from existing information to solve problems.
5Ps5	Use ordered lists and tables to help to solve problems systematically.
5Ps9	Explain methods and justify reasoning orally and in writing; make hypotheses and test them out.
5Ps10	Solve a larger problem by breaking it down into sub-problems or represent it using diagrams.

Measure – Time

Unit overview

This unit reviews time and the relationship between different standard units. Telling and comparing the time in the 24-hour clock format is introduced and learners practise converting 12-hour analogue to 24-hour digital time. Learners study transport timetables and complete, read and interpret data in timetables, using 24-hour notation. They explore calculation strategies, including 'bridging', to find intervals between two times. Finally, calendars are introduced and learners calculate time intervals (in days and weeks) within and across consecutive months.

Common difficulties and remediation

When reading digital time, look out for learners who mix up the hours and minutes, for example, reading 10:12 as 'ten minutes past 12'. For others, the format of the 24-hour clock may cause confusion, particularly the absence of 'a.m.' and 'p.m.' Reading 24-hour times, for example, 'nineteen twenty-five' for 19:25, presents unusual language for young learners. In addition, use of the term 'hundred' in reading times such as '15:00' ('fifteen hundred') can cause confusion. Provide direct instruction on how to read and write digital time to help learners overcome such difficulties. Guide learners in writing time using a colon as it is written on a digital clock and modelling how to write the time on a digital clock when the hour is a single digit and a double digit.

Look out for learners who apply whole-number knowledge to time, which uses a different base. For example, some learners will calculate the time interval between 10:45 and 11:25 as 80 minutes, not 40 minutes. It is important that learners understand that there are 60 seconds in one minute and 60 minutes in one hour. This also applies to calculations with larger units of time: days, weeks, months and years, where knowledge of time unit relationships is essential.

In addition, be aware of learners who apply other incorrect conversion factors to calculations, for example assuming four weeks is equal to one month. When working with calendars, encourage learners to count the right number of days in a particular month.

Promoting and supporting language

Whenever a new key word is introduced, ask learners to write a definition in their own words, for example: 'The 24-hour clock always uses 4 digits with numbers less than 12 for morning, and numbers greater than 12 for afternoon.'

Encourage frequent usage of mathematical language to help embed vocabulary. For example, modelling use of the question 'What is the time interval?' when learners ask 'How much time goes by between …?' as they make the transition to mathematically precise language.

Lesson 1: **Telling the time**

Measure – Time

Learning objectives

Code	Learning objective
5Mt1	Recognise and use the units for time (seconds, minutes, hours, days, months and years).
5Mt2	Tell and compare the time using digital and analogue clocks using the 24-hour clock.

Strand 5: Problem solving

5Pt1 / 5Ps1	Understand everyday systems of measurement in [length, weight, capacity, temperature and] time and use these to perform simple calculations.
5Pt2	Solve single and multi-step word problems (all four operations); represent them, e.g. with diagrams or a number line.
5Pt7	Consider whether an answer is reasonable in the context of a problem.
5Ps2	Choose an appropriate strategy for a calculation and explain how they worked out the answer.
5Ps4	Deduce new information from existing information to solve problems.
5Ps5	Use ordered lists and tables to help to solve problems systematically.
5Ps10	Solve a larger problem by breaking it down into sub-problems or represent it using diagrams.

Prerequisites for learning

Learners need to:
- name and use the units of time; use a.m. and p.m.
- read and write time to the nearest minute on 12-hour digital and analogue clocks.

Success criteria

Learners can:
- identify similarities between a 12-hour and 24-hour clocks

- recognise 12-hour a.m. times and 24-hour times as the same
- convert 12-hour p.m. times to 24-hour times.

Vocabulary

second (s), minute (min), hour (h), analogue, digital, 24-hour

Refresh

- Choose an activity from Measure – Time in the Refresh activities.

Discover 🆂🅱

- Discuss the text and image in the Student's Book. Ask: **What would happen if a 12-hour clock timetable was published with 'a.m.' and 'p.m.' missing from each time?** Discuss possible confusion.

Teach and Learn 🆂🅱 🖥 📊

- Ask: **How many seconds are there in 4/10 minutes?** (240/600) **How did you work it out?** (multiply by 60) Continue questions for conversion factors: hours to minutes; days to hours; weeks to days; years to days.
- Display and discuss **Slide 1**. The two sets of numbers refer to morning/afternoon times. Display the **Clock tool**, set to analogue (24-hour). Show 2:45 p.m. This converts to 14:45 in 24-hour time. Ask: **What time will a 24-hour clock show for 7:37 p.m./11:03 p.m?** (19:37, 23:03) Ask: **How did you work it out?** (add 12 hours) Repeat for 2:19 a.m./9:54 a.m. (02:19, 09:54) Ask: **How did you work it out?** (the 24-hour clock uses four digits so place a zero before all times before 10:00)
- Read the Example text together. Ask: **What time will a 24-clock show for 12:31 a.m?** (00:31) Switch the **Clock tool** to digital (24-hour), and set it at 00:17,

11:01, 5:23 and 23:46. Ask learners to convert the times to analogue.

Practise 🆆🅱

- Workbook: Telling the time page 210

Apply 👥

- Ask: **On a 24-hour digital clock, over a whole day, how many times does a seven appear?** (in units of minutes each hour (six times), multiply by 24 (144); in minutes in 24 hours (24); in hours (2); total = 170)

Review 📊

- Revise relationships between different time units. Ask: **How can you find the number of seconds in seven minutes/hours in 4·5 days?** Write the date and time in digital notation. Ask: **What will the time be 40 minutes/40 hours from now?** Ask: **What will the date be 30 hours/30 days/30 years from now?**
- Display the **Clock tool** showing the 12-hour analogue clock alongside the 24-hour digital clock. Ask learners to convert analogue times 12:13 a.m./06:31 a.m./10:55 p.m. to 24-hour clock times.

Assessment for learning

- Demonstrate how you would convert 12:33 a.m./11:11 a.m./11:11 p.m. to 24-hour time. What rules did you use?

Lesson 2: **Reading timetables**

Learning objectives

Code	Learning objective
5Mt1	Recognise and use the units for time (seconds, minutes, hours, days, months and years).
5Mt3	Read timetables using the 24-hour clock.
Strand 5: Problem solving	
5Pt1 / 5Ps1	Understand everyday systems of measurement in [length, weight, capacity, temperature and] time and use these to perform simple calculations.
5Pt2	Solve single and multi-step word problems (all four operations); represent them, e.g. with diagrams or a number line.
5Pt7	Consider whether an answer is reasonable in the context of a problem.
5Ps2	Choose an appropriate strategy for a calculation and explain how they worked out the answer.
5Ps4	Deduce new information from existing information to solve problems.
5Ps5	Use ordered lists and tables to help to solve problems systematically.
5Ps10	Solve a larger problem by breaking it down into sub-problems or represent it using diagrams.

Prerequisites for learning

Learners need to:

- read, write and convert time between analogue and digital 12-hour and 24-hour clocks
- convert from hours to minutes and vice versa.

Success criteria

Learners can:

- complete, read and interpret data in a timetable in 24-hour notation

- use more than one calculation strategy to find the interval between two given times.

Vocabulary

timetable, 24-hour clock

Resources

paper (per pair); ruler (per pair)

Refresh

- Choose an activity from Measure – Time in the Refresh activities.

Discover 🆂🅱

- Discuss the text and images in the Student's Book. Ask: **Where have you seen a timetable? Have you ever missed an appointment because you misread a timetable? What error did you make?**

Teach and Learn 🆂🅱 🖥

- Refer learners to the text and Example in the Student's Book. Ask: **Why do you think the times are shown in 24-hour clock format?** (24-hour clock times show time of day without the need for a.m./p.m.) Ask questions to practise reading and interpreting 24-hour clock times.
- Display **Slide 1**. Ask, for example: **What time does the train leaving [station] at [time] arrive at [station]? To arrive at [station] at [time] what time would you need to leave [station]? Which train is the slowest? How did you work out the answer?** Ask learners to make up one question about the timetable for their partner to answer.

Practise 🆆🅱

- Workbook: Reading timetables page 212
- Refer to Activity 1 from the Additional practice activities.

Apply 👥

- Say to pairs of learners: **Plan and draw a bus timetable. Draw five rows with destinations and four columns for different departure times. Journeys between bus stations take the same time. Erase some times and swap with another pair to complete the missing times.**

Review 🖥

- Say: **My bus leaves at 16:49. The journey takes 1 hour 25 minutes. When will I arrive?** Discuss strategies to bridge the hour, for example, 25 minutes = (11 + 14) minutes. Add the hour first (17:49), then minutes (17:49 + 11 min = 18:00. 18:00 + 14 min = 18:14).
- Display **Slide 1** again. Say: **You are on the [time] train to [destination]. The time is [time].** Ask: **Where are you on your journey? In how many minutes will you reach [destination]? How did you work out your answer?**

Assessment for learning

- Convert the times on this train timetable to 12-hour clock format.
- How long does it take for the train to travel from A to B? How do you know?

Measure – Time

Lesson 3: **Time intervals**

Learning objectives

Code	Learning objective
5Mt1	Recognise and use the units for time (seconds, minutes, hours, days, months and years).
5Mt4	Calculate time intervals in seconds, minutes and hours using digital or analogue formats.

Strand 5: Problem solving

5Pt1 / 5Ps1	Understand everyday systems of measurement in [length, weight, capacity, temperature and] time and use these to perform simple calculations.
5Pt2	Solve single and multi-step word problems (all four operations); represent them, e.g. with diagrams or a number line.
5Pt7	Consider whether an answer is reasonable in the context of a problem.
5Ps2	Choose an appropriate strategy for a calculation and explain how they worked out the answer.
5Ps4	Deduce new information from existing information to solve problems.
5Ps5	Use ordered lists and tables to help to solve problems systematically.
5Ps9	Explain methods and justify reasoning orally and in writing; make hypotheses and test them out.
5Ps10	Solve a larger problem by breaking it down into sub-problems or represent it using diagrams.

Prerequisites for learning

Learners need to:

- read, write and convert time between analogue and digital 24-hour clocks.

Success criteria

Learners can:

- count in small unit intervals, such as five minutes, to calculate larger time intervals

- use the 'bridging to the next 60' method to calculate a time interval.

Vocabulary

minute (min), hour (h), 'counting on', 'bridging through 60'

Refresh

- Choose an activity from Measure – Time in the Refresh activities.

Discover [SB]

- Discuss the text and the image in the Student's Book. Ask: **What time do you leave for school in the morning? What time do you get to school? How long does the journey take you?**

Teach and Learn [SB] [bar chart icon]

- Discuss the text in the Student's Book. Display the **Clock tool** (analogue) in two windows, one clock set at 08:05, the other at 09:20. Ask: **How would you find the time interval between the two clock times?** Elicit the two strategies: (1) 'counting on' and (2) 'bridging through 60'. Count in 5 min/10 min intervals from 08:05 to 09:20. Ask a volunteer to write the units on the board, add them and say the total. Draw a timeline on the board from 08:05 to 09:20. Demonstrate, using arrows, 'bridging through 60' in intervals (55 min + 20 min). Emphasise the need to count up to 60 minutes to the next hour.

- Set the two clocks at 07:11 and 09:08. Ask: **Which strategy is better to calculate this time interval? Why?** Elicit that when the units are not multiples of 5 or 10 minutes the 'bridging' strategy is easier.

Demonstrate: 9 min to 07:20; 40 min to 08:00; 60 min to 09:00; 8 min to 09:08: total = 117 min. Ask: **How will you convert to hours/minutes?** (count multiples of 60, then subtract, 1 h 57 min)

Practise [WB]

- Workbook: Time intervals page 214

Apply [pair icon]

- Pairs of learners plan a three-part journey that involves different modes of transport, for example, taxi, train and flight, writing the departure and arrival times. They swap plans with another pair and ask them to calculate the total journey time.

Review

- Say: **A swimmer begins swimming at 11:11 and finishes at 13:10.** Ask: **How long was his swim?** Discuss strategies used and ask learners to comment on effectiveness of methods.

Assessment for learning

- Here are two times displayed on analogue clocks. Show me how you calculate the time interval.
- How would you solve the problem for two digital 24-hour times?

Lesson 4: **Calendars**

Learning objectives

Code	Learning objective
5Mt1	Recognise and use the units for time (seconds, minutes, hours, days, months and years).
5Mt5	Use a calendar to calculate time intervals in days and weeks (using knowledge of days in calendar months).
5Mt6	Calculate time intervals in months or years.

Strand 5: Problem solving

5Pt1 / 5Ps1	Understand everyday systems of measurement in [length, weight, capacity, temperature and] time and use these to perform simple calculations.
5Pt2	Solve single and multi-step word problems (all four operations); represent them, e.g. with diagrams or a number line.
5Pt7	Consider whether an answer is reasonable in the context of a problem.
5Ps2	Choose an appropriate strategy for a calculation and explain how they worked out the answer.
5Ps4	Deduce new information from existing information to solve problems.
5Ps5	Use ordered lists and tables to help to solve problems systematically.
5Ps9	Explain methods and justify reasoning orally and in writing; make hypotheses and test them out.
5Ps10	Solve a larger problem by breaking it down into sub-problems or represent it using diagrams.

Prerequisites for learning

Learners need to:

- read simple timetables and use a calendar.

Success criteria

Learners can:

- calculate time intervals in 'bridging' stages
- convert from weeks to days by multiplying by 7, and from years to months by multiplying by 12.

Vocabulary

day, week, month, year, 'bridging through the next month'

Resources

current year's calendar (per class – keep on display for Workbook Practise – Challenge 3)

Refresh

- Choose an activity from Measure – Time in the Refresh activities.

Discover [SB] 🖥

- Display **Slide 1**. Discuss the text and image in the Student's Book. Ask: **Can you think of any other reasons why you might want to use a calendar to calculate time intervals?** Discuss responses.

Teach and Learn [SB]

- Display a current calendar. Ring today's date and say the day of the week. Ask: **What will the date be on this day next week/in two weeks?** Ask a volunteer to locate today's date. Ask them to choose a date later this month on a different day. Trace a finger down columns (weeks) and rows (days) to demonstrate the time interval calculation.

- Ring a date next month. Demonstrate two strategies: (1) column-row strategy; (2) bridging strategy: calculating dates to the end of the month then adding days from following month.

- Ask: **How would you calculate time intervals without a calendar?** You could use knowledge of days in calendar months. Say: **A family is on holiday from 13th July to 14th August.** Ask: **How many days is that?** Learners share solutions:

$((31 - 13) + 1^*) + 14 = 33$ (*add 1 to include July 13th – contrast with a question in which the given date is not included in the calculation, such as 'days before' or 'days after' a given date).

Practise 🔲

- Workbook: Calendars page 216
- Refer to Activity 2 from the Additional practice activities.

Apply 👥

- Learners plan an adventure holiday involving visits to several locations over a period of six months or more. The dates for each visit should be given. They swap plans with another pair and ask them to draw and complete a table with visits and time intervals.

Review

- Say: **A long cycle race begins on 10th June and finishes on August 5th. Over how many days does the race take place?** Discuss strategies.

Assessment for learning

- Here are two dates. Show how you calculate the time interval.

Measure – Time

Measure – Time

Additional practice activities

Activity 1 Challenge 2

Learning objectives

- Read and use a timetable to calculate elapsed time.
- Choose an appropriate strategy for a calculation involving time units and explain how they worked out the answer.

Resources

large sheet of paper (per group)

What to do

- Learners plan and draw a timetable for a coach journey. They draw five rows with headings for destinations and four columns for different departure times. Journeys between coach stations can take different times.
- They decide on a unit cost per minute of travel and use this to calculate passenger prices for journeys made at different departure times.

Variation

Challenge 1 Reduce the number of destinations to three.

Activity 2 Challenge 2

Learning objectives

- Read and use a timetable to calculate elapsed time.
- Choose an appropriate strategy for a calculation involving time units and explain how they worked out the answer.

Resources

calendars from different years (per group)

What to do

- Select two dates that are more than a year apart and write them on the board. Ask each group to work out the time interval between the two dates. Encourage them to split the task into different stages worked on by different individuals/teams.
- Learners discuss solutions, strategies used and the efficiency of each strategy.

Variation

Challenge 3 Each date is also given a time in hours, minutes and seconds. The group has to calculate the time interval in seconds, minutes, hours, days (or days and weeks).

Unit 22: Area and perimeter

Learning objectives

Code	Learning objective
5Ma1	Measure and calculate the perimeter of regular and irregular polygons.
5Ma2	Understand area measured in square centimetres (cm²).
5Ma3	Use the formula for the area of a rectangle to calculate the rectangle's area.
Strand 5: Problem solving	
5Pt1	Understand everyday systems of measurement in length [, weight, capacity, temperature and time] and use these to perform simple calculations.
5Pt2	Solve single and multi-step word problems (all four operations); represent them, e.g. with diagrams or a number line.
5Pt7	Consider whether an answer is reasonable in the context of a problem.
5Ps1	Understand everyday systems of measurement in length [, weight, capacity, temperature and time] and use these to perform simple calculations.
5Ps2	Choose an appropriate strategy for a calculation and explain how they worked out the answer.
5Ps4	Deduce new information from existing information to solve problems.
5Ps5	Use ordered lists and tables to help to solve problems systematically.
5Ps9	Explain methods and justify reasoning orally and in writing; make hypotheses and test them out.
5Ps10	Solve a larger problem by breaking it down into sub-problems or represent it using diagrams.

Unit overview

This unit reviews application of the property of 'opposite sides equal' to find the perimeter of a rectangle. Learners explore the relationship between the number of sides of a regular polygon and its perimeter and are encouraged to find the perimeter of a regular polygon by using a range of methods including formula, measurement and squared paper. The concept of area and square units is reviewed and the relationship between rectangles and arrays is explored. Learners measure and calculate the area of rectangles, using the formula $A = L \times W$ and solve problems involving missing lengths.

Common difficulties and remediation

Learners often get confused between area and perimeter and their units of measure. Misunderstanding arises because both terms are concerned with the boundary of a shape – perimeter being the distance around it and area being the amount of surface within it. Some learners believe that area is equivalent to the 'inside of an object' and perimeter is equivalent to the 'outside of an object'. Teachers should bring the concepts together, as teaching area and perimeter in isolation does not encourage learners to create connections. Learners may have a superficial understanding of the two concepts and can use the formula for calculation, but struggle to answer questions involving both concepts.

Learners who are introduced to the formula too early may develop procedural understanding of area rather than a conceptual and relational understanding. Therefore, it is important for learners to discover the formula for themselves. Computational algorithms should be regarded as quick methods of finding an answer, but should be introduced only after learners realise as they count squares that it is simpler to find the number of squares in one row and multiply this by the number of rows.

Look out for learners who measure or calculate (using an area formula) area correctly, but supply either no units, or incorrect units. Teachers should provide opportunities for learners to arrange square units, for example, tiles in rows and columns, to completely cover a surface and show how arrangements provide a way to measure area.

Promoting and supporting language

At every stage, learners require a mathematical vocabulary to access questions and problem-solving exercises. You may find it useful to refer to the audio glossary on Collins Connect. If appropriate, when a new key word is introduced, ask learners to write a definition in their books, drawing a box around it for emphasis. Encourage learners to write the definition in their own words, for example: 'Perimeter is the length of the outline of a shape.' In addition, be aware of terms that may have other mathematical meanings, for example some learners may believe that 'square centimetres' are units that only apply to squares.

Encourage frequent usage of mathematical language to help embed vocabulary. For example, modelling use of the question 'What is the perimeter?' when learners ask 'What is the distance around the edge of this object?' as they make the transition to mathematically precise language.

Measure – Area and perimeter

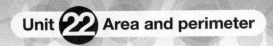

Unit **22** Area and perimeter

Lesson 1: **Perimeter of regular polygons**

Learning objectives

Code	Learning objective
5Ma1	Measure and calculate the perimeter of regular [and irregular] polygons.

Strand 5: Problem solving

5Pt1 / 5Ps1	Understand everyday systems of measurement in length [, weight, capacity, temperature and time] and use these to perform simple calculations.
5Pt2	Solve single and multi-step word problems (all four operations); represent them, e.g. with diagrams or a number line.
5Ps2	Choose an appropriate strategy for a calculation and explain how they worked out the answer.
5Ps4	Deduce new information from existing information to solve problems.

Prerequisites for learning

Learners need to:

- understand and use the vocabulary of perimeter
- draw rectangles and measure and calculate their perimeters.

Success criteria

Learners can:

- work out and express in words a formula for finding the perimeter of a regular polygon
- find the perimeter of a regular polygon by using a range of methods including formula, measurement and grid paper.

Vocabulary

perimeter, centimetre (cm), metre (m)

Resources

regular polygons (per pair); squared paper (per learner); ruler (per learner); classroom table, book, whiteboard, window (per class); large 2D equilateral triangle, regular pentagon, regular hexagon and regular octagon (per learner)

Refresh

- Use the activity *Geoboard Shapes* from the Refresh activities.

Discover [SB]

- Discuss the text and image in the Student's Book. Talk about the shape of the different panes of glass. How would you measure the perimeter of one of them? Say: **A piece of glass is to be inserted into a regular octagonal window frame. The sides of the frame are 20 cm long.** Ask: **What is the perimeter of the glass?** Discuss possible strategies.

Teach and Learn [SB] [chart]

- Revise perimeter in the context of boundaries, for example, a fence that is to be built to hold sheep in a field.
- Draw several squares on the page and label one side only (use a variety of mm, cm, m, km). Ask learners to calculate the perimeter of each square.
- Display the **Geometry set tool** with a square. Ask a volunteer to use a ruler and a formula to calculate the perimeter.
- Distribute regular polygons, rulers and squared paper if necessary. Say: **Take a shape and draw around it.** Ask: **How do you find the perimeter?** Elicit that you would measure the length of the sides and find the sum. Ask: **How did you work out the perimeter? Did anyone use a different method?**

Because the sides of a regular polygon are the same length, the perimeter can be found by multiplying the length of one side by the number of sides.

Practise [WB]

- Workbook: Perimeter of regular polygons page 218. Note: each learner will need a large 2D equilateral triangle, regular pentagon, regular hexagon and regular octagon.

Apply [group] [screen]

- Display **Slide 1**. Give learners squared paper and ask them to construct the sequence of squares shown. They investigate what happens to the perimeter as the terms increase (multiples of 4). Show **Slide 2** and ask learners to investigate an increasing sequence of hexagons.

Review

- Ask: **What is the perimeter of a pentagon with sides of 6 cm?** (30 cm) **Of 9 cm?** (45 cm) Ask: **How did you work it out?** Say: **A game is played on an octagonal pitch with sides of 15 metres.** Ask: **What is the perimeter of the field?** Discuss solutions.

Assessment for learning

- Demonstrate how you would find the perimeter of a square, pentagon, hexagon and octagon with sides of 7 cm.

Lesson 2: **Perimeter of irregular polygons**

Learning objectives

Code	Learning objective
5Ma1	Measure and calculate the perimeter of [regular and] irregular polygons.

Strand 5: Problem solving

5Pt1 / 5Ps1	Understand everyday systems of measurement in length [, weight, capacity, temperature and time] and use these to perform simple calculations.
5Pt2	Solve single and multi-step word problems (all four operations); represent them, e.g. with diagrams or a number line.
5Ps2	Choose an appropriate strategy for a calculation and explain how they worked out the answer.
5Ps4	Deduce new information from existing information to solve problems.
5Ps5	Use ordered lists and tables to help to solve problems systematically.
5Ps10	Solve a larger problem by breaking it down into sub-problems or represent it using diagrams.

Prerequisites for learning

Learners need to:

- understand and apply the term 'perimeter' to the sides of a polygon
- measure and calculate the perimeter of an irregular polygon.

Success criteria

Learners can:

- use a ruler to measure and find the perimeter of an irregular polygon.

Vocabulary

perimeter, centimetre (cm), metre (m)

Resources

regular and irregular polygons (per pair); protractor (per pair); paper (per pair); ruler (per learner); circles (per pair)

Refresh

- Use the activity *Measure perimeter* from the Refresh activities.

Discover SB

- Discuss the text and images in the Student's Book. Say: **A field has straight sides but is irregular in shape. A farmer wants to construct a fence at the boundary.** Ask: **Is it possible for the farmer to calculate the perimeter accurately, or does it have to be measured?** Accept comments and discuss.

Teach and Learn SB 📊

- Discuss the text and image in the Student's Book. Display the **Geoboard tool**. Construct an irregular pentagon. Ask: **How do you find the perimeter?** Elicit that you would measure the length of the sides and find the sum.
- Distribute regular and irregular polygons, rulers, protractors and paper if necessary. Say: **Work in pairs. Take two shapes – one you consider to be regular, the other irregular – and draw round them. All regular polygons have equal sides while all irregular polygons have at least two sides or two angles that are different. Prove that one shape is regular and the other is irregular.** Learners measure sides and angles (if necessary). Ask learners to share their solutions and discuss.
- Ask them to solve problems, for example: 'A lawn border is an irregular hexagon with three sides of

6·5 m, two sides of 7·5 m and one side of 10 m. What is its perimeter? (44·5 m).
- Prior to the Workbook activities, give learners large 2D shapes to measure.

Practise WB

- Workbook: Perimeter of irregular polygons page 220
- Refer to Activity 1 from the Additional practice activities.

Apply 👥

- Distribute circles and rulers. Present learners with a challenge: Draw round a circle, mark any four points on the circumference and join them to form a quadrilateral. What is the largest perimeter you can achieve? Repeat for five- and six-sided polygons.

Review

- Ask: **What is the perimeter of an irregular pentagon with sides of 6 cm, 7 cm, 8 cm, 9 cm and 10 cm. How did you work it out?**
- Ask learners to work in pairs. They each draw an irregular hexagon (with sides of whole or half centimetres), swap papers and measure to find the perimeter of the shape.

Assessment for learning

- Demonstrate how you would find the perimeter of an irregular polygon where the adjacent sides increase by 1 cm.

Measure – Area and perimeter

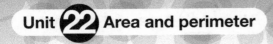
Lesson 3: **Area of a rectangle (1)**

Measure – Area and perimeter

Learning objectives

Code	Learning objective
5Ma2	Understand area measured in square centimetres (cm²).
5Ma3	Use the formula for the area of a rectangle to calculate the rectangle's area.

Strand 5: Problem solving

5Pt1 / 5Ps1	Understand everyday systems of measurement in length [, weight, capacity, temperature and time] and use these to perform simple calculations.
5Pt2	Solve single and multi-step word problems (all four operations); represent them, e.g. with diagrams or a number line.
5Pt7	Consider whether an answer is reasonable in the context of a problem.
5Ps2	Choose an appropriate strategy for a calculation and explain how they worked out the answer.
5Ps10	Solve a larger problem by breaking it down into sub-problems or represent it using diagrams.

Prerequisites for learning

Learners need to:

- understand that area is measured in square units
- find the area of rectilinear shapes drawn on a square grid by counting squares.

Success criteria

Learners can:

- calculate the area of a rectangle by multiplying

the number of squares in a row by the number of columns.

Vocabulary

array, area, row, column, centimetre (cm), square centimetre (cm²)

Resources

squared paper (per learner); ruler (per learner)

Refresh

- Use the activity *Count the squares* from the Refresh activities.

Discover ⬛

- Discuss the text and images in the Student's Book. Say: **You must make a curtain to cover a window two by three metres.** Ask: **What is the smallest amount of fabric required? How would you work this out?** Discuss strategies.

Teach and Learn ⬛ 🖥 📊

- Discuss the text and images in the Student's Book. Display **Slide 1** and discuss area and square centimetres.
- Display the **Geoboard tool**. Construct a rectangle 2 × 4. Say: **The side of each unit square is one centimetre.** Ask: **What is the area of the rectangle? How do you know?** Write the answer: 8 square centimetres. Recap: 'square cm' and the use of the abbreviations cm² and 'sq. cm'. Say: **You have found area by counting squares, but using arrays is a more efficient way.** Discuss real world arrays, for example, ice cubes in a tray.
- Construct a 4 × 5 cm rectangle on the **Geoboard tool.** Ask: **How many rows are there?** (5) **How many squares are in each row?** (4) **How many squares are there altogether?** (20) **Each square measures 1 cm across. What is the area of the rectangle?** (20 square centimetres) Say: **With the array method,**

you count the rows and squares in each row, then multiply the numbers.

- Repeat for 6 × 3 and 8 × 4 rectangles. Distribute squared paper. Learners shade squares to form rectangles and use the array method to calculate area, checking by counting squares.

Practise 📒

- Workbook: Area of a rectangle (1) page 222

Apply 👥 🖥

- Display **Slide 2**. Distribute squared paper. Ask: **How many rectangles can you make where the perimeter is numerically equal to the area?** Learners investigate the problem and share results.

Review

- On the board, draw a rectangle, 8 × 4 cm. Ask: **What is the area of this rectangle?** (32 sq. cm) **How did you work it out?** Ask: **How could you make a rectangle that has double the area?** (double the length of one side). Say: **Imagine a rectangle has an area of six square centimetres.** Ask: **What might be its perimeter?** (1 × 6: 1 + 1 + 6 + 6 = 14 cm; 2 × 3: 2 + 2 + 3 + 3 = 10 cm)

Assessment for learning

- Demonstrate how you would find the area of three rectangles in which the length is double the width.

Lesson 4: **Area of a rectangle (2)**

Learning objectives

Code	Learning objective
5Ma2	Understand area measured in square centimetres (cm^2).
5Ma3	Use the formula for the area of a rectangle to calculate the rectangle's area.

Strand 5: Problem solving

5Pt1 / 5Ps1	Understand everyday systems of measurement in length [, weight, capacity, temperature and time] and use these to perform simple calculations.
5Pt2	Solve single and multi-step word problems (all four operations); represent them, e.g. with diagrams or a number line.
5Pt7	Consider whether an answer is reasonable in the context of a problem.
5Ps2	Choose an appropriate strategy for a calculation and explain how they worked out the answer.
5Ps4	Deduce new information from existing information to solve problems.
5Ps5	Use ordered lists and tables to help to solve problems systematically.
5Ps9	Explain methods and justify reasoning orally and in writing; make hypotheses and test them out.
5Ps10	Solve a larger problem by breaking it down into sub-problems or represent it using diagrams.

Prerequisites for learning

Learners need to:

- measure the area of a shape in cm^2 or m^2
- find the area of rectilinear shapes drawn on a square grid by counting squares
- calculate the area of a rectangle by representing its dimensions as an array.

Success criteria

Learners can:

- calculate the area of a rectangle, given its dimensions
- compute the missing dimension of a rectangle, given the area and the other dimension.

Vocabulary

array, area, row, column, centimetre (cm), square centimetre (cm^2); square metre (m^2); squared paper, ruler, pencil (per learner)

Resources

ruler (per learner); rectangles (per pair)

Refresh

- Use the activity *Largest area* from the Refresh activities.

Discover SB

- Discuss the text and images in the Student's Book. Ask: **How would you find the area of a football pitch?** Discuss strategies.

Teach and Learn SB

- Discuss the text and images in the Student's Book. Display the **Geometry set tool** with the red rectangle. Ask: **Who can think of a way to measure the area of the rectangle without using a grid?** Ask a volunteer to use a ruler to measure the length and width of the rectangle. Say: **To find the area of a surface you multiply the length by the width, giving the answer in square units.** Write: $A = L \times W$. Say: **For the rectangle this is 10 cm × 6·5 cm = 65 cm^2.** Repeat for the blue square.
- Draw a large rectangle with dimensions 12 m × 9 m. Say: **The rectangle represents a field.** Ask: **What is its area? What unit will you use?** (108 square metres) Ask: **What else is measured in square metres?** Draw another rectangle and write inside: A = 40 m^2. Ask: **If the length of the rectangle is 8 m, how can you use a formula to find the width?** To find the missing side, divide the area by the given length (5 m).

Practise WB

- Workbook: Area of a rectangle (2) page 224
- Refer to Activity 2 from the Additional practice activities.

Apply 👥

- Distribute squared paper. Each learner draws three rectangles, measures the sides and calculates the area, writing the area inside the rectangle with the correct unit. They label only one side with the length. Learners swap papers and determine the missing lengths.

Review

- On the board, draw a 3 cm × 11 cm rectangle and label the sides. Ask: **What is the area?** (33 cm^2) **If you double the length of the sides, what happens to the area?** (132 cm^2; area quadrupled) Draw a rectangle. Write A = 63 m^2 in the centre and label one side 9 m. Ask: **What is the length of the other side?** (7 m)

Assessment for learning

- How would you use a formula to calculate the area of a classroom table?

Measure – Area and perimeter

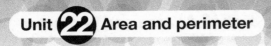

Measure – Area and perimeter

Additional practice activities

Activity 1 👥 Challenge **2**

Learning objectives
- Calculate the perimeter of a rectangle.
- Choose an appropriate strategy for a calculation involving perimeter and explain how they worked out the answer.

Resources
1–6 dice, alternatively use Resource sheet 4: 1–6 spinners (per pair)

What to do
- Learners roll a dice twice. Each roll represents the length of opposite sides of a rectangle.
- They have 15 seconds to state the perimeter of the rectangle. If they cannot answer after 15 seconds the turn passes to their partner, who has five seconds to respond.
- The winner is the learner with the higher score after five rounds.

Variation
Challenge **3** Include polygons of three, five, six and eight sides in the game. Learners roll the dice the number of sides and have 15 seconds to answer. They score points equivalent to the number of sides.

Activity 2 👥 Challenge **2**

Learning objectives
- Use the formula to calculate the area of a rectangle.
- Choose an appropriate strategy for a calculation involving area and explain how they worked out the answer.

Resources
ruler (per pair); large piece of paper (per pair); 1 cm squared paper (per pair)

What to do
- Say: **Draw a rectangle 2 cm by 3 cm, calculate its area using the formula and write it inside the rectangle.**
- Say: **Double the length of each side of the rectangle.** Ask: **What happens to the area?**
- Learners continue to double dimensions and calculate area, then share their findings.
- Ask: **Do you get the same type of pattern if you begin with a different sized rectangle?**

Variation
Challenge **1** Learners draw rectangles on 1 cm squared paper and use the grid to determine the area.

Unit 23: Handling data

Learning objectives

Code	Learning objective
5Dh1	Answer a set of related questions by collecting, selecting and organising relevant data; draw conclusions from their own and others' data and identify further questions to ask.
5Dh2	Draw and interpret frequency tables, pictograms and bar line charts, with the vertical axis labelled for example in twos, fives, tens, twenties or hundreds. Consider the effect of changing the scale on the vertical axis.
5Dh3	Construct simple line graphs, e.g. to show changes in temperature over time.
5Dh4	Understand where intermediate points have and do not have meaning, e.g. comparing a line graph of temperature against time with a graph of class attendance for each day of the week.
5Dh5	Find and interpret the mode of a set of data.
5Db1	Describe the occurrence of familiar events using the language of chance or likelihood.
Strand 5: Problem solving	
5Pt2	Solve single and multi-step word problems (all four operations); represent them, e.g. with diagrams or a number line.
5Pt7	Consider whether an answer is reasonable in the context of a problem.
5Ps4	Deduce new information from existing information to solve problems.
5Ps5	Use ordered lists and tables to help to solve problems systematically.
5Ps8	Investigate a simple general statement by finding examples which do or do not satisfy it, [e.g. the sum of three consecutive whole numbers is always a multiple of three].
5Ps9	Explain methods and justify reasoning orally and in writing; make hypotheses and test them out.

Unit overview

This unit reviews the use of frequency tables and tally charts to record data and the types of tables and graphs used to present data. Learners gain practice in interpreting information and answering questions from pictograms and bar line charts. They explore the effect a change in scale has on a graph and investigate the plotting of data on graphs of different scales. They learn to construct line graphs, plotting points accurately and joining them using a pencil and ruler. The differences between discrete and continuous data are discussed, together with the types of graphs best suited to represent each type. Learners are encouraged to plan their own investigations, suggesting hypotheses to test in the classroom. They devise their own data collection methods and construct graphs to best present the data, drawing conclusions regarding the validity of the statement that prompted the enquiry. In addition, they discover how to find the mode of a set of data and are introduced to probability.

Common difficulties and remediation

Some learners struggle to interpret a graph when the scale is not 1 : 1; others overlook the scale altogether. Teachers should highlight scales when many-to-one correspondence is featured. Learners should be encouraged to look for a key to any graph or chart before they begin interpretation.

Learners should be reminded that tally marks rely on groupings of five. A common mistake is for students to draw five vertical lines, then draw a sloping line across them to show the group.

Learners need to understand that individual bars on a bar line chart are the same width and do not touch each other, emphasising that the values in-between have no meaning. Equally the tops of the bars should not be joined.

Some learners require support interpreting category data where the responses are numbers. Teachers must address this directly, helping learners differentiate between the set of numbers that represents categories and the set that records frequency.

Teachers should encourage the construction of tables and graphs on squared paper. The use of squared graph paper encourages students to create regular-sized symbols, occupying the same space. This prevents learners from mistakenly interpreting the size of a graph and its features and symbols as significant, instead of focusing on the plotted data.

Promoting and supporting language

Encourage frequent usage of mathematical language to help embed vocabulary. For example, model use of the question 'What is the probability of the event taking place?' when learners ask 'How likely is it …?' as they make the transition to mathematically precise language.

Lesson 1: **Interpreting graphs and tables**

Handling data – Organising, categorising and representing data

Learning objectives

Code	Learning objective
5Dh2	Draw and interpret frequency tables, pictograms and bar line charts, with the vertical axis labelled for example in twos, fives, tens, twenties or hundreds. Consider the effect of changing the scale on the vertical axis.

Strand 5: Problem solving

Code	Learning objective
5Pt2	Solve single and multi-step word problems (all four operations); represent them, e.g. with diagrams or a number line.
5Pt7	Consider whether an answer is reasonable in the context of a problem.
5Ps4	Deduce new information from existing information to solve problems.
5Ps5	Use ordered lists and tables to help to solve problems systematically.

Prerequisites for learning

Learners need to:
- use tables and tally charts to record data
- know that data can be presented in pictograms and bar charts.

Success criteria

Learners can:
- construct a pictogram to represent data and explain the value of the pictorial symbol used

- interpret information and answer questions from simple pictograms and bar line charts
- read frequency values from a frequency table and find the greatest and least values.

Vocabulary

tally chart, frequency table, pictogram, bar line chart

Resources

squared paper (per pair); ruler (per learner); Resource sheet 20: Bar line template (per learner)

Refresh

- Use the activity *Tally charts* from the Refresh activities.
- In this lesson, learners will be refreshing their knowledge of tables, tally charts and graphs. The Student's Book focuses on bar and line graphs and the Workbook focuses on pictograms and tallies. This is intentional and is designed to cover a broad range of familiar data handling methods.

Discover SB 🖵

- Display **Slide 1.** Say: **Think of a table or graph you have used.** Ask: **How did it help organise the information? What did the information tell you?** Discuss the text and image in the Student's Book. Discuss examples and alternative formats for organising and representing such data.

Teach and Learn SB 🖵 📊

- Discuss Learn in the Student's Book and display **Slide 2** of the graph from the Example. Ask questions about the graph.
- On the board, construct a tally and frequency table with five rows. Ask: **What type of data do we use a frequency table to collect?** Elicit that the table is for collecting discrete data. Add five fruit headings to the rows and collect tally data from learners, asking: **What is your favourite fruit?** Ask learners to count the tally marks and complete the frequency column.
- Display the **Pictogram tool** with the fruit symbol. Demonstrate how to represent collected data as a

pictogram, labelling categories and adding a title. Ask questions that progress from reading single items of data to comparison of two categories. **What is the most/least popular fruit? How many more/fewer learners voted for banana than orange?**

Practise WB

- Workbook: Interpreting graphs and tables page 226. Learners will each need a copy of Resource sheet 20: Bar line template for Challenges 2 and 3.

Apply 👥

- Provide squared paper and rulers. Discuss how best to construct data collection and frequency tables, and the types of question that data can be used to answer. Learners collect data on a survey of favourite pets and draw a chart or graph.

Review 📊

- Invent and display the results of a survey that asked learners to name their favourite sea creature in the **Pictogram tool.** Without learners seeing, remove key pieces of information, such as title, key, labels and ask them to identify the missing features. Prompt them to explain the importance of such information. Ask questions that progress from reading single items of data to comparison of two categories of data.

Assessment for learning

- Demonstrate how you would use a pictogram to find the most/least popular category of data.

Lesson 2: **Changing the scale**

Learning objectives

Code	Learning objective
5Dh2	Draw and interpret frequency tables, pictograms and bar line charts, with the vertical axis labelled for example in twos, fives, tens, twenties or hundreds. Consider the effect of changing the scale on the vertical axis.

Strand 5: Problem solving

5Pt2	Solve single and multi-step word problems (all four operations); represent them, e.g. with diagrams or a number line.
5Pt7	Consider whether an answer is reasonable in the context of a problem.
5Ps4	Deduce new information from existing information to solve problems.
5Ps5	Use ordered lists and tables to help to solve problems systematically.

Prerequisites for learning

Learners need to:

* use tables and tally charts to record data
* answer questions about the data in frequency tables, pictograms and bar line charts.

Success criteria

Learners can:

* read single pieces of information and compare category data for frequency tables, pictograms and bar line charts with the vertical axis labelled in intervals greater than 1

* comment on the effect on a bar line chart where the information is presented in the same way, but the scale on the vertical axis is changed to intervals greater than 1.

Vocabulary

tally chart, frequency table, pictogram, bar line chart, scale, interval

Resources

squared paper (per pair); ruler (per learner)

Refresh

* Use the activity *Pictogram goals* from the Refresh activities.

Discover SB 🖥

* Display **Slide 1**. Discuss the text and images in the Student's Book. Say: **Look at the two graphs.** Ask: **How does changing the scale affect the look of the graph? Why might a car salesman who is under pressure to keep sales similar each month prefer to show his bosses the second graph?**

Teach and Learn SB 🖥

* Discuss the text in the Student's Book. Display **Slide 2** showing weather graphs of the numbers of sunny and rainy days over five years. Point to each graph in turn. Ask: **What does the key tell you? Which year had most/least sun? Which year had most/least rain?**

* Display **Slide 3**, which shows a similar graph of sunny days with the pictogram scale changed to one sun equals 20 sunny days. Ask: **How does this change the look of the pictogram? Which of the scales gives a more accurate picture of weather changes over the years?**

Practise WB

* Workbook: Changing the scale page 229
* Refer to Activity 1 from the Additional practice activities.

Apply 👥

* Use the internet to present weather data for the past six months. Ask learners to represent the data in two pictograms or bar line charts. The two charts should differ in pictogram scale, or vertical axis increments. Discuss how the scale changes the look of the graph.

Review 🖥

* Display **Slide 4**, a frequency table showing data for daily sales of TVs at a store and a frequency graph of the data. Ask: **What is the missing scale on the graph? How did you work it out?** Ask: **If the scale units are increased/decreased, how do you think this will change the look of the graph?** Accept comments and discuss.

Assessment for learning

* What affect does changing the scale on a graph have on the data?

Handling data – Organising, categorising and representing data

Lesson 3: **Line graphs**

Handling data – Organising, categorising and representing data

Learning objectives

Code	Learning objective
5Dh3	Construct simple line graphs, e.g. to show changes in temperature over time.

Strand 5: Problem solving	
5Pt2	Solve single and multi-step word problems (all four operations); represent them, e.g. with diagrams or a number line.
5Pt7	Consider whether an answer is reasonable in the context of a problem.
5Ps4	Deduce new information from existing information to solve problems.
5Ps5	Use ordered lists and tables to help to solve problems systematically.
5Ps9	Explain methods and justify reasoning orally and in writing; make hypotheses and test them out.

Prerequisites for learning

Learners need to:

- be able to answer questions about the data in frequency tables, pictograms and bar line charts
- be able to plot data points on a graph.

Success criteria

Learners can:

- draw a line graph to represent a set of data with an appropriate scale, plotting points accurately and joining them using a pencil and ruler.

Vocabulary

horizontal axis, vertical axis, line graph, scale

Resources

squared paper (per learner and per pair); ruler (per learner)

Refresh

- Use the activity *Plotting points* from the Refresh activities.

Discover SB 🖥

- Display **Slide 1**. Say: **Look at the graph. What does it tell you about temperature over this period?** Ask: **How might a line graph like this help you plan a day out in the near future?**

Teach and Learn SB 📊

- Discuss the text and Example in the Student's Book. On the board, draw the table below:

The Castle Museum – daily visits							
Day	Mon	Tues	Weds	Thurs	Fri	Sat	Sun
Number of visits	20	40	30	45	60	90	95

Use the data to construct a graph using the **Line grapher tool**. Ask: **What would be a suitable scale? Why?** (increments of 10 or 20) Show how to plot the data and draw lines between points. Ask: **What does the line graph show?** (the number of daily visits over one week) **What do the numbers on the vertical and horizontal axes represent?** (number of visits/ days) **How many visits were made on Tuesday?** (40) **How did you work out the answer? On what day were the highest/lowest visits recorded?** (Sunday/Monday) **Why do you think the results for Sunday and Monday are so different?**

Practise 📘

- Workbook: Line graphs page 233. Each learner will need some squared paper for Challenge 3.

Apply 👥

- On the board, draw the table below, showing the distance a toy car travelled down a ramp. Learners use the data to construct a line graph. Ask: **How is the distance travelled by the toy car affected by the height of the ramp? How do you know?**

Height of ramp (cm)	10	20	30	40	50
Distance travelled (cm)	45	60	75	85	100

Review

- Discuss the purpose and uses of a line graph. Ask why it is important to choose the correct scale and plot points accurately. Refer to the line graph from Teach and Learn. Ask: **How many visits were made on Saturday?** (90) **Which day had 30 more visits than Wednesday?** (Friday) **Which two consecutive days showed the greatest increase in visitors?** (Friday to Saturday)

Assessment for learning

- Demonstrate how you would use a line graph to show an increase in the quantity of something over time

Lesson 4: Intermediate points

Learning objectives

Code	Learning objective
5Dh4	Understand where intermediate points have and do not have meaning, e.g. comparing a line graph of temperature against time with a graph of class attendance for each day of the week.

Strand 5: Problem solving

Code	Learning objective
5Pt2	Solve single and multi-step word problems (all four operations); represent them, e.g. with diagrams or a number line.
5Pt7	Consider whether an answer is reasonable in the context of a problem.
5Ps4	Deduce new information from existing information to solve problems.
5Ps5	Use ordered lists and tables to help to solve problems systematically.
5Ps9	Explain methods and justify reasoning orally and in writing; make hypotheses and test them out.

Prerequisites for learning

Learners need to:

• show and interpret data in a bar or bar line chart.

Success criteria

Learners can:

• interpret and present discrete data in a line graph
• interpret line graphs and decide whether intermediate points have meaning

• read intermediate points and use the data displayed to make predictions.

Vocabulary

discrete data, continuous data, intermediate point

Resources

squared paper (per pair); ruler (per learner)

Refresh

• Use the activity *Tide times* from the Refresh activities.

Discover 〔SB〕 🖥

• Display **Slide 1**. Discuss the text and images in the Student's Book. Ask: **How might a line graph like this help you plan a day out? Could you plot the data represented by Graph type B on Graph type A. Why, or why not?**

Teach and Learn 〔SB〕 🖥 📊

• Display **Slide 2**. Discuss the text and images in the Student's Book. Provide examples of discrete and continuous data – discrete: savings/month, cost/ supermarket visit; continuous: time/speed, age/ height. Ask learners to explain in their own words why each type of data is discrete or continuous.

• Display the **Line grapher tool**. Plot data for the cost of multiple ice creams priced at $4 each. Ask: **Do the points between the plotted points have any meaning?** They have no meaning, because ice creams cannot be split up.

• Use real data to plot a graph of outside temperature against time, showing rise and fall. Ask: **Do the points between the plotted points have meaning?** They have meaning, because the temperature can be a reading at any point on the scale.

Practise 〔WB〕

• Workbook: Intermediate points page 236

• Refer to Activity 2 from the Additional practice activities.

Apply 👥 🖥

• Display **Slide 3** which shows two data tables: number of learners in five different classes (left); time it takes for meat to cook (right). Learners plot both graphs using squared paper and rulers. Ask: **Which data is continuous** (right) **and which is discrete** (left)?

Class	Number of learners	Weight (kg)	Time (hours)
A	45	1	1·5
B	60	2	3
C	75	3	4·5
D	85	4	5·5
E	100	5	6·5

Review

• Discuss the differences between discrete and continuous data with the class. Learners give examples of each, discussing whether points in between plotted points have any meaning.

Assessment for learning

• Here are two graphs. Which one shows continuous data and which one shows discrete? How do you know?

Lesson 5: **Mode**

Learning objectives

Code	Learning objective
5Dh5	Find and interpret the mode of a set of data.

Strand 5: Problem solving

5Pt7	Consider whether an answer is reasonable in the context of a problem.
5Ps4	Deduce new information from existing information to solve problems.
5Ps5	Use ordered lists and tables to help to solve problems systematically.
5Ps9	Explain methods and justify reasoning orally and in writing; make hypotheses and test them out.

Prerequisites for learning

Learners need to:
- put a set of numbers in ascending or descending order.

Success criteria

Learners can:
- find the mode of a set of data by identifying the value that occurs most frequently in the set

Vocabulary

set, mode

Resources

paper (per pair); ruler (per pair)

Refresh

- Use the activity *Colour modes* from the Refresh activities.

Discover [SB]

- Discuss text and questions in the Student's Book. Elicit that it is important to manufacture more of the most popular books or the most common size of shoe. This is because making things in equal numbers would cause a shortage of some and an overstocking of others.

Teach and Learn [SB]

- Discuss the text and images in the Student's Book. Explain to learners how to find the mode of a set of data. Provide other illustrations, for example, the mode of a set of 2D shapes, or the mode of a set of different coloured counters. Establish that the mode is the item that appears most often in a set.
- On the board, write: 170 cm, 167 cm, 173 cm, 171 cm, 167 cm, 170 cm, 171 cm, 167 cm. Say: **The figures are the distances recorded in a standing jump competition.** Ask: **What is the mode of the set of distances?** (167 cm) Discuss strategies. **Remind learners that the mode is the item, or value, within a set of data that occurs the most frequently.**
- Demonstrate and complete a frequency table for each measurement. Say: **An easy way to remember the meaning is that it has a similar beginning to 'most'.**

- Explain that the mode can be used for any type of data, not just numbers. Draw a frequency table that records the frequency of colour of parked cars: green (12); blue (15); white (9); red (14). Ask: **What is the mode for this set?** Elicit that the mode is a category, in this case colour (blue), not frequency.

Practise [WB]

- Workbook: Mode page 238

Apply 👥💻

- Display **Slide 1**. Ask: **What is the mode of the set of colours?** Learners draw and complete their own frequency table for colours and find the mode. Ask: **Why might this information be useful to a clothes manufacturer?**

Review

- Invite ten learners to each think of a random number from 0 to 9. In turn, they write their numbers on the board. Ask the class to find the mode of the set of numbers. Repeat for a set of letters from A to J.

Assessment for learning

- Throw a dice ten times and record the results. What is the mode of this set of data? How do you know?

Lesson 6: **Collecting data**

Learning objectives

Code	Learning objective
5Dh1	Answer a set of related questions by collecting, selecting and organising relevant data; draw conclusions from their own and others' data and identify further questions to ask.

Strand 5: Problem solving	
5Pt7	Consider whether an answer is reasonable in the context of a problem.
5Ps4	Deduce new information from existing information to solve problems.
5Ps5	Use ordered lists and tables to help to solve problems systematically.
5Ps8	Investigate a simple general statement by finding examples which do or do not satisfy it, [e.g. the sum of three consecutive whole numbers is always a multiple of three].
5Ps9	Explain methods and justify reasoning orally and in writing; make hypotheses and test them out.

Prerequisites for learning

Learners need to:

- interpret information and answer questions from simple pictograms and bar line charts.

Success criteria

Learners can:

- collect data in the form of frequency tables with a range of suitable data categories and justify the form of representation.

Vocabulary

data, hypothesis, collect

Resources

ruler (per pair); 1–6 dice, alternatively use Resource sheet 4: 1–6 spinners (per pair); wooden plank (per pair); books (per pair); toy car (per pair)

Refresh

- Use the activity *Favourite fruits* from the Refresh activities.

Discover [SB] 🖥

- Display **Slide 1** and discuss the text and image in the Student's Book. Say: **Think of a hypothesis that you could test in the classroom.** Ask: **What sort of data would you need to collect?** Ask learners to share their hypotheses and comment on the feasibility of testing them.

Teach and Learn [SB]

- Discuss the text and example in the Student's Book.
- On the board, write: Most children in the class travel to school by car. Say: **This is a hypothesis yet to be proven.** Ask: **How can we test this statement?** Agree a survey should be conducted. Ask: **What possible answers might we get?** Agree the answer categories: walk, cycle, car, bus, train. Ask: **How should we collect and record the data?** Elicit that the best format is a frequency table.
- With help from volunteers, construct the table on the board with columns headed Tally and Frequency and rows labelled with transport type. Ask learners to come up in turn to add a tally mark to the relevant row. Complete the frequency column and ask learners to comment on the results.

Practise [WB]

- Workbook: Collecting data page 240
- Refer to Activity 3 from the Additional practice activities.

Apply 👥

- On the board write: Boys watch more TV than girls. Discuss the hypothesis and how best to test it. Learners plan a survey of class members asking the question: 'How many hours of TV on average do you watch per week?' They construct frequency tables, deciding on the categories, and collect data. Lead a discussion of results.

Review

- Ask learners to comment on the data collected for the survey in Teach and Learn. Ask: **Is the data continuous or discrete? Why?** (discrete because each category represents a single mode of transport)
- On the board, write: Girls can run faster than boys. Ask: **How would you test this statement to see if it is true? What data would you collect and how would you record it?**

Assessment for learning

- If you were collecting data to test the statement, 'Most books have 100 pages', what data would you collect?

Handling data – Organising, categorising and representing data

Lesson 7: **Presenting data**

Learning objectives

Code	Learning objective
5Dh1	Answer a set of related questions by collecting, selecting and organising relevant data; draw conclusions from their own and others' data and identify further questions to ask.

Strand 5: Problem solving

5Pt7	Consider whether an answer is reasonable in the context of a problem.
5Ps4	Deduce new information from existing information to solve problems.
5Ps5	Use ordered lists and tables to help to solve problems systematically.
5Ps8	Investigate a simple general statement by finding examples which do or do not satisfy it, [e.g. the sum of three consecutive whole numbers is always a multiple of three].
5Ps9	Explain methods and justify reasoning orally and in writing; make hypotheses and test them out.

Prerequisites for learning

Learners need to:

- interpret information and answer questions from simple pictograms and bar line charts
- draw a line graph to represent a set of data with an appropriate scale, plotting points accurately and joining them using a pencil and ruler
- interpret line graphs and decide whether intermediate points have meaning.

Success criteria

Learners can:

- represent information in graphs, charts or tables, and justify the form of representation and check the plausibility of their conclusions
- use the tools within data handling software to present information in an appropriate format.

Vocabulary

data, hypothesis, conclusion

Resources

squared paper (per learner (for Workbook) and per pair); ruler (per learner)

Refresh

- Use the activity *Shoe size* from the Refresh activities.

Discover [SB]

- Discuss the text and image in the Student's Book. Ask: **What is the best graph to represent a set of data?** Say: **Think of a hypothesis that could be tested and make suggestions for the graph best suited to represent the data.**

Teach and Learn [SB] [image]

- Discuss the text and example in the Student's Book. Ask learners to look at the data they collected in the Lesson 6. The data is discrete, not continuous. Ask: **How does knowing the type of data influence which graph we choose to present it?** Line graphs work best for continuous data, whereas bar charts, bar line charts and pictograms work best for categorical, or discrete, data. Ask: **Which type of graph best suits the data we collected?** Agree a bar chart or pictogram. Display the **Bar charter tool**. Ask learners to look at their results and make a suggestion for the vertical scale. Ask volunteers to set up category groups and input data. Ask: **What do the results show? Do they support the hypothesis? Why?**

Practise [WB]

- Workbook: Collecting data page 243
- Refer to Activity 3 from the Additional practice activities.

Apply [icon]

- Discuss the type of data collected in Lesson 6 Apply, and how best to represent it. Learners construct a graph on squared paper from the data. Ask: **What do the results show? Do they support the hypothesis that boys watch more TV than girls? Why?**

Review

- Review the difference between discrete and continuous data. Write: The taller the person, the heavier the person will be. Ask: **What data would you collect to test this statement? What type would it be?** (continuous) **What is the best graph to represent this data?** (line graph) **Why? Discuss whether intermediate points on the graph would have meaning.** (yes)

Assessment for learning

- You are collecting data to test the statement: 'The most frequent number of words spelled correctly in any spelling test of 20 words is 15.' What data would you collect and how would you present it?

Handling data – Organising, categorising and representing data

Lesson 8: **Probability**

Learning objectives

Code	Learning objective
5Db1	Describe the occurrence of familiar events using the language of chance or likelihood.

Strand 5: Problem solving

Code	Learning objective
5Pt7	Consider whether an answer is reasonable in the context of a problem.
5Ps4	Deduce new information from existing information to solve problems.
5Ps8	Investigate a simple general statement by finding examples which do or do not satisfy it, [e.g. the sum of three consecutive whole numbers is always a multiple of three].
5Ps9	Explain methods and justify reasoning orally and in writing; make hypotheses and test them out.

Prerequisites for learning

Learners need to:
* understand the rules of simple board games
* use basic vocabulary associated with the chance events of game play.

Success criteria

Learners can:
* use the vocabulary that describes the probability of events: impossible, unlikely, even chance, likely, certain (no chance, poor chance, good chance)

* show probabilities on a probability scale.

Vocabulary

likelihood, chance, impossible, certain, even chance

Resources

Resource sheet 21: blank four sector spinner (per pair); scissors (per pair);

Refresh

* Use the activity *Likely thumbs* from the Refresh activities.

Discover [SB]

* Discuss the text and image in the Student's Book. Ask: **What other examples can you think of where you are told the likelihood of something happening?** Say: **Sometimes you are told that an event is likely to happen.** Ask: **Does this mean it WILL happen?** Discuss.

Teach and Learn [SB] 🖥

* Display **Slides 1 and 2** and discuss the text and images in the Student's Book.
* Draw a probability scale on the board. Say: **Think of an event that is certain to happen.** Lead a discussion where learners share events. Add two suggestions to the 'certain' end of the probability scale. Say: **Think of an event that is impossible.** Accept suggestions, then add two events to the scale at the 'impossible' end.
* Explain that the middle of the scale is labelled 'even chance' and describes events that have two equally likely outcomes, such as tossing a coin to get a heads.
* Ask: **What other events have an even chance of happening?** Elicit that rolling a 1–6 dice and getting an even number is an example.

Practise [WB]

* Workbook: Probability page 245
* Refer to Activity 4 from the Additional practice activities.

Apply 👥

* Learners use Resource sheet 21, to make two four-sector spinners and test them, according to the probability rules: (1) even chance of spinning a 3 or 4; (2) likely to spin red, unlikely to spin blue. They spin 50 times, recording the outcomes. Ask them to check that outcomes match predictions, confirming the design of each spinner.

Review

* Review different points and sections of the probability scale. Discuss with the class the likelihood of rolling (a) a 1 (b) a 6 (c) an even number (d) an odd number (e) a number greater than 1 (f) a letter, on a 1–6 dice?
* Draw a probability scale labelled 'certain' at one end and 'impossible' at the other end, and 'even chance' in the middle. The sections split by the midpoint 'even chance' are labelled 'likely' and 'unlikely'. As a class, work together to label the events at appropriate points on the scale.

Assessment for learning

* You drop a piece of toast. How likely is it that the toast will fall butter side down? Butter side up? On its side? Fly up in the air?

Handling data – Probability

231

Additional practice activities

Activity 1

Learning objectives

- Collect and organise data to find out about a subject and present the data as a pictogram.
- Construct and interpret a simple pictogram.

Resources

squared paper (per pair); ruler (per pair)

What to do

- Learners collect and record data for a class survey, 'What is your favourite pizza topping?'
- They create five categories and construct tally and frequency tables. They present the survey results in a pictogram on squared paper, with a symbol representing one pizza.
- Learners write five statements commenting on what the graph reveals.
- They construct a second pictogram, this time with a symbol representing more than one pizza, and comment on how this changes the look of the graph.

Variation

Challenge 1 Learners use the **Pictogram tool** to record the results of the survey and to change the scale.

Activity 2

Learning objectives

- Collect and organise data to investigate a general statement and present the data using a suitable graph.
- Identify whether data is continuous or discrete and say whether intermediate points on a graph have meaning or not.

Resources

squared paper (per pair); metre ruler (per pair)

What to do

- On the board, write: The older the learner, the greater their height.
- Learners investigate whether this statement is true by collecting relevant data. They plot the data on a suitable graph on squared paper and comment on the results.
- Learners share conclusions, saying whether the data is continuous or discrete, and commenting on the meaning of intermediate points.

Variation

Challenge 3 Learners extrapolate the data, extending the graph line and make predictions about height at different ages.

Additional practice activities

Activity 3 Challenge 2

Learning objectives

- Collect and organise data to investigate a general statement and present the data using a suitable graph.
- Draw conclusions from the data plotted on a graph and say whether or not it supports a general statement.

Resources

squared paper (per pair); ruler (per pair)

What to do

- On the board, write: Most learners in the class have two brothers or sisters.
- Ask learners to investigate whether this statement is true by conducting a class survey.
- They prepare a frequency table for number of siblings, deciding on a sensible range.
- They complete the survey, plot the data on a suitable graph and write a report of the results.

Variation

Challenge 3 Learners include a conclusion in their report stating whether the results support the statement that led to the enquiry.

Activity 4 Challenge 2

Learning objectives

- Describe the probability of a number being spun on a spinner using the language of chance or likelihood.
- Add an event to a scale to indicate the likelihood of the event happening.

Resources

Resource sheet 21: Blank four sector spinner (one spinner per pair); scissors (per pair); paper (per pair)

What to do

- Ask learners to label the four sectors of a blank spinner: 1, 2, 3 and 4.
- Learners draw a probability scale and consider the likelihood of the following items being spun: a number; a letter; a 1; a 3; a 1 or a 2; an even number; an odd number.
- Learners add each event to the scale to indicate its likelihood of happening.
- They spin the spinner 30 times and record the results. For each event, they check how close their prediction is to the actual outcome.

Variation

Challenge 1 Use a two-sector spinner that learners label 1 and 2. Adapt the questions accordingly.

Resource sheet 1: Thermometer template

Cut along grey line and join at the black dotted line.

Resource sheet 2: 0–100 number cards

6	13	20	27
5	12	19	26
4	11	18	25
3	10	17	24
2	9	16	23
1	8	15	22
0	7	14	21

34	41	48	55
33	40	47	54
32	39	46	53
31	38	45	52
30	37	44	51
29	36	43	50
28	35	42	49

Resource sheet 2: 0–100 number cards 3 of 4

62	69	76	83
61	68	75	82
60	67	74	81
59	66	73	80
58	65	72	79
57	64	71	78
56	63	70	77

Resource sheet 2: 0–100 number cards

88	93	98	100
87	92	97	99
86	91	96	
85	90	95	
84	89	94	

Resource sheet 3: Blank place value grid

U	
T	
H	
Th	
TTh	
HTh	

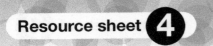

Resource sheet 4: 1–6 spinner

How to use the spinner

Hold the paper clip in the centre of the spinner using the pencil and gently flick the paper clip with your finger to make it spin.

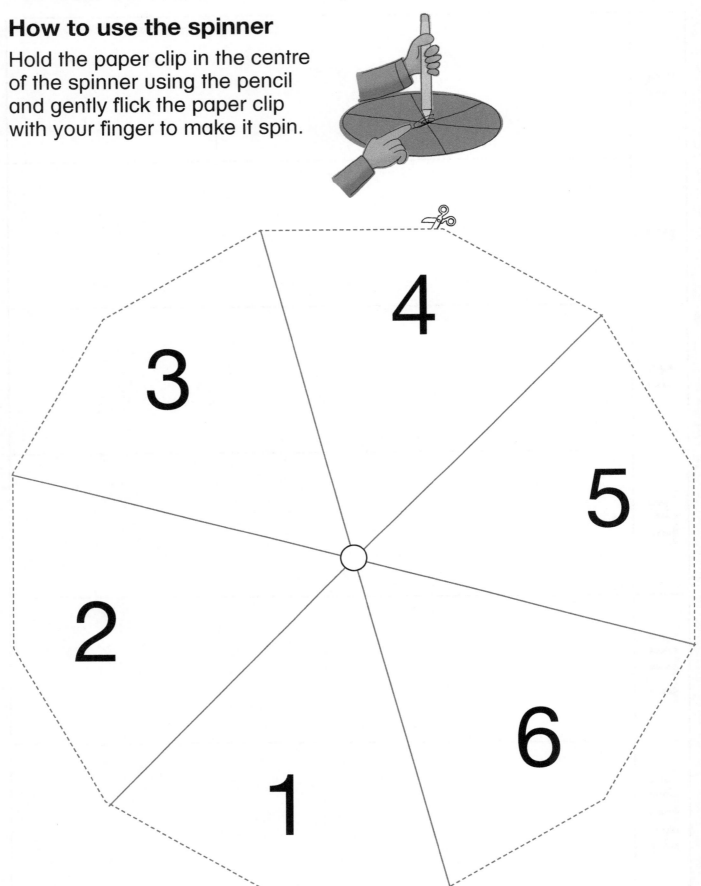

Resource sheet 5: Arrow cards

100	000
200	006
300	008
400	007
500	009

10	001	1	0
20	062	2	6
30	083	3	8
40	074	4	7
50	095	5	6

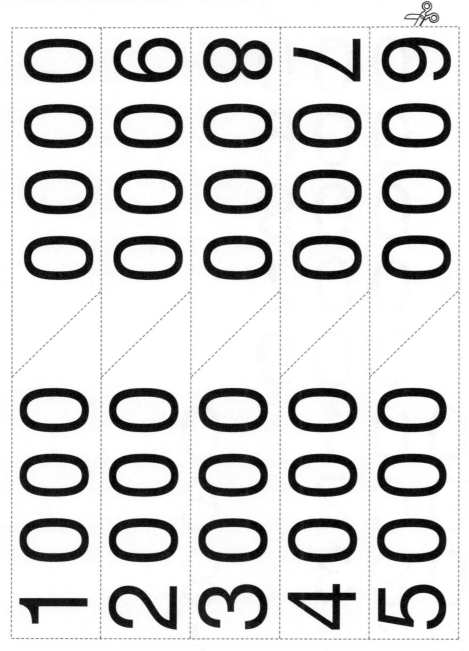

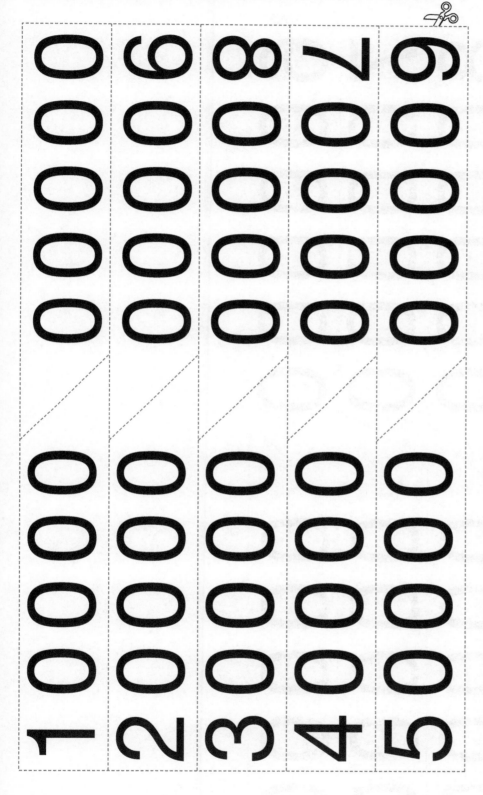

Resource sheet 5: Arrow cards

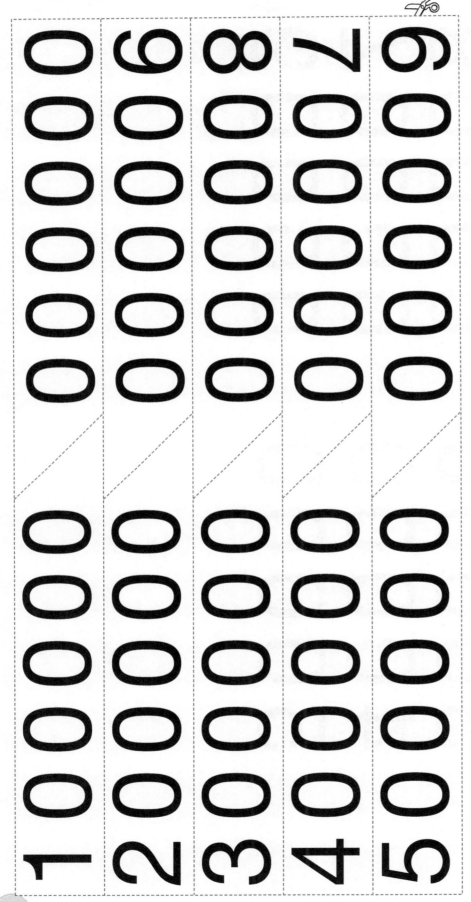

Resource sheet 6: Squared paper

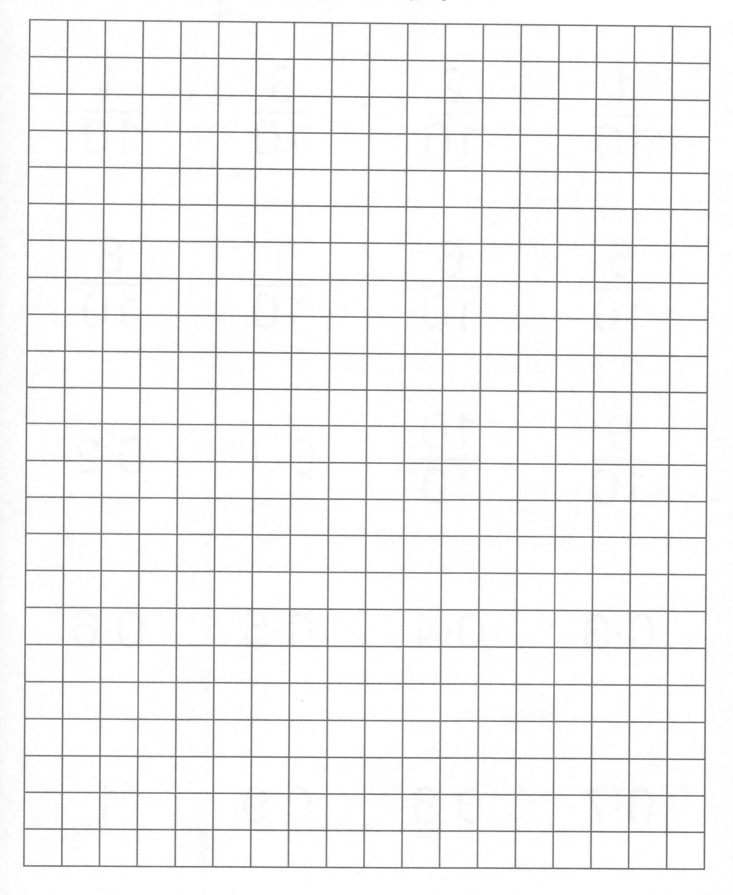

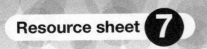

Resource sheet 7: Equivalence snap

$\dfrac{1}{10}$	$\dfrac{2}{10}$	$\dfrac{3}{10}$	$\dfrac{4}{10}$
$\dfrac{5}{10}$	$\dfrac{6}{10}$	$\dfrac{1}{10}$	$\dfrac{8}{10}$
$\dfrac{9}{10}$	$\dfrac{10}{10}$	0·1	0·2
0·3	0·4	0·5	0·6
0·7	0·8	0·9	1

Resource sheet 8: Blank place value grid (decimals)

h	
t	
•	
U	
T	
H	
Th	

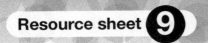

Resource sheet 9: 100 square

1	2	3	4	5	6	7	8	9	10
11	12	13	14	15	16	17	18	19	20
21	22	23	24	25	26	27	28	29	30
31	32	33	34	35	36	37	38	39	40
41	42	43	44	45	46	47	48	49	50
51	52	53	54	55	56	57	58	59	60
61	62	63	64	65	66	67	68	69	70
71	72	73	74	75	76	77	78	79	80
81	82	83	84	85	86	87	88	89	90
91	92	93	94	95	96	97	98	99	100

Resource sheet 10: Fraction spinner

How to use the spinner

Hold the paper clip in the centre of the spinner using the pencil and gently flick the paper clip with your finger to make it spin.

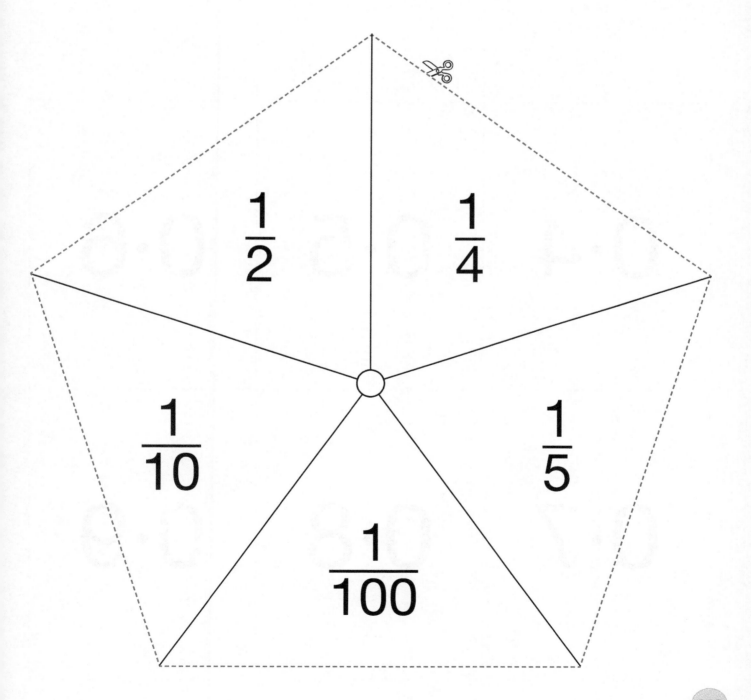

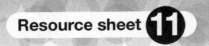

Resource sheet 11: One-place decimal digit cards

0·1	0·2	0·3
0·4	0·5	0·6
0·7	0·8	0·9

Resource sheet 12: Blank number lines

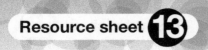

Resource sheet 13: Addition strategies

Name	Description	Example
Counting on	Continue counting from any number in steps	Calculate 86 + 40 Strategy: Count on in 10s, i.e. 96, 106, 116, 126
'Friendly' numbers (compatible numbers)	Numbers easily recognised as adding to a multiple of 10, 100, …	Calculate 425 + 375 Split: 400 + 25 + 300 + 75 25 + 75 = 100 400 + 300 + 100 = 800
Partioning	Mentally splitting one or both numbers into place values before adding.	Calculate 237 + 324 Split: 237 into 200, 30 and 7; 324 into 300, 20 and 4 Add hundreds: 200 + 300 = 500 Add tens: 30 + 20 = 50 Add ones: 7 + 4 = 11 237 + 324 = 500 + 50 + 11 = 561
Bridging	Using multiples of 10/100 … as 'landing points' for adding.	Calculate 477 + 127 Split 127: 23 + 104 Bridge to 500, with 477 + 23 = 500, then 500 + 104 = 604
Compensation	Rounding to a multiple of 10, 100, …, adding second number to rounded number, then making adjustment to compensate for rounding.	Calculate 98 + 85 Bridge 98 to 100 100 + 85 = 185 Adjust 185 down 2 to compensate for rounding up by 2 185 − 2 = 183

Resource sheet 14: Subtraction strategies

Name	Description	Example
Counting on	Continue counting back from any number in steps.	Calculate 97 – 30 Count back in 10s, i.e. 87, 77, 67
Partioning	Mentally split one or both numbers into place values before subtracting.	Calculate 744 – 431 Split 744: 700, 40 and 4; 431: 400, 30 and 1 Subtract hundreds: 700 – 400 = 300 Subtract tens: 40 – 30 = 10 Subtract ones: 4 – 1 = 3 300 + 10 + 3 = 313
Bridging	Use multiples of 10, 100, … as 'landing points' for subtraction.	Calculate 814 – 367 Split 367: 314 + 53 Bridge to 500: 814 – 314 = 500, then 500 – 53 = 447
Compensation	Round the second number to a multiple of 10, 100, …, subtract the number, then make adjustment to compensate for rounding.	Calculate 587 – 298 Bridge 298 to 300 587 – 300 = 287 Adjust 287 up 2 to compensate for taking 2 too many 287 + 2 = 289
'Friendly' numbers	Numbers are easily recognised as subtracting to give a multiple of 10, 100, …	Calculate 925 – 175 Split 175: 125 + 50 925 – 125 = 800 800 – 50 = 750

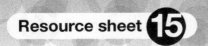

Resource sheet 15: 1–3 spinner

How to use the spinner

Hold the paper clip in the centre of the spinner using the pencil and gently flick the paper clip with your finger to make it spin.

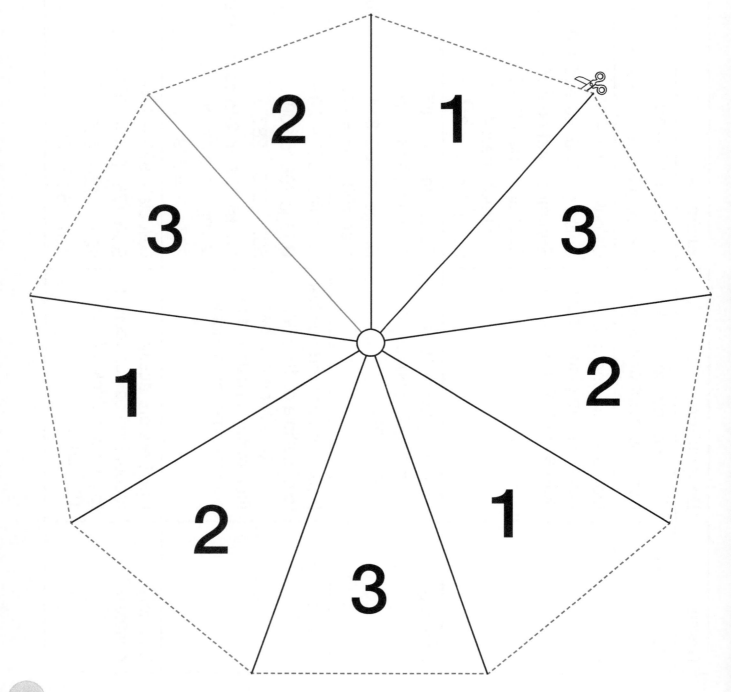

Resource sheet 16: 0–3 spinner

How to use the spinner

Hold the paper clip in the centre of the spinner using the pencil and gently flick the paper clip with your finger to make it spin.

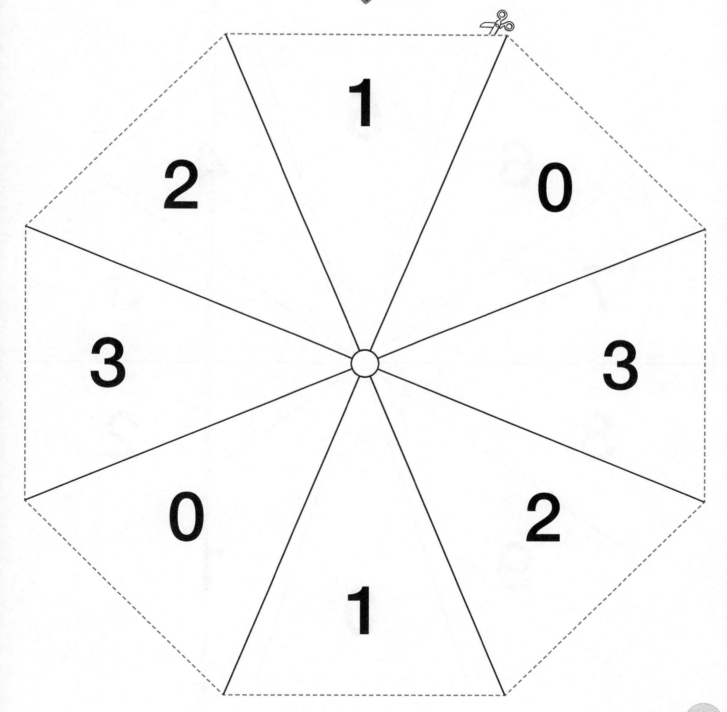

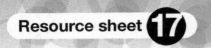

Resource sheet 17: 0–9 spinner

How to use the spinner

Hold the paper clip in the centre of the spinner using the pencil and gently flick the paper clip with your finger to make it spin.

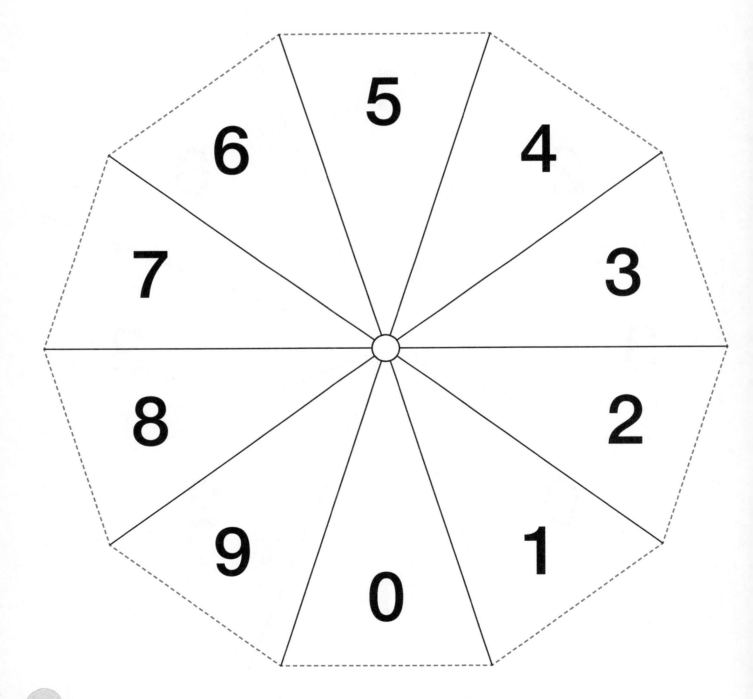

Resource sheet 18: Sorting triangles

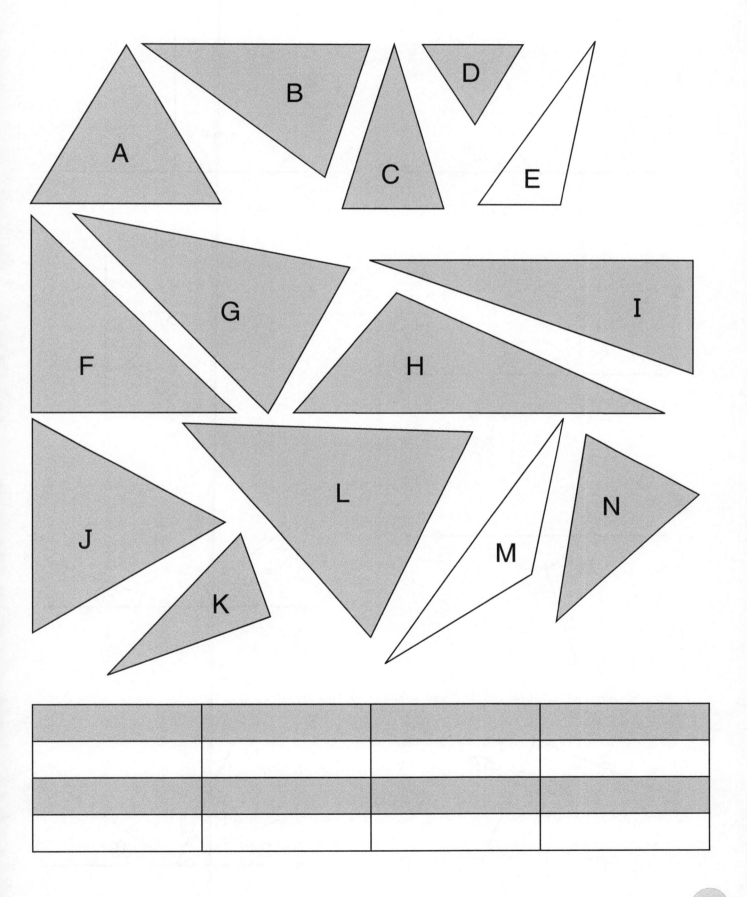

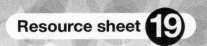

Resource sheet 19: Angles on a straight line

1

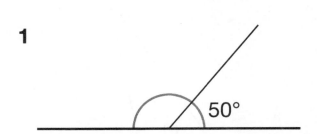

50°

2

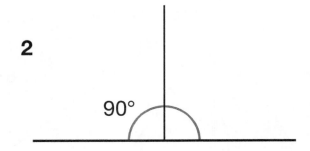

90°

3

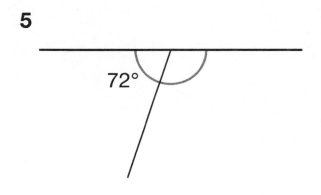

35°

4

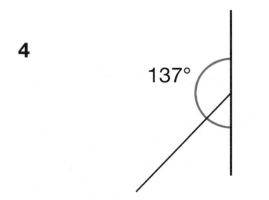

137°

5

72°

6

35°   20°

7

90°   67°

8

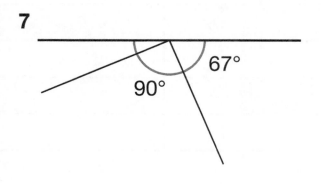

41°

24°   35°

Resource sheet 20: Bar line template

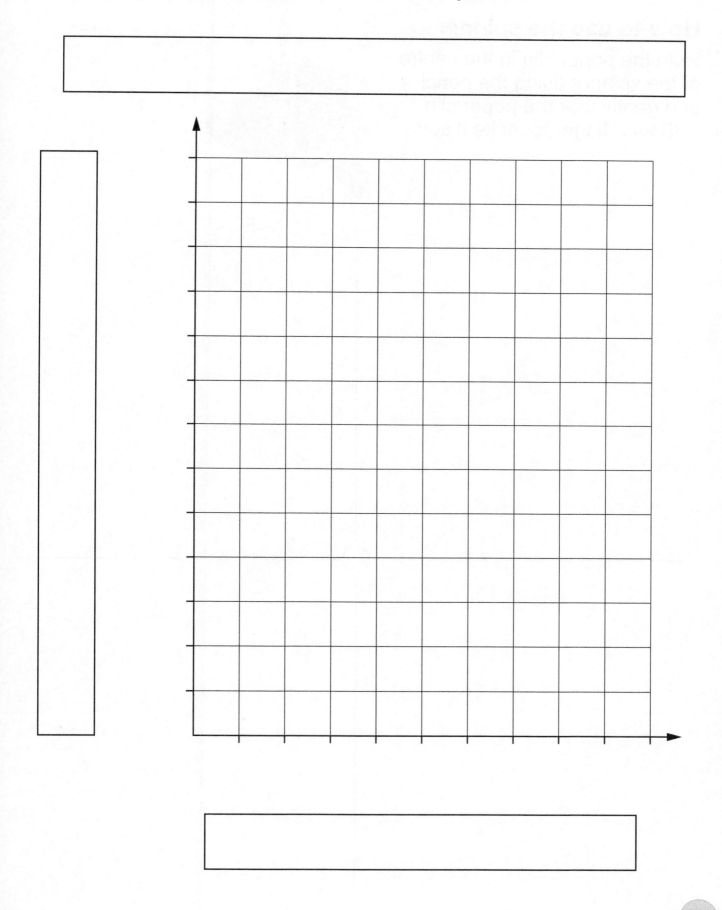

Resource sheet 21: Blank four sector spinner

How to use the spinner

Hold the paper clip in the centre of the spinner using the pencil and gently flick the paper clip with your finger to make it spin.

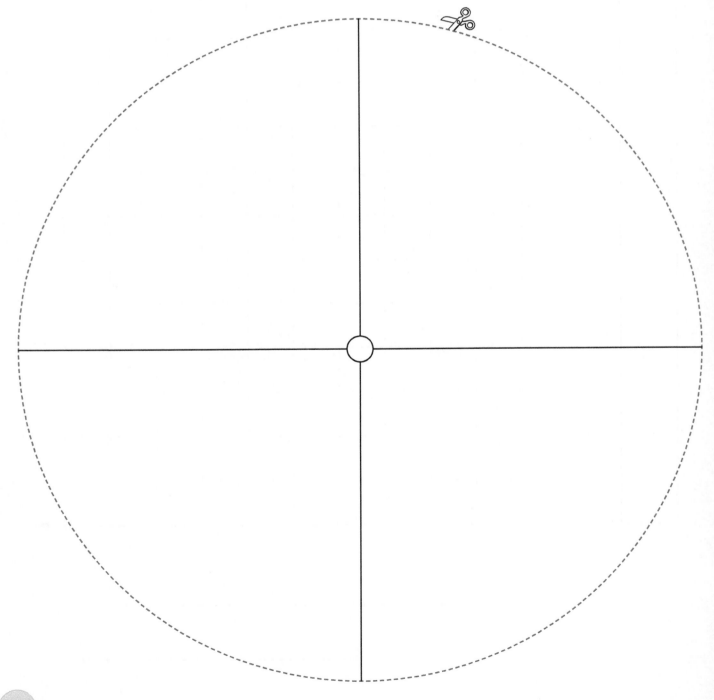

Unit 1

Lesson 1: Counting on and back

Challenge 1
a 18, **20, 22, 24**, 26, **28, 30**, 32, **34, 36**
b 125, **120, 115, 110**, 105, **100, 95**, 90, **85, 80**
c 31, **34, 37, 40**, 43, **46, 49**, 52, **55, 58**
d 290, **280, 270, 260**, 250, **240, 230**, 220, **210, 200**

Challenge 2
a 46, **55, 64, 73**, 82, **91, 100**, 109, **118, 127**
b 83, **76, 69, 62**, 55, **48, 41**, 34, **27, 20**
c 13, **21, 29, 37**, 45, **53, 61**, 69, **77, 85**
d 61, **57, 53, 49**, 45, **41, 37**, 33, **29, 25**
e 210, **200, 190, 180**, 170, **160, 150**, 140, **130, 120**

Challenge 3
a 17 is not part of the sequence
b 55 is not part of the sequence
c 29 circled
d 75 circled

Lesson 2: Number sequences

Challenge 1
a 40, 45, 50   **b** 17, 20, 23
c 40, 30, 20   **d** 11, 9, 7

Challenge 2
1 a 47, 39, 31 Subtract 8
  b 1, 7, 13 Add 6
2 a −7, −14, −21, −28
  b 17, 22, 27 52

Challenge 3
Learners' exercise: answers may vary

Lesson 3: Place value (1)

Challenge 1
6 – hundred thousands (grey)
3 – ten thousands (yellow)
1 – thousands (pink)
7 – hundreds (blue)
0 – tens (red)
9 – units (green)

Challenge 2
1 b 80   **2** 87 421, 12478
  c 8000   96320, 20369
  d 0   875 310, 103 578
  e 400 000   866 420, 204 668
  f 8000   988 332, 233 889

Challenge 3
Answers may vary

Lesson 4: Place value (2)

Challenge 1
a 7000   **b** 50 000, 9
c 60 000, 300   **d** 400 000, 90

Challenge 2
1 b Seventy thousand, nine hundred and thirty-three
  c Sixty thousand, seven hundred and seventeen
  d Forty-four thousand, four hundred and forty-four
  e Eight hundred and twenty-nine thousand, six hundred and forty-five
  f Five hundred and five thousand, two hundred and two
  g Seven hundred and eleven thousand and ten
  h Three hundred and thirty-three thousand and thirty-three

2 a 47 235, forty-seven thousand, two hundred and thirty-five
  b 83 041, eighty-three thousand and forty-one
  c 921 623, nine hundred an twenty-one thousand, six hundred and twenty-three
  d 504 598, five hundred and four thousand, five hundred and ninety-eight
  e 170 806, one hundred and seventy thousand, eight hundred and six
  f 600 167, six hundred thousand, one hundred and sixty-seven

Challenge 3
b 8000 (8 thousands)
c 300 000 (3 hundred thousands)
d 200 (2 hundreds)
e 10 000 (1 ten thousand)

Lesson 5: Multiplying and dividing by 10 or 100

Challenge 1
Numbers correctly placed in the grid:

a 3260   **b** 62 700   **c** 234   **d** 908

Challenge 2
Numbers correctly placed in the grid:

1 a 497 300   **b** 282 900
  c 809 000   **d** 79
  e 38   **f** 99
2 35 790
  4370
  93 500
  40 400

Challenge 3
$4.20, $42
$52.50, $525
$17, $1700
$31.30, $313

Lesson 6: Rounding

Challenge 1
b 4560   **c** 1230   **d** 7810
e 6340   **f** 9980

Challenge 2

1 b 1500   **c** 3000   **d** 2700   **e** 8000
   f 7800   **g** 4200   **h** 5000

2 b 4000   **c** 7000   **d** 9000   **e** 6000
   f 6000   **g** 4000   **h** 9000   **i** 10 000
   j 7000

Challenge 3

Maya: 645, Sarah: 4500, Kyle: 7949, Puneet: 4499

Lesson 7: Comparing and ordering

Challenge 1

a >   **b** >   **c** >   **d** <

Challenge 2

1 a 24 281, 24 734, 35 748, 27 201, 27 369
   b 52 799, 53 463, 53 491, 54 627, 54 634

2 a 323 567, 428 139, 428 913, 673 656, 789 452
   b 102 255, 102 522, 107 656, 117 566, 117 656
   c 123 909, 123 990, 124 099, 124 909, 125 909
   d 827 679, 827 696, 827 697, 872 770, 872 707
   e 305 008, 305 080, 305 800, 308 505, 308 550

Challenge 3

Learners' exercise: answers may vary

Lesson 8: Odds, evens and multiples (1)

Challenge 1

23, 99, 337 circled blue
48, 340, 450, 862, 990 circled green
340, 450, 990 circled red and orange

Challenge 2

1 957 − 478, 779 + 436, 836 − 463 coloured blue
   243 + 349, 552 + 638, 782 − 356 coloured green
   552 + 638, 779 + 436 coloured red

2 9 × 9, 99 ÷ 3, 49 ÷ 7 coloured blue
   24 ÷ 6, 5 × 14, 4 × 12 coloured green
   5 × 14 coloured red

Challenge 3

Answers may vary to find that there are 125

Unit 2

Lesson 1: Whole numbers

Challenge 1

a −6, **−4**, **−2**, 0, 2, **4**, **6**, 8, **10**
b 10, **5**, **0**, **−5**, −10, **−15**, **−20**, −25, **−30**
c −60, **−50**, **−40**, **−30**, −20, **−10**, **0**, 10, **20**
d 20, **10**, **0**, **−10**, −20, **−30**, **−40**, −50, **−60**

Challenge 2

a 12, **6**, **0**, **−6**, −12, **−18**, **−24**, −30, **−36**
b −0.9, **−0.8**, **−0.7**, **−0.6**, −0.5, **−0.4**, **−0.3**, −0.2, **−0.1**
c −20, **−12**, **−4**, 4, 12, **20**, **28**, 36, **44**
d 15, **8**, **1**, **−6**, −13, **−20**, **−27**, −34, **−41**
e −14, **−11**, **−8**, **−5**, −2, 1, **4**, 7, **10**
f 12, **3**, **−6**, **−15**, −24, **−33**, **−42**, −51, **−60**

Challenge 3

a **−21**, −27, −33, −39, −45, **−51, −57, −63**
b **41, 33**, 25, 17, 9, 1, **−7, −15**
c **−14, −11, −8, −5**, −2, 1, 4, **7**
d **−12, −7, −2, 3**, 8, 13, 18, 23

Lesson 2: Positive and negative numbers

Challenge 1

Temperatures correctly matched to the scale
Moscow, Wellington

Challenge 2

a −11, −3, 7, 8   **b** −17, −14, 1, 8
c −10, −4, 12, 15   **d** −19, −5, 0, 13

Challenge 3

Louis Oosthuizen, Branden Grace, Jason Day, Bubba
Watson, Tiger Woods, Sergio Garcia, Rory McIlroy,
Emiliano Grillo

Lesson 3: Calculating temperature change

Challenge 1

Day 2: rise of 14 degrees   Day 3: fall of 9 degrees
Day 4: rise of 14 degrees   Day 5: fall of 7 degrees
Day 6: rise of 15 degrees   Day 7: rise of 15 degrees

Challenge 2

1 a 14 degrees   **b** 26 degrees   **c** 29 degrees
2 Day 1: 12°C, 23   Day 2: 9°C, 35   Day 3: 17°C, 16
   Day 4: 20°C, 15   Day 5: 9°C, 37   Day 6: 24°C, 11
   Day 7: 3°C, 36

Challenge 3

a 12   **b** 13 degrees
c 23 degrees   **d** 23 degrees

Lesson 4: Odds, evens and multiples (2)

Challenge 1

even + even = even
even − even = even
even × even = even

even + odd = odd
even − odd = odd
even × odd = even

odd + even = odd
odd − even = odd
odd × even = even

odd + odd = even
odd − odd = even
odd × odd = odd

Challenge 2

Learners' exercise: answers may vary

Challenge 3

Learners' exercise

Unit 3

Lesson 1: Tenths

Challenge 1
a $\frac{3}{10}$, 0.3   **b** $\frac{7}{10}$, 0.7   **c** $\frac{5}{10}$, 0.5
d $\frac{9}{10}$, 0.9   **e** $\frac{1}{10}$, 0.1   **f** $\frac{10}{10}$, 1

Challenge 2
1 0.3, 0.6, 0.9, 1.3, 1.5, 1.9
2 a 0.3   **b** 0.9   **c** 0.1   **d** 0.6
   e 0.4   **f** 0.8   **g** 1   **h** 0.7
3 a 0.4, 0.5   **b** 0.9, 1   **c** 0.5, 0.4
   d 0.2, 0.1   **e** 1.5, 1.6   **f** 1.9, 2

Challenge 3
b 2.2 (2.3)   **c** 4.3 (4.4)   **d** 3.5 (3.6)
e 7.6 (7.7)   **f** 18.8 (18.9)   **g** 10.4 (10.5)
h 20.2 (20.3)   **i** 17.2 (17.3)   **j** 13.4 (13.5)

Lesson 2: Hundredths

Challenge 1
a $\frac{66}{100}$, 0.66   **b** $\frac{84}{100}$, 0.84
c $\frac{28}{100}$, 0.28   **d** $\frac{43}{100}$, 0.43

Challenge 2
1 0.07, 0.13, 0.25, 0.41, 0.49, 0.77, 0.83, 0.99
2 a 0.33   **b** 0.47   **c** 0.79   **d** 0.11
   e 0.7   **f** 0.01   **g** 0.61   **h** 0.5
3 a 0.26, 0.27   **b** 0.69, 0.7   **c** 0.90, 0.91
   d 0.4, 0.39   **e** 2, 2.01   **f** 0.02, 0.01

Challenge 3
H, U, N, D, R, E, D, P, A, R, T, S

Lesson 3: Multiplying by 10 or 100

Challenge 1
870, 9340, 6120
400, 6700, 4800, 21 700

Challenge 2
1 9, 237, 708, 22, 303, 4043 coloured yellow
   55, 484, 123, 6, 5054, 99, 106, 404, 10 000, 1, 666,
   1111, 37, 8089 coloured blue
2 300, 4500, 72 500, 115 000, 190 300, 400 400, 700 700

Challenge 3
a $20   **b** $300   **c** $500
d $120   **e** $2500   **f** $99 900

Lesson 4: Dividing by 10 or 100

Challenge 1
5.5, 38.1, 88.8, 190.2
2.12, 6.06, 23.45, 99.99

Challenge 2
1 11, 51, 3330, 71.7, 77 777, 6006, 2, 5505, 44, 299,
   9191, 7272 coloured green
   437, 2302, 606, 1010, 3, 8100, 99 009, 51 005
   coloured red
2 0.04, 0.70, 0.58, 2.04, 30.09, 240.89, 5324.88

Challenge 3
a $10.25   **b** $27.25   **c** $788.90
d $876.40   **e** $8734.55   **f** $93.30

Unit 4

Lesson 1: Tenths and hundredths

Challenge 1
b four tenths   **c** two ones   **d** zero tenths
e five tenths   **f** three tenths   **g** six tens
h seven hundredths

Challenge 2
1 a 0.6 + 0.07   **b** 0.3 + 0.03
   c 0.5 + 0.01   **d** 4 + 0.8 + 0.09
   e 9 + 0.1 + 0.01   **f** 6 + 0.06
   g 5 + 0.5 + 0.05
2 a 0.43   **b** 0.72   **c** 9.61
   d 6.01   **e** 24.78   **f** 69.05
   g 800.08   **h** 9140.52   **i** 6000.66

Challenge 3
a 9.37   **b** 54.82   **c** 235.1
d 577.26   **e** 4846   **f** 10 000.18

Lesson 2: Comparing decimals

Challenge 1
0.3 – 0.7 – 0.23 – 0.26; 0.26 on the box

Challenge 2
1 a <   **b** >   **c** <   **d** >   **e** <   **f** >
   g >   **h** <   **i** <   **j** <
2 a >   **b** >   **c** >   **d** <   **e** >   **f** <
   g >   **h** >   **i** <   **j** >

Challenge 3
Answers may vary

Lesson 3: Ordering decimals

Challenge 1
1 a 0.4, 1, 1.4   **b** 2, 3.2, 3.6   **c** 2.4, 4.2, 4.8
2 a 1.5   **b** 3.8   **c** 0.5   **d** 2.9

Challenge 2
1 a 4.12, 4.17, 4.23, 4.24, 4.25
   b 7.17, 7.25, 7.26, 7.32, 7.34
   c 2.86, 2.87, 2.91, 2.94, 2.99
   d 9.01, 9.02, 9.51, 9.56, 9.58
2 Answers may vary, for example:
   a 23.42, 23.4   **b** 50.09, 50.11
   c 80, 79.9   **d** 33.34, 33.37
   e 92.11, 92.15   **f** 111.92, 111.95

Challenge 3
Answers may vary

Lesson 4: Rounding decimals
Challenge 1
3, 3, 4, 4, 4, 5

Challenge 2
1 a 16, 17   b 5, 6   c 12, 13
  d 3, 4   e 24, 25   f 50, 51
  g 100, 101 h 136, 137 i 998, 999
2 a 1   b 6   c 12
  d 32   e 30   f 86
  g 0   h 150
3 a $7   b $21   c $60
  d $100   e $237   f $556

Challenge 3
a 25.71   b 210.19 or 210.11
c 539.79

Unit 5

Lesson 1: Equivalent fractions
Challenge 1
One $\frac{1}{2}$ bar shaded, two $\frac{1}{4}$ bars shaded,
four $\frac{1}{8}$ bars shaded
$\frac{1}{2} = \frac{2}{4} = \frac{4}{8}$

Challenge 2
1 a One $\frac{1}{5}$ bar shaded, two $\frac{1}{10}$ bars shaded
    $\frac{1}{5} = \frac{2}{10}$
  b One $\frac{1}{3}$ bar shaded, two $\frac{1}{6}$ bars shaded
    $\frac{1}{3} = \frac{2}{6}$
2 $\frac{1}{2} - \frac{2}{4}$
  $\frac{2}{8} - \frac{1}{4}$
  $\frac{8}{4} - \frac{3}{4}$
  $\frac{6}{10} - \frac{2}{5}$
  $\frac{4}{10} - \frac{2}{5}$

Challenge 3
$\frac{1}{2}, \frac{2}{4}, \frac{4}{8}, \frac{5}{10}$ coloured red
$\frac{1}{3}, \frac{2}{6}$ coloured yellow
$\frac{1}{5}, \frac{2}{10}$ coloured blue

Lesson 2: Fraction and decimal equivalents
Challenge 1
1 $\frac{7}{10} - 0.7$
  $\frac{1}{4} - 0.25$
  $\frac{3}{10} - 0.3$
  $\frac{1}{10} - 0.1$
  $\frac{3}{4} - 0.75$
  $\frac{9}{10} - 0.9$
  $\frac{1}{2} - 0.5$
2 $\frac{1}{10}, \frac{1}{4}, \frac{3}{10}, \frac{1}{2}, \frac{7}{10}, \frac{3}{4}, \frac{9}{10}$

Challenge 2
1 a 4   b 2   c 1   d 10
  e 75   f 10   g 75   h 25
2 a 0.75, $\frac{3}{4}$ b 0.9, $\frac{9}{10}$ c 0.5, $\frac{1}{2}$
  d 0.75, $\frac{3}{4}$ e 0.25, $\frac{1}{4}$ f 0.3, $\frac{3}{10}$
3 a 0.1, $\frac{1}{4}$, 0.3, $\frac{1}{2}$
  b 0.2, $\frac{1}{4}$, 0.7, $\frac{3}{4}$
  c $\frac{25}{100}, \frac{3}{4}, \frac{76}{100}$, 0.8

Challenge 3
a 2.75   b 3.5   c 7.25   d 5.75
e 9.7   f 12.3   g 15.09   h 17.29
i 16.67   j 20.99   k 35.07   l 100.01

Lesson 3: Mixed numbers
Challenge 1
Proper fractions: $\frac{3}{5}, \frac{7}{10}$
Mixed numbers: $6\frac{2}{3}, 4\frac{1}{6}, 11\frac{5}{8}, 8\frac{1}{4}$
Improper fractions: $\frac{9}{4}, \frac{10}{3}, \frac{11}{9}$

Challenge 2
1 a $3\frac{1}{4}$ b $3\frac{1}{3}$ c $5\frac{2}{5}$ d $5\frac{7}{8}$
2 a $3\frac{1}{8}, 3\frac{3}{8}, 3\frac{7}{8}, 4\frac{3}{8}, 4\frac{5}{8}$ b $7\frac{1}{10}, 7\frac{3}{10}, 7\frac{7}{10}, 8\frac{1}{5}, 8\frac{4}{5}$
3 $2\frac{1}{5}, 2\frac{2}{5}, 2\frac{3}{5}, 3\frac{2}{5}, 3\frac{4}{5}$

Challenge 3
a $\frac{11}{3}$ b $\frac{5}{2}$ c $\frac{19}{4}$ d $\frac{42}{5}$ e $\frac{97}{10}$ f $\frac{71}{12}$

Lesson 4: Fractions of quantities
Challenge 1
a 10   b 20   c 8   d 12
e 6   f 11   g 5   h 9

Challenge 2
1 a 20, 3, 60   b 9, 2, 18
  c 9, 5, 45   d 7, 5, 35
2 a 30   b 36   c 56
  d 27   e 48   f 60

Challenge 3
$330, $220
$1330, $570
$303, $404

Unit 6

Lesson 1: Per cent symbol
Challenge 1
a 9   b 49   c 27   d 73

Challenge 2
1 a 19 squares shaded   b 27 squares shaded
  c 55 squares shaded   d 73 squares shaded
2 a 13   b 29   c 76   d 81

Challenge 3
Learners' exercise: answers may vary

Lesson 2: Expressing fractions as percentages

Challenge 1
a $\frac{9}{100} = 9\%$ **b** $\frac{33}{100} = 33\%$ **c** $\frac{77}{100} = 77\%$

Challenge 2
1 a 50 squares shaded, 50%
  b 10 squares shaded, 10%
  c 1 square shaded, 1%

2 a 50 **b** 23 **c** 10
  d 61 **e** 1 **f** 93

Challenge 3
1 a $\frac{23}{100}$ **b** $\frac{67}{100}$ **c** $\frac{59}{100}$ **d** $\frac{1}{10}$
  e $\frac{1}{100}$ **f** $\frac{50}{100}$ or $\frac{1}{2}$

2 a 0.5, 50% **b** 0.25, 25%
  c 0.75, 75% **d** 0.7, 70%
  e 0.4, 40% **f** 0.75, 75%

Lesson 3: Percentages of quantities

Challenge 1
a 1 **b** $150 **c** 20 km **d** 73 m
e 46 g **f** $100 **g** 200 litres **h** 350 years

Challenge 2
1 a $100 **b** 87.5 km **c** $721
  d 94 g **e** $13.45 **f** 684 hours
2 a 32 **b** 192 **c** 156

Challenge 3
a $6 **b** 15 g **c** $1350
d 690 kg **e** 2820 hours **f** $1071
g $1942.50 **h** $6939

Lesson 4: Percentage problems

Challenge 1
a 27 birds **b** 43 sheep
c 284 tiles **d** 23 balloons

Challenge 2
1 a $21 **b** $2400 **c** 1700 **d** $1050
2 a 5
  b 35
  c 333
  d 975

Challenge 3
a 2161 **b** $1251 **c** $1080 **d** 1302

Unit 7

Lesson 1: Proportion

Challenge 1
a Pattern continued, 2
b Pattern continued, 4
c Pattern continued, 5

Challenge 2
1 b Angela has $\frac{1}{3}$ as many biscuits as Toby, Angela has 3 biscuits
  c Angela has $\frac{1}{5}$ as many biscuits as Toby, Angela has 2 biscuits

2 a 2 in every 3 tiles is spotted, $\frac{2}{3}, \frac{1}{3}$
  b 1 in every 4 tiles is spotted, $\frac{1}{4}, \frac{3}{4}$

Challenge 3

a	b	c
$\frac{2}{3}$	$\frac{3}{4}$	$\frac{5}{6}$
$\frac{1}{3}$	$\frac{1}{4}$	$\frac{1}{6}$
16	18	20
8	6	4
24	27	30
12	9	6

Lesson 2: Proportion problems

Challenge 1
a 20 **b** 36 **c** 100

Challenge 2
1 a 15, 10 **b** 24, 42 **c** 70, 56
2 a 360, 540 **b** 400, 720

Challenge 3
a 50, 100 **b** 100, 200 **c** 150, 300

Lesson 3: Ratio

Challenge 1
1 b 2:3 **c** 4:3 **d** 2:7

Challenge 2
1 a 1:3 **b** 5:1 **c** 3:1
2 a 3 **b** 12 **c** 3 **d** 10

Challenge 3
a 36 **b** 24
c No. There will be a total of 40 wheels.

Lesson 4: Ratio problems

Challenge 1
Oats: 40, 80, 160
Butter: 20, 40, 80
Golden syrup: 4, 8, 16
Sugar: 10, 20, 40
Ginger: 2, 4, 8

Challenge 2

1 a Daffodils: 4, 6, 24, 60
Crocuses: 6, 9, 36, 90
Roses: 10, 15, 60, 150
Pansies: 2, 3, 12, 30
Hyacinths: 8, 12, 48, 120

b No. $\frac{1}{3}$ of 22 is not a whole number

2 Longs: 50, 100, 800, 1800
Shorts: 32, 64, 512, 1152
Cogs: 90, 180, 1440, 3240
Wheels: 72, 144, 1152, 2592
Bases: 46, 92, 736, 1656

Challenge 3

Side	Size 1	Size 2	Size 3	Size 4	Size 5
Side A	3	6	18	30	54
Side B	7	14	42	70	126
Side C	4	8	24	40	72
Side D	2	4	12	20	36
Side E	9	18	54	90	162

Unit 8

Lesson 1: Counting on or back (2)

Challenge 1

a 273, 283, 293, 303, 323, 333, 343
b 4969, 5069, 5169, 5369, 5469, 5569, 5769
c 4306, 5306, 6306, 8306, 9306

Challenge 2

1 a 298, 308, 318, 328, 348, 358, 368
b 2183, 2193, 2203, 2223, 2233, 2243, 2263
c 46 629, 47 629, 48 629, 49 629, 50 629, 52 629, 53 629, 54 629

2 3768, 4068, 4368, 5468, 10 468
5007, 5037, 5067, 5287, 5587, 5887, 6987, 11 987
6229, 6259, 6289, 6509, 6809, 7109, 8209, 13 209
4243, 4213, 4183, 4053, 3753, 3453, 3253, 1253
5122, 5092, 5062, 4932, 4632, 4332, 4132, 2132
8066, 8036, 8006, 7876, 7576, 7276, 7076, 5076

Challenge 3

b +100 **c** −1000 **d** +100
e +10 **f** −5000 **g** −600
h +100 **i** −3000 **j** −600

Lesson 2: Adding 2- and 3-digit numbers

Challenge 1

a 133 **b** 168 **c** 213 **d** 327
e 432 **f** 566 **g** 694 **h** 989
i 1206 **j** 1241 **k** 1223 **l** 1708

Challenge 2

1 a i 25 **ii** 50 **iii** 220 **iv** 540
b i 43 **ii** 534 **iii** 326 **iv** 113
c i 96 **ii** 516 **iii** 284 **iv** 470
d i 391 **ii** 590 **iii** 395 **iv** 297

2 a Answers may vary
b i 106 **ii** 325 **iii** 276

Challenge 3

a 70 **b** 160 **c** 474 **d** 579 **e** 428 **f** 388
g 624 **h** 325 **i** 298 **j** 796 **k** 243 **l** 965

Lesson 3

Challenge 1

a 133 **b** 168 **c** 213 **d** 58 **e** 33 **f** 237
g 506 **h** 415 **i** 717 **j** 209 **k** 112 **l** 198
m 187 **n** 384 **o** 93

Challenge 2

1 a i 25 **ii** 50 **iii** 220 **iv** 540
b i 43 **ii** 534 **iii** 326 **iv** 113
c i 96 **ii** 516 **iii** 284 **iv** 470
d i 391 **ii** 590 **iii** 395 **iv** 297

2 a Answers may vary
b i 106 **ii** 325 **iii** 276

Challenge 3

a 70 **b** 160 **c** 476 **d** 579 **e** 428 **f** 386
g 624 **h** 325 **i** 298 **j** 796 **k** 243 **l** 965

Lesson 4: Adding more than two numbers

Challenge 1

a 98 **b** 195 **c** 493 **d** 593 **e** 967 **f** 380
g 373 **h** 2171 **i** 2426 **j** 2946

Challenge 2

1 a $931 **b** $1626 **c** $2559
d $3040 **e** $3263 **f** $3935

2 Red: 2853, Blue: 2879, Green: 2848, Yellow: 2848
Blue wins a trophy

Challenge 3

A: 3357 B: 2478
C: 2340 D: 3482
E: 2872
Order: 2340, 2478, 2872, 3357, 3482

Unit 9

Lesson 1: Near multiples of 10 or 100

Challenge 1

a 38, 58 **b** 87, 107 **c** 144, 164
d 203, 223 **e** 468, 488 **f** 2573, 2593

Challenge 2

1 696, 648, 738, 599, 976, 475, 1175, 279
970, 922, 1012, 873, 1250, 749, 1449, 553
2636, 2588, 2678, 2539, 2916, 2415, 3115, 2219
3440, 3392, 3482, 3343, 3720, 3219, 3919, 3023
5905, 5857, 5947, 5808, 6185, 5684, 6384, 5488
7062, 7014, 7104, 6965, 7342, 6841, 7541, 6645

2 Learners' exercise

Challenge 3

c +89 **d** −161 **e** +188 **f** −778
g −2302 **h** −958 **i** −539 **j** +298
k −352

Lesson 2: Near multiples of 1000

Challenge 1

a i 1009 **ii** 1016 **iii** 1029
b i 1006 **ii** 1017 **iii** 1026

Challenge 2

1 3004, 3006, 2005, 2002, 1005, 1004, 6, 1
3022, 3024, 2023, 2020, 1023, 1022, 24, 19
4007, 4009, 3008, 3005, 2008, 2007, 1009, 1004
5023, 5025, 4024, 4021, 3024, 3023, 2025, 2020
7014, 7016, 6015, 6012, 5015, 5014, 4016, 4011
8017, 8019, 7018, 7015, 6018, 6017, 5019, 5014

2 Learners' exercise

Challenge 3

a 27 136 **b** 28 419 **c** 24 262 **d** 192 377
e 256 665 **f** 337 008 **g** 175 316 **h** 162 191

Lesson 3: Decimals that total 1 or 10

Challenge 1

a 0.5 **b** 0.7 **c** 0.6 **d** 0.9
e 0.2 **f** 0.4 **g** 0.3 **h** 0.1

Challenge 2

1 81, 53, 42, 76, 34, 23, 39, 17, 5, 49, 67, 1
2 a 7.6 **b** 4.9 **c** 1.7 **d** 2.3
e 5.3 **f** 4.2 **g** 0.1 **h** 8.1
i 3.4 **j** 3.9

3 Learners' exercise: answers may vary

Challenge 3

a 0.22 **b** 0.87 **c** 0.06 **d** 0.43
e 0.35 **f** 0.52 **g** 0.74 **h** 0.69
i 0.12 **j** 0.03 **k** 0.98 **l** 0.91

Lesson 4: Near multiples of 1

Challenge 1

a 0.7 **b** 0.5 **c** 1.3 **d** 2.2
e 1.7 **f** 1.8 **g** 1.9 **h** 1
i 1.8 **j** 1.8 **k** 0.9 **l** 0.6

Challenge 2

1 13.3, 12.5, 9.4, 7.3, 3.4, 1.5
15.4, 14.6, 11.5, 9.4, 5.5, 3.6
17.2, 16.4, 13.3, 11.2, 7.3, 5.4
19.5, 18.7, 15.6, 13.5, 9.6, 7.7
24.6, 23.8, 20.7, 18.6, 14.7, 12.8
27.7, 26.9, 23.8, 21.7, 17.8, 15.9

2 Answers may vary
3 Learners' exercise: answers may vary

Challenge 3

a 19.5 **b** 16.9 **c** 18.9 **d** 46.5
e 41.9 **f** 26.5 **g** 32.9 **h** 27.9
i 56.7 **j** 78.9

Lesson 5: Adding and subtracting 2- and 3-digit numbers

Challenge 1

Learners' exercise

Challenge 2

1 a 41 **b** 50 **c** 119 **d** 39
e 175 **f** 34 **g** 109 **h** 682
i 700 **j** 145 **k** 235 **l** 521
m 830 **n** 626 **o** 1796

2 75 − 50 = 25 807 − 649 = 158
733 − 498 = 235 96 − 68 = 28
37 + 198 = 235 255 + 179 = 434
87 − 59 = 28 682 − 524 = 158
561 − 536 = 25 473 − 445 = 28

Challenge 3

Learners' exercise: answers may vary

Lesson 6: Adding more than two numbers

Challenge 1

1 a 96 **b** 98 **c** 99 **d** 100

Challenge 2

1 a 111, 484, 967 **b** 114, 508, 986
2 a 1703 **b** 2442 m **c** 2842 ml

Challenge 3

Learner's exercise: answers may vary

Lesson 7: Adding decimals

Challenge 1

1 a 6.6 **b** 98.9 **c** 5.79 **d** 79.89
2 a 8.5 **b** 53.7 **c** 11.64 **d** 97.29

Challenge 2

1 a 14.4 **b** 128.1 **c** 16.79 **d** 76.98
e 19.1 **f** 94.3 **g** 16.21 **h** 126.23
2 W, R, I, T, T, E, N M, E, T, H, O, D

Challenge 3

1 $101.72, $120.86, $102.65, $129.32, $127.30
2 $116.07, $120.74, $151.86, $136.23, $143.44

Lesson 8: Subtracting decimals

Challenge 1

1 a 4.2 **b** 22.3 **c** 1.51 **d** 55.63
2 a 3.6 **b** 37.3 **c** 4.16 **d** 42.84

Challenge 2

1 a 2.6 **b** 32.7 **c** 4.72 **d** 44.67
e 2.6 **f** 11.6 **g** 4.07 **h** 5.44
2 D, E, C, I, M, A, L P, L, A, C, ES

Challenge 3

1 $33.74, $74.60, $56.69, $61.30, $70.25
2 $30.13, $41.78, $50.89, $57.79, $56.71

Unit 10

Lesson 1: Adding and subtracting decimals mentally

Challenge 1

a 6.5 b 6.8 c 5.8 d 8.9 e 9.9 f 9.7
g 10 h 9 i 2.1 j 2.1 k 3.2 l 2.2
m 1.5 n 1.1 o 3 p 1

Challenge 2

1 a 19.2 b 48.1 c 70.2 d 42.2
  e 79.3 f 92.4 g 86.5 h 96.8
  i 2.4 j 17.4 k 28.8 l 38.6
  m 13.7 n 48.8 o 61.8 p 53.8

2 a 181.5
    90.2, 91.3
    41.9, 48.3, 43
    18, 23.9, 24.4, 18.6
    7.3, 10.7, 13.2, 11.2, 7.4
    3.1, 4.2, 6.5, 6.7, 4.5, 2.9
    302.4
    151.2, 151.2
    73, 78.2, 73
    33.9, 39.1, 39.1, 33.9
    15.1, 18.8, 20.3, 18.8, 15.1
    6.5, 8.6, 10.2, 10.1, 8.7, 6.4
  b 177.2
    89.2, 88
    42.5, 46.7, 41.3
    18.9, 23.6, 23.1, 18.2
    7.9, 11, 12.6, 10.5, 7.7
    3.4, 4.5, 6.5, 6.1, 4.4, 3.3
    1.8, 1.6, 2.9, 3.6, 2.5, 1.9, 1.4
    187.6
    93.2, 94.4
    44.3, 48.9, 45.5
    20, 24.3, 24.6, 20.9
    8.9, 11.1, 13.2, 11.4, 9.5
    4.3, 4.6, 6.5,6.7, 4.7, 4.8
    2.4, 1.9, 2.7, 3.8, 2.9, 1.8, 3

Challenge 3

a 13.9 b 26.6 c 28.6 d 19.8
e 45.7 f 56.7 g 19.9 h 64.9
i 16.8 j 49.3 k 18.8 l 79.5
m 19.7 n 64.2 o 49.5 p 68.1

Lesson 2: Adding and subtracting near multiples mentally

Challenge 1

a 27 b 48 c 76 d 86
e 412 f 555 g 813 h 935
i 2452 j 4683 k 7366 l 9628
m 74 n 63 o 45 p 73
q 223 r 583 s 537 t 206
u 3535 v 5884 w 4034 x 3922

Challenge 2

1 a 18, 37, 65, 114
  b 305, 277, 238, 180
  c 375, 773, 1071, 1268
  d 897, 600, 502, 303
  e 1022, 4020, 6019, 7016
  f 8235, 5236, 3238, 2241

2 Learners' exercise

Challenge 3

a 29, 29, 35, 47
b 39, 39, 59, 63
c 89, 49, 84, 88
d 59, 59, 39, 93
e 539, 449, 1435, 1687
f 1649, 789, 2209, 2213

Lesson 3: Adding more than two numbers

Challenge 1

1 a 136 b 126 c 162 d 226
  e 194 f 165 g 200 h 435

Challenge 2

1 a 1342 b 1590 c 2046
2 a $1942 b 3207 kg c 928 minutes

Challenge 3

Learners' exercise: answers may vary

Lesson 4: Adding and subtracting decimals

Challenge 1

1 a 7.5 b 68.2 c 9.58 d 118.47
2 a 5.9 b 25.3 c 4.06 d 15.26

Challenge 2

1 a 15.7 b 130.5 c 14.26 d 135.65
  e 2.7 f 22.5 g 3.61 h 26.26
2 a $8.70 b $140.50 c $14.63 d $148.73
  e $11 f $119.90 g $11.77 h $99.32

Challenge 3

a $22.41 b $31.33 c $17.52 d $10.65
e $31.53

Unit 11

Lesson 1: Multiples

Challenge 1

a 45, 70, 155, 230, 680, 795 circled
b 180, 270, 410, 700, 2920 circled
c 600, 1100, 2400, 6000, 8500 circled

Challenge 2

1 Divisible by 2: 6, 788
  Divisible by 5: 5, 725, 2485
  Divisible by 10: none
  Intersection of divisible by 2 and divisible by 5: none
  Intersection of divisible by 2 and divisible by 10: none
  Intersection of divisible by 5 and divisible by 10: none
  Intersection of all three: 930
  In the rectangle outside the circles: 9, 11, 23
  All multiples of 10 are also multiples of 2 and 5

2

	80 or less but greater than 59	40 or less but greater than 19	60 or less but greater than 39	100 or less but greater than 79	20 or less
Multiple of 6	60, 72	24, 30, 36	48, 42, 54		12, 6, 18
Multiple of 7	70, 63	28, 21, 35	49, 56, 42		14, 7
Multiple of 8	64, 72	24, 32	40, 48, 56	80	16, 8
Multiple of 9	72, 63	36, 27	45, 54	81	9, 18

72 appears three times

Challenge 3

	Multiple of 6	Multiple of 7
Multiple of 8	24, 48, 72, 96, 120, 144, 168, 192	56, 112, 168
Multiple of 9	18, 36, 54, 72, 90, 108, 126, 144, 162, 180, 198	63, 126, 189

a 168
b 126

Lesson 2: Factors

Challenge 1

Answers may vary, for example:

a $5 \times 3 = 15$, $15 \div 3 = 5$
  $8 \times 3 = 24$, $24 \div 8 = 3$
  $3 \times 10 = 30$, $30 \div 3 = 10$
  $4 \times 3 = 12$, $12 \div 3 = 4$
  $9 \times 3 = 27$, $27 \div 3 = 9$

b $8 \times 5 = 40$, $40 \div 5 = 8$
  $8 \times 4 = 32$, $32 \div 4 = 8$
  $8 \times 7 = 56$, $56 \div 8 = 7$
  $8 \times 2 = 16$, $16 \div 2 = 8$
  $8 \times 9 = 72$, $72 \div 8 = 9$

Challenge 2

1 a 4   b 6   c 3   d 8   e 10   f 6
  g 4   h 3   i 10   j 5   k 3   l 10
  m 6   n 6   o 9

2 1, 18 2, 9 3, 6
  1, 45 3, 15 5, 9
  1, 81 3, 27 9, 9
  1, 22 2, 11
  1, 40 2, 20 4, 10 5, 8

Challenge 3

a 18: 2, 6, 18
  45: 5, 15, 45
  both: 1, 3, 9
c 36: 4, 12, 36
  54: 27, 54
  both: 1, 2, 3, 6, 9, 18

b 27: 9, 27
  30: 2, 5, 6, 10, 15, 30
  both: 1, 3
d 22: none
  44: 4, 44
  both: 1, 2, 11, 22

Lesson 3: Multiples and factors

Challenge 1

Learners' exercise: answers may vary

Challenge 2

1 a 1, 32 2, 16 4, 8
  b 1, 28 2, 14 4, 7
  c 1, 30 2, 15 3, 10 5, 6

2 a 6, 13 circled
  b 4, 10, 12 circled
  c 18 circled
  d 7, 16 circled

Challenge 3

a Three: 1, 68 2, 34, 4, 17
b Four: 1, 54 2, 27 3, 18 6, 9
c Six: 1, 42 2, 42, 3, 28 4, 21 6, 14 7, 12
d Four: 1, 88 2, 44 4, 22 8, 11

Lesson 4: Multiplying a 3-digit number by a single-digit number (1)

Challenge 1

a 400, 20, 7   b 200, 40, 9   c 500, 8
d 700, 20, 6   e 900, 10, 5   f 800, 30, 8

Challenge 2

1 a 874   b 2792   c 4848
2 a 591   b 6712

Challenge 3

a 978   b 2870   c 5467   d 7264

Lesson 5: Multiplying a 2-digit number by a 2-digit number (1)

Challenge 1

b 136   c 117   d 174

Challenge 2

1 a 667   b 1666   c 2898
2 a 2736   b 3498   c 6141

Challenge 3

a 754   b 2183   c 4221   d 9114

Lesson 6: Multiplying a 2-digit number by a 2-digit number (2)

Challenge 1

12, 48	19, 76	23, 92
8, 32	13, 52	14, 56
18, 72	9, 36	17, 68
11, 44	28, 112	24, 96
22, 88	16, 64	

Challenge 2

1 a 576   **b** 1200   **c** 432   **d** 1280
   e 672   **f** 1104   **g** 400   **h** 720

2 Learners' exercise

Challenge 3

a 40   **b** 55   **c** 36   **d** 42   **e** 330   **f** 270

Lesson 7: Dividing a 3-digit number by a single-digit number (1)

Challenge 1

1 a 3	**b** 30	**c** 300
2 a 4	**b** 40	**c** 400
3 a 4	**b** 40	**c** 400
4 a 5	**b** 50	**c** 500

Challenge 2

1 a 50   **b** 102   **c** 130   **d** 130
   e 27   **f** 82   **g** 51   **h** 52
   i 62   **j** 323   **k** 118   **l** 83

2 M, E, N, T, A, L S, T, R, A, T, E, G, Y

Challenge 3

186, 98 172, 89
672, 68 837, 46

Lesson 8: Dividing a 3-digit number by a single-digit number (2)

Challenge 1

1 8, 16, 24, 32, 40, 48, 56, 64, 72, 80
   a 24   **b** 16   **c** 72
   d 48   **e** 64   **f** 32

Challenge 2

1 a 43   **b** 77
2 a 23 remainder 3   **b** 27 remainder 6

Challenge 3

1 89   **2** 69   **3** 68, £3   **4** 5

Unit 12

Lesson 1: Multiplication and division facts

Challenge 1

a $2 \times 3 = 6$, $3 \times 2 = 6$, $6 \div 3 = 2$, $6 \div 2 = 3$
b $6 \times 7 = 42$, $7 \times 6 = 42$, $42 \div 6 = 7$, $42 \div 7 = 6$

Challenge 2

1 a 3   **b** 4   **c** 12   **d** 8   **e** 8   **f** 49
   g 9   **h** 1   **i** 36
2 a 2   **b** 9   **c** 12   **d** 4   **e** 4   **f** 9
   g 60   **h** 9   **i** 100
3 Learners' exercise

Challenge 3

a 1750   **b** 740   **c** 1290

Lesson 2: Multiplying multiples of 10 and 100 (1)

Challenge 1

a 70   **b** 10   **c** 90   **d** 8   **e** 7   **f** 300
g 100   **h** 8   **i** 5   **j** 10   **k** 40   **l** 2
m 500   **n** 6   **o** 100

Challenge 2

1 120, 210, 270, 900, 1500, 2400
   280, 490, 630, 2100, 3500, 5600
   200, 350, 450, 1500, 2500, 4000
   360, 630, 810, 2700, 4500, 7200
   80, 140, 180, 600, 1000, 1600
   160, 280, 360, 1200, 2000, 3200
2 a £6710   **b** $8020   **c** $7970   **d** $11 080

Challenge 3

a 3, 4
b 15, 1 or 10, 4 or 5, 7
c Any combination of 1, 2, 3, 4, 5, 6, 7, 8 laptops and respectively 121, 105, 89, 73, 57, 41, 25, 9 cars

Lesson 3: Multiplying by 19, 21 or 25 (1)

Challenge 1

6 – 120, 3 – 60, 8 – 160, 5 – 100, 9 – 180, 4 – 80

Challenge 2

1 a 57   **b** 168   **c** 125   **d** 133   **e** 84   **f** 200
   g 114   **h** 189   **i** 150
2 a $284   **b** $326   **c** $443   **d** $615

Challenge 3

a 19, 25   **b** 21, 19 or 25, 16
c 25, 21   **d** 19, 23 or 25, 19

Lesson 4: Multiplication by factors (1)

Challenge 1

Learners' exercise

Challenge 2

1 a 255   **b** 585   **c** 810
   d 1026   **e** 1044   **f** 2268
2 Answers may vary, for example:
   $23 \times 24 = 552$, $19 \times 15 = 285$, $37 \times 36 = 1332$,
   $27 \times 18 = 486$, $49 \times 12 = 588$

Challenge 3

a 1296   **b** 1728   **c** 4096   **d** 4608

Lesson 5: Doubles and halves (1)

Challenge 1
a 28 **b** 13 **c** 16 **d** 17

Challenge 2
1 22, 88   230, 920   4850, 19 700
220, 880   2300, 9200   44.5, 178
2200, 8800   39, 156   445, 1780
31, 124   390, 1560   4450, 17 800
310, 1240   3900, 15 600   49.5, 198
3100, 12 400   48, 192   495, 1980
19, 76   480, 1920   4950, 19 800
190, 760   4800, 19 200
1900, 7600   48.5, 197
23, 92   485, 1970

2 a 39, 69, 615, 8350, 174, 1360, 16 800
b 72, 2700, 2720, 6150, 3200, 5800, 15 400

Challenge 3
a 514 **b** 2810

Lesson 6: Multiplying a 3-digit number by a single-digit number (2)

Challenge 1
a 468 **b** 1464 **c** 3222

Challenge 2
1 a 3522 **b** 5551 **c** 8514
2 a 2898 **b** 6104

Challenge 3
Learners' exercise

Lesson 7: Multiplying a 2-digit number by a 2-digit number (3)

Challenge 1
b 129 **c** 177 **d** 222

Challenge 2
1 a 1269 **b** 2537 **c** 4368
2 a 3087 **b** 5244 **c** 8742

Challenge 3
a 1221 **b** 2646 **c** 6319 **d** 9118

Lesson 8: Dividing a 3-digit number by a single-digit number (3)

Challenge 1
a 110 **b** 105 **c** 74 **d** 140 **e** 103 **f** 54
g 210 **h** 104 **i** 64

Challenge 2
1 a 95 **b** 93 **c** 29 r 3 **d** 16 r 5
2 a 27 **b** 59 **c** 74

Challenge 3
1 346 **2** 196 **3** 3

Unit 13

Lesson 1: Multiplying multiples of 10 and 100 (2)

Challenge 1
210, 420, 560, 1400, 2800, 6300
270, 540, 720, 1800, 3600, 8100
60, 120, 160, 400, 800, 1800
120, 240, 320, 800, 1600, 3600

Challenge 2
1 a 5470 **b** 4950 **c** 14 370
2 a 188 **b** 392 **c** 498 **d** 408
e 752 **f** 693 **g** 1585 **h** 1152

Challenge 3
a 100 800 **b** 66 800 **c** 69 200 **d** 150 700

Lesson 2: Multiplying by 19, 21, or 25 (2)

Challenge 1
a $76 **b** $63 **c** $175 **d** $95
e $126 **f** $200 **g** $152 **h** $189
i $150 **j** $171 **k** $168 **l** $225

Challenge 2
1 40 × 21 = 840   **2 a** 1710 m
70 × 19 = 1330   **b** 1470 m
50 × 21 = 1050   **c** 10 000 m
80 × 19 = 1520   **d** 15 200 m
60 × 21 = 1260
90 × 19 = 1710

Challenge 3
a $2850 **b** $3990 **c** $5750
d $4940 **e** $6720 **f** $9500

Lesson 3: Multiplication by factors (2)

Challenge 1
a 2 **b** 4 **c** 2 **d** 8 **e** 2 **f** 9
g 2 **h** 15

Challenge 2
1 a 102 **b** 182 **c** 228 **d** 378
2 6 × 19 = 114   17 × 8 = 136
13 × 12 = 156   9 × 14 = 126

Challenge 3
a 1380 **b** 1820 **c** 1360
d 3780 **e** 2880 **f** 6720

Lesson 4: Doubles and halves (2)

Challenge 1
b Double 90 is 180
Double 900 is 1800
Double 9000 is 18 000
d Double 60 is 120
Double 600 is 1200
Double 6000 is 12 000
f Half 180 is 90
Half 1800 is 900
Half 18 000 is 9000

c Half 160 is 80
Half 1600 is 800
Half 16 000 is 8000
e Half 80 is 40
Half 800 is 400
Half 8000 is 4000

Challenge 2
1 60 – 30
12 000 – 6000
4000 – 2000
180 – 90

18 000 – 9000
14 000 – 7000
80 – 40
240 – 120

2 a 110 **b** 960 **c** 8400 **d** 440
e 1180 **f** 11 000 **g** 700 **h** 1320
i 12 400 **j** 880 **k** 1540 **l** 13 200
3 a 60 **b** 360 **c** 1600 **d** 180
e 170 **f** 2700 **g** 120 **h** 80
i 3800 **j** 240 **k** 290 **l** 1700

Challenge 3
a 245 **b** 1355 **c** 3317 **d** 185
e 3675 **f** 4189 **g** 305 **h** 4485
i 9482 **j** 425 **k** 2631 **l** 5688

Lesson 5: Doubles and halves (3)

Challenge 1
b Double 4.8 is 9.6
d Double 1.9 is 3.8
f Half 9.6 is 4.8

c Half 7.4 is 3.7
e Half 3.6 is 1.8

Challenge 2
1 9.6 – 4.8
3.8 – 1.9
7.2 – 3.6
13.2 – 6.6

11 – 5.5
6.4 – 3.2
2.8 – 5.6
2.4 – 1.2

2 a 0.8 **b** 3.4 **c** 9.6 **d** 1.8
e 5.2 **f** 11.8 **g** 2.6 **h** 6.8
i 13.4 **j** 17.2 **k** 6.22 **l** 4.54
m 2.66 **n** 8.9 **o** 4.96
3 a 1.1 **b** 2.8 **c** 6.23 **d** 0.7
e 3.9 **f** 1.64 **g** 3.1 **h** 2.18
i 3.86 **j** 0.8 **k** 3.24 **l** 4.97
m 1.9 **n** 4.17

Challenge 3
4.37 – 2.185
17.61 – 3.805
3.43 – 1.715
5.29 – 2.645

4.19 – 2.095
7.77 – 3.885
3.57 – 1.785
5.33 – 2.665

Lesson 6: Multiplying a decimal by a single-digit number

Challenge 1
b $7.3 \times 7 = 51.1$
d $3.7 \times 4 = 14.8$
f $9.5 \times 9 = 85.5$

c $7.8 \times 8 = 62.4$
e $6.2 \times 9 = 55.8$

Challenge 2
1 a 10.8 **b** 28.8 **c** 46.2 **d** 71.2
2 b 14.8 **c** 54.6 **d** 84.6
3 a 13.8 **b** 78.3 **c** 48.3
4 a 40.2 **b** 80 **c** 102.6

Challenge 3
Dalia: 11.1, 19.6, 37.1, 67.8
Jasper: 33.3, 14.7, 21.2, 69.2
Carina: 29.6, 9.8, 26.5, 65.9

Lesson 7: Writing a remainder as a fraction

Challenge 1
a $6\frac{1}{6}, 7\frac{1}{2}, 9\frac{1}{3}$ **b** $5\frac{1}{8}, 8\frac{3}{8}, 10\frac{3}{8}$ **c** $4\frac{1}{3}, 5\,4/9, 8\frac{1}{3}$

Challenge 2
1 a $10\frac{2}{3}$ **b** $8\frac{1}{2}$ **c** $13\frac{2}{5}$
d $10\frac{2}{3}$ **e** $9\frac{2}{7}$ **f** $8\frac{1}{2}$
2 a $31\frac{2}{3}$ ml **b** $16\frac{1}{3}$ ml **c** $12\frac{1}{4}$

Challenge 3
Learners' exercise: answers may vary

Lesson 8: Order of operations

Challenge 1
a 14 **b** 2 **c** 37 **d** 23
e 10 **f** 38 **g** 11 **h** 115

Challenge 2
1 a 43 **b** 76 **c** 9 **d** 74 **e** 1 **f** 8
g 68 **h** 45 **i** 162 **j** 104 **k** 77 **l** 288
2 A, D, D, I, T, I, O, N B, E, F, O, R, E S, U, B, T, R, A, C, T, I, O, N

Challenge 3
a $(88 - 8) \div 2$
c $81 \div (53 - 44)$
e $38 + 10 - 2 \times 6$
g $8 + 8 \times 9 - 11$

b $68 \div 2 + 2 \times 8 + 31 - 42$
d $(9 \times 48) \div (3 + 5) - 13$
f $(64 - 16) \div 3 + 2 \times 6$
h $64 \div (16 \times 4) + 12 - (7 + 5)$

Unit 14

Lesson 1: Triangles

Challenge 1
Triangles correctly coloured

Challenge 2

1 Equilateral triangles: A, H, E
Isosceles triangles: B, F, I,
Scalene triangles: C, D, G, J

2

Triangle	Equilateral	Isosceles	Scalene
all sides equal	✓	✗	✗
two sides equal, two angles equal	✓	✓	✗
no sides equal, no angles equal	✗	✗	✓
can have a right angle	✗	✓	✓
can have an angle greater than 90°	✗	✓	✓
has two or more angles less than 90°	✓	✓	✓
regular shape	✓	✗	✗
irregular shape	✗	✓	✓

Challenge 3
Square: isosceles
Rectangle: scalene
Rhombus: isosceles
Trapezium: scalene
Kite: isosceles or scalene

Lesson 2: Symmetry in regular polygons

Challenge 1
a ✓   **b** ✗   **c** ✓
✗   ✓   ✓

Challenge 2
1 a 4   **b** 5   **c** 8
2 a 3   **b** 8   **c** 6
3 Answers may vary

Challenge 3
Learners' exercise: answers may vary

Lesson 3: Symmetrical patterns

Challenge 1
Learners' exercise: patterns completed in given line of symmetry

Challenge 2
1, 2 Learners' exercise: patterns completed in given line(s) of symmetry

3 Learners' exercise: answers may vary

Challenge 3
a, b Learners' exercise: patterns completed in given line(s) of symmetry which will give rotational symmetry order 2

c Learners' exercise: answers may vary

Lesson 4: Perpendicular and parallel lines

Challenge 1
a Yes   **b** No   **c** No   **d** No
e No   **f** Yes **g** No   **h** No

Challenge 2
1 a parallel   **b** perpendicular
  c parallel   **d** neither
  e perpendicular   **f** parallel
2 Lines identified as described

Challenge 3
Lines drawn as described

Unit 15

Lesson 1: Visualising 3D shapes

Challenge 1
a Cube   **b** Square-based triangle
c Cylinder   **d** Hexagonal prism

Challenge 2
1 Hexagonal pyramid, cone, tetrahedron, pentagonal prism
2 Learners' exercise

Challenge 3
Learners' exercise

Lesson 2: Nets

Challenge 1
a Top row, second net   **b** Second row, second net

Challenge 2
1 A ✗   B ✓   C ✓   D ✗
   E ✓   F ✓   G ✓
2 Learners' exercise

Challenge 3
Net C

Lesson 3: Constructing 3D shapes

Challenge 1
a 20   **b** 27   **c** 36

Challenge 2
1 A: 7   B: 17   C: 136
2 Square, 6, 8, 12
   Triangle, 4, 4, 6
   Square, triangle, 5, 5, 8

Challenge 3
Pentagon, rectangle, 7, 10, 15
Hexagon, rectangle, 8, 12, 18
Triangle, 8, 6, 12

Lesson 4: Relationships between 3D shapes

Challenge 1
Learners' exercise

Challenge 2
a Pentagonal prism **b** Square-based pyramid
c Hexagonal prism **d** Triangular-based pyramid

Challenge 3

Shape	Number of faces	Number of vertices	Number of edges
triangular-based pyramid	4	4	6
square-based pyramid	5	5	8
triangular prism	5	6	9
pentagonal prism	7	10	15
hexagonal prism	8	12	18
tetrahedron	4	4	6
octahedron	8	6	12

Unit 16

Lesson 1: Measuring angles

Challenge 1
a 180° **b** 360° **c** 90°

Challenge 2
1 a 20° **b** 55° **c** 115° **d** 160°
2 a 70° or 75° **b** 135°
  c 175° **d** 30°

Challenge 3
a 52° **b** 106° **c** 142° **d** 99°

Lesson 2: Angle size

Challenge 1
Learners' exercise

Challenge 2
E, C, B, A, D
D, A, B, C, E

Challenge 3
Learners' exercise

Lesson 3: Classifying angles

Challenge 1
a Obtuse **b** Right angle **c** Acute angle

Challenge 2
1 a Acute **b** Obtuse **c** Right angle
  d Right angle **e** Acute **f** Obtuse
2 Learners' exercise

Challenge 3
Learners' exercise

Lesson 4: Angles on a straight line

Challenge 1
a 85° **b** 70° **c** 105°

Challenge 2
1 a 77° **b** 63° **c** 121° **d** 21°

Challenge 3
a 85° **b** 67° **c** 73° **d** 67°

Unit 17

Lesson 1: Reading and plotting coordinates

Challenge 1
a (1, 3) **b** (3, 3) **c** (0, 1)
d (2, 2) **e** (4, 0)

Challenge 2
1 Learners' exercise
2 a Hut, mountain, cave
  b Starfish, mountain
  c Hut, cave
  d Answers may vary, for example: they have the same y-co-ordinate
  e Answers may vary, for example: they have the same x-co-ordinate
  f Answers may vary

Challenge 3
a Octopus at (1, 4) **b** Lake at (2, 2)
c Boat at (2, 5) **d** Dolphin at (1.5, 4.5)
e Answers may vary

Lesson 2: Shapes from co-ordinates

Challenge 1
Square, right-angled triangle, rectangle

Challenge 2
1 Answers may vary, for example: (5, 2), (5, 4), (1, 4)
Answers may vary, for example: (7, 7), (7, 4), (4, 4), (4, 7)
Answers may vary, for example: (0, 6), (3, 6), (3, 0), (0, 0)
2 (7, 2), answers may vary, for example: (2, 3), (3, 7)

Challenge 3
Answers may vary, for example: (8, 2), (2, 5), (2, 2)

Lesson 3: Testing for symmetry

Challenge 1
Rectangle, isosceles triangle, trapezium circled

Challenge 2
1 None: H

1: A, B, C	2: E	3: D
4: I	5: F	6: G

2 Shapes completed symmetrically

Challenge 3
answers may vary

Lesson 4: Reflection (1)
Challenge 1
Shapes completed symmetrically

Challenge 2
Learners' exercises

Challenge 3
Learners' exercise

Lesson 5: Reflection (2)
Challenge 1
Learners' exercise

Challenge 2
Learners' exercises

Challenge 3
Learners' exercises

Lesson 6: Understanding translation
Challenge 1
A cross drawn through B, D and E

Challenge 2
1 a 4, up   **b** 3, right   **c** 5, down
2 Learners' exercise

Challenge 3
Learners' exercise: answers may vary

Lesson 7: Shape translation (1)
Challenge 1
Right 2 squares, up 4 squares
Down 4 squares, right 1 square
Left 2 squares, up 4 squares
Down 4 squares, left 1 square

Challenge 2
1 Learners' exercise
2 a Right 3 squares, down 5 squares
  b Right 5 squares, up 5 squares
  c Down 5 squares, left 2 squares

Challenge 3
Learners' exercise

Lesson 8: Shape translation (2)
Challenge 1
Learners' exercise

Challenge 2
1 a 3 squares right
    3 squares down
    3 squares left
  b 3 squares right
    3 squares down
    3 squares left
  c 3 squares right
    3 squares down
    3 squares left
2 a, b A cross drawn through both diagrams

Challenge 3
Learners' exercise

Unit 18

Lesson 1: Measuring length
Challenge 1
Screw: 17 mm
Crayon: 110 mm
Pencil: 160 mm

Challenge 2
1 97
    107 (circled)
    107 (circled)
    117
    127
    107 (circled)
2 Learners' exercise

Challenge 3
Learners' exercise

Lesson 2: Converting units
Challenge 1
3 km – 3000 m, 23 m – 2300 cm, 77 cm – 770 mm, 8.4 km – 8400 m, 6.7 m – 670 cm

Challenge 2

1 a 6000	**b** 800	**c** 90
d 46 000	**e** 1900	**f** 830
g 5200	**h** 910	**i** 357

2 a 6 m – 600 cm – 6000 mm
  b 0.4 km – 400 m – 40 000 cm
  c 3.9 m – 390 cm – 3900 mm
  d 4000 cm – 40 m – 0.04 km

Challenge 3
1 a 147 cm **b** 1250 m   **c** 31 cm   **d** 1750 m
2 30, 90
   44.6, 133.8
   54.8, 164.4
   62.8, 188.4

Lesson 3: Ordering and rounding length
Challenge 1
a 730, 800, 817, 870   **b** 89, 91, 95.9, 96

Challenge 2
1 a Answers may vary, for example:
   600 mm, 659 mm, 660 mm, 661 mm, 690 mm
  b Answers may vary, for example:
   400 m, 469.09 m, 470.01 m, 4690 m, 4694 m
2 A: 10
   B: 12
   C: 11

Challenge 3
A: 370.5, C: 370.6, B: 371.0, D: 371.1

Lesson 4: Measuring lines
Challenge 1
Learners' exercise

Challenge 2
Learners' exercise

2 a i 8   **ii** 10   **iii** 7
  b i 117   **ii** 45   **iii** 98

Challenge 3
a Learners' exercise
b Learners' exercise: answers may vary
c Learners' exercise

Unit 19

Lesson 1: Measuring mass
Challenge 1
a 2   **b** 3.5   **c** 4.5   **d** 1.5

Challenge 2
1 a 100   **b** 290   **c** 330   **d** 480
2 a 0.8   **b** 2.4   **c** 4.9   **d** 3.1

Challenge 3
Learners' exercise

Lesson 2: Converting units
Challenge 1
4 kg – 4000 g, 16 kg – 16 000 g, 55 kg – 5500 g,
7.9 kg – 7900 g, 5.5 kg – 5500 g

Challenge 2
1 21 g, 3.5 kg, 90 g, 1800 g
2 a 2000   **b** 7000   **c** 10 000   **d** 31 000
  e 3100   **f** 4300   **g** 43 300   **h** 99 900

Challenge 3
A: 1500, B: 140, C: 1500

Lesson 3: Ordering and rounding mass
Challenge 1
3900, 3999, 4000, 4001

Challenge 2
1 a i 100   **ii** 200   **iii** 400   **iv** 500
  b i 2   **ii** 7   **iii** 6   **iv** 3
2 a E, A, C, D, B   **b** D, B, A, C, E

Challenge 3
a 32.0   **b** 31.8   **c** 31.7   **d** 37.9
Lucy, Gopal, Becky, Jackson

Lesson 4: Reading and weighing scales
Challenge 1
a 4.75   **b** 6.5   **c** 270

Challenge 2
1 a 2750   **b** 1250
  c 205 to 210   **d** 710 to 715
2 a 2500   **b** 600   **c** 5225

Challenge 3
a 760   **b** 1800

Unit 20

Lesson 1: Measuring capacity
Challenge 1
350 millilitres, 500 millilitres, 8 *l*, 5 ml

Challenge 2
1 A: 150, B: 600, C: 50, D: 700
2 Learners' exercise

Challenge 3
a 550 ml marked   **b** 9 *l* marked

Lesson 2: Converting units
Challenge 1
3 *l* – 3000 ml, 7 *l* – 7000 ml, 4.3 *l* – 4300 ml,
19 *l* – 19 000 ml, 43 *l* – 43 000 ml

Challenge 2

1 Less than 250 ml: H
Between 250 and 950 ml: A, B, E
Between 950 and 8000 ml: C, D, F, G
Between 8000 ml and 15 000 ml: I, J

2 a 700 ml marked
b 1300 ml marked
c 5500 ml marked
d 35 000 ml marked

Challenge 3

Answers may vary, for example:
250 + 250 = 0.5 *l*
750 + 200 + 50 = 1 *l*
1250 + 100 + 150 = 1.5 *l*

Lesson 3: Ordering and rounding capacity

Challenge 1

4100, 4900, 4909, 4990, 4999

Challenge 2

1 a 100 **b** 300 **c** 400
d 900 **e** 4400 **f** 600
2 a B, A, E, D, C **b** H, G, F, J, I

Challenge 3

1 12 430, 12 450, 12 400, 12
16 950, 16 950, 17 000, 17
18 090, 18 100, 18 100, 18
2 C, E, B, A, D

Lesson 4: Reading capacity scales

Challenge 1

1 a 220 **b** 108 **c** 285 **d** 1700
2 a 710 **b** 1600 **c** 3250 **d** 6750

Challenge 3

Anwar: C
Samina: D
Harry: B
Ella: A
Rajesh: E

Unit 21

Lesson 1: Telling the time

Challenge 1

a > **b** > **c** > **d** <
e > **f** = **g** < **h** <

Challenge 2

1 00:05, 01:37, 04:35, 06:10, 09:15, 11:22
12:19, 14:58, 17:04, 20:31, 22:23, 23:46
2 2:33 a.m., 11:54 a.m., 12:07 p.m.
2:41 p.m., 5:26 p.m., 11:16 p.m.
3 a 07:07 **b** 05:37 **c** 01:58
d 20:13 **e** 14:18 **f** 18:03

Challenge 3

11:09 a.m – 11:09
11:29 a.m. – 11:29
11:49 a.m. – 11:49
12:09 p.m. – 12:09
12:29 p.m. – 12:29
12:49 p.m. – 12:49
1:09 p.m. – 13:09
1:29 p.m. – 13:29
1:49 p.m. – 13:49
2:09 p.m. – 14:09
2:29 p.m. – 14:29
2:49 p.m. – 14:49

Lesson 2: Reading timetables

Challenge 1

a 08:10 **b** 16:50 **c** 05:55 **d** 13:15

Challenge 2

1 9:35 a.m., 12:05 p.m., 1:10 p.m., 2:15 p.m., 3:00 p.m.

2

Station	Arrival time		
	Bus 1	Bus 2	Bus 3
Central station	07:15	10:45	14:55
Market square	07:30	11:00	15:10
Church lane	07:45	11:15	15:25
The Bank	08:00	11:30	15:40
High street	08:15	11:45	15:55

Challenge 3

a 21:01 **b** 38 minutes
c 20:28 **d** 21:01

Lesson 3: Time intervals

Challenge 1

a 45 minutes **b** 50 minutes
c 50 minutes **d** 1 hour 15 minutes

Challenge 2

1 a 47 minutes **b** 1 hour 5 minutes
c 36 minutes **d** 52 minutes
2 9 hours 10 minutes, 550 minutes
10 hours 45 minutes, 645 minutes
10 hours 28 minutes, 628 minutes
9 hours 33 minutes, 573 minutes
3 a 3 hour 41 minutes
b 7 hours 39 minutes
c 11 hours 38 minutes

Challenge 3

Clara: 09:12
Freddy: 08:51
Jay: 07:52
Antonio: 07:36

Lesson 4: Calendars
Challenge 1
a 16 days   **b** 23 days

Challenge 2
1 44 days, 6 weeks 2 days
   39 days, 7 weeks 4 days
   62 days, 8 weeks, 6 days
2 a 28   **b** 7 weeks 5 days

Challenge 3
Proposals A and D

Unit 22

Lesson 1: Perimeter of regular polygons
Challenge 1
a 120   **b** 100   **c** 48   **d** 360

Challenge 2
1 a 45   **b** 150   **c** 240
2 Learners' exercise: answers may vary

Challenge 3
a 80   **b** 600

Lesson 2: Perimeter of irregular polygons
Challenge 1
a 14   **b** 55   **c** 63

Challenge 2
a 14   **b** 12   **c** 15.5
d 15   **e** 17   **f** 22

Challenge 3
a 3.5   **b** 1.5   **c** 2.5

Lesson 3: Area of a rectangle (1)
Challenge 1
a 4   **b** 12   **c** 8

Challenge 2
1 A: 10, B: 18, C: 44
2 A: 18, B: 32, C: 20, D: 12
   D, A, C, B

Challenge 3
0.2, 18, 2.4. All three walls circled

Lesson 4: Area of a rectangle (2)
Challenge 1
a 20   **b** 48   **c** 72

Challenge 2
1 a 4.2   **b** 13.5   **c** 17.6
2 a 2000   **b** 600   **c** 6000
   d 5500   **e** 61.5

Challenge 3
a 9   **b** 13   **c** 31

Unit 23

Lesson 1: Interpreting tables and graphs
Challenge 1
Frequency: green: 16, white: 14, red: 10,
black: 12, blue: 8
Pictogram: green: 16 circles, white: 14 circles,
red: 10 circles, black: 12 circles, blue: 8 circles

a 12
b green
c blue

Challenge 2
1 Frequency: history: 6, space: 12, sport: 4, art and
   crafts: 14, nature: 16
   Bar line chart constructed
   a nature   **b** sport   **c** 10   **d** 52
2 a 18   **b** 2   **c** 68

Challenge 3
Learners' exercise

Lesson 2: Changing the scale
Challenge 1
Learners' exercise: bar line graph

Challenge 2
1 Answers may vary, for example: it makes the bars
   taller and the graph
   can be read more accurately
2 Answers may vary, for example: it makes the bars
   taller and the graph
   can be read more accurately

Challenge 3
Learners' exercise: bar line graphs

a Answers may vary, for example: it makes the bars
   taller and the graph can be read more accurately

Lesson 3: Line graphs
Challenge 1
Learners' exercise: Line graph

a Weeks 1 and 3   **b** Week 5

Challenge 2
Learners' exercise: Line graph

1 a 6
   b They increased
   c Answers may vary, for example: the time goes
      from winter to spring
2 Answers may vary, for example: It makes the graph
   smaller and more difficult to read accurately

Challenge 3
a It increased
b Between 21 and 28 days
c Between 7 and 14 days

Lesson 4: Intermediate points
Challenge 1
Answers may vary

Challenge 2
1 a Answers may vary
  b Yes
  c Answers may vary, for example: they show her height at intermediate ages
  d 125 cm
  e 150 cm

Challenge 3
a Yes
b Answers may vary, for example: they show intermediate values of data
c 20 km/h, 40 km/h, 40 km/h, 62 km/h, 80 km/h
d A: The car is accelerating (increasing speed)
  B: The car is travelling at constant speed
  C: The car is accelerating (increasing speed)

Lesson 5: Mode
Challenge 1
a Cookie dough   b Cookie dough

Challenge 2
1 a 3   b 12   c apple
  d pink   e R   f 9
2 G

Challenge 3
a High: 61   Low: 50
b The mode would change to 28

Lesson 6: Collecting data
Challenge 1
Learners' exercise: answers may vary

Challenge 2
Learners' exercises: answers may vary

Challenge 3
Learners' exercise: answers may vary

Lesson 7: Presenting data
Challenge 1
Learners' exercise: answers may vary

Challenge 2
Learners' exercises: answers may vary

Challenge 3
Learners' exercise: answers may vary

Lesson 8: Probability
Challenge 1
Impossible, certain, possible, certain

Challenge 2
1 B: possible
  C: possible
  E: impossible
  A letter: certain
2 a Labelled 2, 3, 4
  b Labelled 4, 4, 4, 4
  c Labelled 2, 2, 3, 3 in any order

Challenge 3
1 You toss a coin and it lands on heads – even chance
  You roll a 1–6 dice and get a 6 – unlikely (between even chance and impossible)
  You flap your arms and fly up in the sky – impossible
  Your age will increase each year – certain
  The next vehicle you see will have 4 wheels – likely (between even chance and certain)
2 >   <   >   =

Strand: Number | Sub-strand: Numbers and the number system

Stage 1	Stage 2	Stage 3	Stage 4	Stage 5	Stage 6
1Nn4 Count on in tens from zero or a single-digit number to 100 or just over.	**2Nn3** Count on in ones and tens and two-digit numbers and back again.	**3Nn3** Count on and back in ones, tens and hundreds from two- and three-digit numbers.	**4Nn2** Count on and back in ones, tens, hundreds and thousands from four-digit numbers.	**5Nn1** Count on and back in steps of constant size, extending beyond zero.	**6Nn1** Count on and back in fractions and decimals, e.g. $\frac{1}{3}$s, 0.1s, and repeated steps of whole numbers (and through zero).
1Nn6 Begin partitioning two-digit numbers into tens and ones and reverse.	**2Nn6** Know what each digit represents in two-digit numbers; partition into tens and ones.	**3Nn5** Understand what each digit represents in three-digit numbers and partition into hundreds, tens and units.	**4Nn3** Understand what each digit represents in a three- or four-digit number and partition into thousands, hundreds, tens and units.	**5Nn2** Know what each digit represents in five- and six-digit numbers.	**6Nn2** Know what each digit represents in whole numbers up to a million.
				5Nn3 Partition any number up to one million into thousands, hundreds, tens and units.	
			4Nn4 Use decimal notation and place value for tenths and hundredths in context, e.g. order amounts of money; convert a sum of money such as $13.25 to cents, or a length such as 125 cm to metres; round a sum of money to the nearest pound.	**5Nn4** Use decimal notation for tenths and hundredths and understand what each digit represents.	**6Nn3** Know what each digit represents in one- and two-place decimal numbers.
			4Nn5 Understand decimal notation for tenths and hundredths in context, e.g. length.		**6Nn16** Recognise and use decimals with up to three places in the context of measurement.
		3Nn7 Multiply two-digit numbers by 10 and understand the effect.	**4Nn7** Multiply and divide three-digit numbers by 10 (whole number answers) and understand the effect; begin to multiply numbers by 100 and perform related divisions.	**5Nn5** Multiply and divide any number from 1 to 10 000 by 10 or 100 and understand the effect.	**6Nn4** Multiply and divide any whole number from 1 to 10 000 by 10, 100 or 1000 and explain the effect.
					6Nn5 Multiply and divide decimals by 10 or 100 (answers up to two decimal places for division).

Strand: Number | Sub-strand: Numbers and the number system Continued

Stage 1	Stage 2	Stage 3	Stage 4	Stage 5	Stage 6
	2Nn8 Round two-digit numbers to the nearest multiple of 10.	3Nn8 Round two-digit numbers to the nearest 10 and round three-digit numbers to the nearest 100.	4Nn9 Round three- and four-digit numbers to the nearest 10 or 100.	5Nn6 Round four-digit numbers to the nearest 10, 100 or 1000.	6Nn8 Round whole numbers to the nearest 10, 100 or 1000.
			4Nn4 Use decimal notation and place value for tenths and hundredths in context, e.g. [order amounts of money; convert a sum of money such as $13.25 to cents, or a length such as 125 cm to metres] round a sum of money to the nearest pound.	5Nn7 Round a number with one or two decimal places to the nearest whole number.	6Nn9 Round a number with two decimal places to the nearest tenth or to the nearest whole number.
1Nn8 Use more or less to compare two numbers, and give a number which lies between them.	2Nn11 Recognise and use ordinal numbers up to at least the 10th number and beyond.	3Nn11 Compare three-digit numbers, use < and > signs, and find a number in between.	4Nn12 Compare pairs of three-digit or four-digit numbers, using the > and < signs, and find a number in between each pair.	5Nn8 Order and compare numbers up to a million using the > and < signs.	6Nn11 Order and compare positive numbers to one million, and negative integers to an appropriate level.
1Nn9 Order numbers to at least 20 positioning on a number track; use ordinal numbers	2Nn12 Order numbers to 100; compare two numbers using the > and < signs.	3Nn12 Order two- and three-digit numbers.			6Nn12 Use the >, < and = signs correctly. 6Nc13 Find the difference between a positive and negative integer, and between two negative integers in a context such as temperature or on a number line.
1Nn10 Use the = sign to represent equality.					
			4Nn13 Use negative numbers in context, e.g. temperature	5Nn9 Order and compare negative and positive numbers on a number line and temperature scale.	
				5Nn10 Calculate a rise or fall in temperature.	

Strand: Number | Sub-strand: Numbers and the number system Continued

Stage 1	Stage 2	Stage 3	Stage 4	Stage 5	Stage 6
			4Nn4 Use decimal notation and place value for tenths and hundredths in context, e.g. order amounts of money [convert a sum of money such as $13.25 to cents, or a length such as 125 cm to metres; round a sum of money to the nearest pound].	5Nn11 Order numbers with one or two decimal places and compare using the $>$ and $<$ signs.	6Nn14 Order numbers with up to two decimal places (including different numbers of places).
1Nn5 Count on in twos, beginning to recognise odd/even numbers to 20 as 'every other number'.	2Nn4 Count in twos, fives and tens, and use grouping in twos, fives or tens to count larger groups of objects. 2Nn5 Begin to count on in small constant steps such as threes and fours.	3Nn4 Count on and back in steps of 2, 3, 4 and 5 to at least 50.	4Nn14 Recognise and extend number sequences formed by counting in steps of constant size, extending beyond zero when counting back.	5Nn12 Recognise and extend number sequences.	6Nn15 Recognise and extend number sequences.
1Nn5 Count on in twos, beginning to recognise odd/even numbers to 20 as 'every other number'. 1Nc21 Try to share numbers to 10 to find which are even and which are odd. 1Nc7 (MS) Begin to recognise multiples of 2 and 10.	2Nn14 Understand even and odd numbers and recognise these up to at least 20. 2Nn15 Sort numbers, e.g. odd/even, multiples of 2, 5 and 10. 2Nc4 (MS) Learn and recognise multiples of 2, 5 and 10 [and derive the related division facts].	3Nc5 (MS) Recognise two- and three-digit multiples of 2, 5 and 10.	4Nn15 Recognise odd and even numbers. 4Nn8 Recognise multiples of 5, 10 and 100 up to 1000. 4Nc5 (MS) Recognise and begin to know multiples of 2, 3, 4, 5 and 10, up to the tenth multiple.	5Nn13 Recognise odd and even numbers and multiples of 5, 10, 25, 50 and 100 up to 1000.	6Nn17 Recognise odd and even numbers and multiples of 5, 10, 25, 50 and 100 up to 1000. 6Nn7 Find some common multiples, e.g. for 4 and 5.
			4Nn16 Make general statements about the sums and differences of odd and even numbers.	5Nn14 Make general statements about sums, differences and multiples of odd and even numbers.	6Nn18 Make general statements about sums, differences and multiples of odd and even numbers.
	2Nn17 Recognise that $\frac{2}{2}$ and $\frac{4}{4}$ make a whole and $\frac{1}{2}$ and $\frac{2}{4}$ are equivalent.	3Nn16 Recognise equivalence between $\frac{1}{2}$, $\frac{2}{4}$, $\frac{4}{8}$ and $\frac{5}{10}$ using diagrams.	4Nn18 Recognise the equivalence between: $\frac{1}{2}$, $\frac{4}{8}$ and $\frac{5}{10}$; $\frac{1}{4}$ and $\frac{2}{8}$; and $\frac{2}{5}$ and $\frac{4}{10}$.	5Nn15 Recognise equivalence between: $\frac{1}{2}$, $\frac{1}{4}$ and $\frac{1}{8}$; $\frac{1}{3}$ and $\frac{1}{6}$; and $\frac{1}{5}$ and $\frac{1}{10}$.	6Nn22 Recognise equivalence between fractions, e.g. between $\frac{1}{100}$ s, $\frac{1}{10}$ s and $\frac{1}{2}$ s.

Strand: Number | Sub-strand: Numbers and the number system Continued

Stage 1	Stage 2	Stage 3	Stage 4	Stage 5	Stage 6
		3Nn18 Order simple [or mixed] fractions on a number line, e.g. using the knowledge that $\frac{1}{2}$ comes half way between $\frac{1}{4}$ and $\frac{3}{4}$, [and that $1\frac{1}{2}$ comes half way between 1 and 2].	4Nn17 Order and compare two or more fractions with the same denominator (halves, quarters, thirds, fifths, eighths or tenths). 4Nn19 Use equivalence to help order fractions, e.g. $\frac{7}{10}$ and $\frac{3}{4}$. 4Nn20 Understand the equivalence between one-place decimals and fractions in tenths. 4Nn21 Understand that $\frac{1}{2}$ is equivalent to 0.5 and also to $\frac{5}{10}$. 4Nn22 Recognise the equivalence between the decimal fraction and vulgar fraction forms of halves, quarters, tenths and hundredths.	5Nn16 Recognise equivalence between the decimal and fraction forms of halves, tenths and hundredths and use this to help order fractions, e.g. 0.6 is more than 50% and less than $\frac{7}{10}$.	6Nn21 Compare fractions with the same denominator and related denominators, e.g. $\frac{3}{4}$ with $\frac{7}{8}$. 6Nn23 Recognise and use the equivalence between decimal and fraction forms. 6Nn27 Begin to convert a vulgar fraction to a decimal fraction using division.
		3Nn17 Recognise simple mixed fractions, e.g. $1\frac{1}{2}$ and $2\frac{1}{4}$. 3Nn18 Order [simple or] mixed fractions on a number line, e.g. using the knowledge that [$\frac{1}{2}$ comes half way between $\frac{1}{4}$ and $\frac{3}{4}$, and that] $1\frac{1}{2}$ comes half way between 1 and 2.	4Nn23 Recognise mixed numbers, e.g. $5\frac{3}{4}$, and order these on a number line.	5Nn17 Change an improper fraction to a mixed number, e.g. $\frac{7}{4}$ to $1\frac{3}{4}$; order mixed numbers and place between whole numbers on a number line.	6Nn24 Order mixed numbers and place between whole numbers on a number line. 6Nn25 Change an improper fraction to a mixed number, e.g. $\frac{17}{8}$ to $2\frac{1}{8}$.
1Nn12 Find halves of small numbers and shapes by folding, and recognise which shapes are halved.	2Nn18 Recognise which shapes are divided in halves or quarters and which are not. 2Nn19 Find halves and quarters of shapes and small numbers of objects.	3Nn19 Begin to relate finding fractions to division. 3Nn14 Find half of odd and even numbers to 40, using notation such as $13\frac{1}{2}$. 3Nn20 Find halves, thirds, quarters and tenths of shapes and numbers (whole number answers).	4Nn24 Relate finding fractions to division. 4Nn25 Find halves, quarters, thirds, fifths, eighths and tenths of shapes and numbers.	5Nn18 Relate finding fractions to division and use to find simple fractions of quantities.	6Nc21 Relate finding fractions to division and use them as operators to find fractions including several tenths and hundredths of quantities.

Strand: Number | Sub-strand: Numbers and the number system Continued

Stage 1	Stage 2	Stage 3	Stage 4	Stage 5	Stage 6
				5Nn19 Understand percentage as the number of parts in every 100 and find simple percentages of quantities.	6Nn28 Understand percentage as parts in every 100 and express $\frac{1}{2}$, $\frac{1}{4}$, $\frac{1}{3}$, $\frac{1}{10}$, $\frac{1}{100}$ as percentages.
				5Nn20 Express halves, tenths and hundredths as percentages.	6Nn29 Find simple percentages of shapes and whole numbers.
			4Nc26 Begin to understand simple ideas of ratio and proportion, e.g. a picture is one fifth the size of the real dog. It is 25 cm long in the picture, so it is 5 × 25 cm long in real life.	5Nn21 Use fractions to describe and estimate a simple proportion, e.g. $\frac{1}{5}$ of the beads are yellow.	6Nn30 Solve simple problems involving ratio and direct proportion.
				5Nn22 Use ratio to solve problems, e.g. to adapt a recipe for 6 people to one for 3 or 12 people.	

Strand: Number | Sub-strand: Calculation – Addition and subtraction, including Mental strategies (MS)

Stage 1	Stage 2	Stage 3	Stage 4	Stage 5	Stage 6
1Nc17 Recognise the use of a sign such as □ to represent an unknown, e.g. 6 + □ = 10.	2Nc3 (MS) Find all pairs of multiples of 10 with a total of 100 and record the related addition and subtraction facts.	3Nc2 (MS) Know the following addition and subtraction facts: – multiples of 100 with a total of 1000 – multiples of 5 with a total of 100	4Nc1 (MS) Derive quickly pairs of two-digit numbers with a total of 100, e.g. 72 + □ = 100.	5Nc1 (MS) Know by heart pairs of one-place decimals with a total of 1, e.g. 0.8 + 0.2.	6Nc1 (MS) Recall [addition and subtraction facts for numbers to 20 and] pairs of one-place decimals with a total of 1, e.g. 0.4 + 0.6.
	2Nc9 Recognise the use of a symbol such as □ or △ to represent an unknown, e.g. △ + □ = 10.	3Nc13 Find complements to 100, solving number equations such as 78 + □ = 100.	4Nc2 (MS) Derive quickly pairs of multiples of 50 with a total of 1000, e.g. 850 + □ = 1000.	5Nc2 (MS) Derive quickly pairs of decimals with a total of 10, and with a total of 1.	6Nc2 (MS) Derive quickly pairs of one-place decimals totalling 10, e.g. 7.8 and 2.2, and two-place decimals totalling 1, e.g. 0.78 + 0.22.
	2Nc10 Solve number sentences such as 27 + □ = 30.		4Nc3 (MS) Identify simple fractions with a total of 1, e.g. $\frac{1}{4}$ + □ = 1.		

Strand: Number | Sub-strand: Calculation – Addition and subtraction, including Mental strategies (MS) Continued

Stage 1	Stage 2	Stage 3	Stage 4	Stage 5	Stage 6
1Nc13 Relate counting on and back in tens to finding 10 more/less than a number (<100). **1Nc18** Begin to add single- and two-digit numbers.	**2Nc6** Relate counting on/back in tens to finding 10 more/less than any two-digit number and then to adding and subtracting other multiples of 10, e.g. 75 – 30. **2Nc11** Add and subtract a single digit to and from a two-digit number.	**3Nc9** Add and subtract 10 and multiples of 10 to and from two- and three-digit numbers. **3Nc10** Add 100 and multiples of 100 to three-digit numbers. **3Nc17** Add/subtract single-digit numbers to/from three-digit numbers. **3Nc18** Find 20, 30, … 90, 100, 200, 300 more/less than three-digit numbers.	**4Nc7 (MS)** Add three two-digit multiples of 10, e.g. 40 + 70 + 50. **4Nc12** Subtract a small number crossing 100, e.g. 304 – 8.	**5Nc8 (MS)** Count on or back in thousands, hundreds, tens and ones to add or subtract.	**6Nc4 (MS)** Use place value and number facts to add or subtract [two-digit whole numbers and to add or subtract] three-digit multiples of 10 [and pairs of decimals], e.g. 560 + 270; [2.6 + 2.7; 0.78 + 0.23].
			4Nc8 (MS) Add and subtract near multiples of 10 or 100 to or from three-digit numbers, e.g. 367 – 198 or 278 + 49. **4Nc11 (MS)** Find a difference between near multiples of 100, e.g. 304 – 296.	**5Nc9 (MS)** Add or subtract near multiples of 10 or 100, e.g. 4387 – 299. **5Nc11 (MS)** Calculate differences between near multiples of 1000, e.g. 5026 – 4998, or near multiples of 1, e.g. 3.2 – 2.6.	**6Nc5 (MS)** Add/subtract near multiples of one when adding numbers with one decimal place, e.g. 5.6 + 2.9; 13.5 – 2.1. **6Nc6 (MS)** Add/subtract a near multiple of 10, 100 or 1000, or a near whole unit of money, and adjust, e.g. 3127 + 4998; 5678 – 1996.
	2Nc12 Add pairs of two-digit numbers. **2Nc13** Find a small difference between pairs of two-digit numbers.	**3Nc14** Add pairs of two-digit numbers. **3Nc15** Add three-digit and two-digit numbers using notes to support.	**4Nc9 (MS)** Add any pair of two-digit numbers, choosing an appropriate strategy. **4Nc10 (MS)** Subtract any pair of two-digit numbers, choosing an appropriate strategy. **4Nc17** Add pairs of three-digit numbers.	**5Nc10 (MS)** Use appropriate strategies to add or subtract pairs of two- and three-digit numbers and numbers with one decimal place, using jottings where necessary.	**6Nc4 (MS)** Use place value and number facts to add or subtract two-digit whole numbers and to add or subtract [three-digit multiples of 10 and] pairs of decimals, e.g. [560 + 270;] 2.6 + 2.7; 0.78 + 0.23.
		3Nc15 Add three-digit and two-digit numbers using notes to support.	**4Nc17** Add pairs of three-digit numbers. **4Nc18** Subtract a two-digit number from a three-digit number. **4Nc19** Subtract pairs of three-digit numbers.	**5Nc18** Find the total of more than three two- or three-digit numbers using a written method. **5Nc19** Add or subtract any pair of three- and/or four-digit numbers, with the same number of decimal places, including amounts of money.	**6Nc11** Add two- and three-digit numbers with the same or different numbers of digits/decimal places. **6Nc12** Add or subtract numbers with the same and different numbers of decimal places, including amounts of money.

Strand: Number | Sub-strand: Calculation – Multiplication and division, including Mental strategies (MS)

Stage 1	Stage 2	Stage 3	Stage 4	Stage 5	Stage 6
1Nn4 Count on in tens from zero [or a single-digit number] to 100 or just over.	2Nc19 Use counting in twos, fives or tens to solve practical problems involving repeated addition.	3Nc3 (MS) Know multiplication/division facts for 2x, 3x, 5x, and 10x tables.	4Nc4 (MS) Know multiplication for 2x, 3x, 4x, 5x, 6x, 9x and 10x tables and derive division facts.	5Nc3 (MS) Know multiplication and division facts for the 2x to 10x tables.	6Nc7 (MS) Use place value and multiplication facts to multiply/divide mentally, e.g. 0.8 × 7; 4.8 ÷ 6.
1Nn5 Count on in twos, [beginning to recognise odd/even numbers to 20 as 'every other number'].	2Nc22 Work out multiplication and division facts for the 3x and 4x tables.	3Nc4 (MS) Begin to know 4x table.	4Nc13 (MS) Multiply any pair of single-digit numbers together.	5Nc6 (MS) Know squares of all numbers to 10 × 10.	6Nc17 Use number facts to generate new multiplication facts, e.g. the 17x table from 10x + 7x table.
		3Nc21 Multiply single-digit numbers and divide two-digit numbers by 2, 3, 4, 5, 6, 9 and 10.			
1Nc7 (MS) Begin to recognise multiples of 2 and 10.	2Nc4 (MS) Learn and recognise multiples of 2, 5 [and 10 and derive the related division facts].	3Nc5 (MS) Recognise two- and three-digit multiples of 2, 5 and 10.	4Nn8 Recognise multiples of 5, 10 and 100 up to 1000.	5Nc4 (MS) Know and apply tests of divisibility by 2, 5, 10 and 100.	6Nn17 Recognise [odd and even numbers and] multiples of 5, 10, 25, 50 and 100 up to 1000.
	2Nn15 Sort numbers, e.g. [odd/even], multiples of 2, 5 and 10.		4Nc5 (MS) Recognise and begin to know multiples of 2, 3, 4, 5 and 10, up to the tenth multiple.	5Nc5 (MS) Recognise multiples of 6, 7, 8 and 9 up to the 10th multiple.	6Nc3 (MS) Know and apply tests of divisibility by 2, 4, 5, 10, 25 and 100.
					6Nn7 Find some common multiples, e.g. for 4 and 5.
				5Nc7 (MS) Find factors of two-digit numbers.	6Nn6 (MS) Find factors of two-digit numbers.
			4Nc21 Multiply multiples of 10 to 90 by a single-digit number.	5Nc12 (MS) Multiply multiples of 10 to 90, and multiples of 100 to 900, by a single-digit number.	6Nc8 (MS) Multiply pairs of multiples of 10, e.g. 30 × 40, or multiples of 10 and 100, e.g. 600 × 40.
					6Nc14
				5Nc13 (MS) Multiply by 19 or 21 by multiplying by 20 and adjusting.	6Nc15 Multiply near multiples of 10 by multiplying by the multiple of 10 and adjusting.
				5Nc14 (MS) Multiply by 25 by multiplying by 100 and dividing by 4.	6Nc16 Multiply by halving one number and doubling the other, e.g. calculate 35 × 16 with 70 × 8.
				5Nc15 (MS) Use factors to multiply, e.g. multiply by 3, then double to multiply by 6.	6Nc17 Use number facts to generate new multiplication facts, e.g. the 17x table from 10x + 7x tables.

Strand: Number | Sub-strand: Calculation – Multiplication and division, including Mental strategies (MS) Continued

Stage 1	Stage 2	Stage 3	Stage 4	Stage 5	Stage 6
1Nc5 (MS) Know doubles to at least double 5.	**2Nc5 (MS)** Find and learn doubles for all numbers up to 10 and also 15, 20, 25 and 50.	**3Nc6 (MS)** Work out quickly the doubles of numbers 1 to 20 and derive the related halves.	**4Nc16 (MS)** Derive quickly doubles of all whole numbers to 50, doubles of multiples of 10 to 500, doubles of multiples of 100 to 5000, and corresponding halves.	**5Nc16 (MS)** Double any number up to 100 and halve even numbers to 200 and use this to double and halve numbers with one or two decimal places, e.g. double 3.4 and half of 8.6	**6Nc9 (MS)** Double quickly any two-digit number, e.g. 78, 7.8, 0.78 and derive the corresponding halves.
1Nc6 (MS) Find near doubles using doubles already known, e.g. 5 + 6.	**2Nc20** Find doubles of multiples of 5 up to double 50 and corresponding halves.	**3Nc7 (MS)** Work out quickly the doubles of multiples of 5 (<100) and derive the related halves.	**4Nc20** Double any two-digit number.	**5Nc17 (MS)** Double multiples of 10 to 1000 and multiples of 100 to 10000, e.g. double 360 or double 3600, and derive the corresponding halves.	**6Nc18** Multiply two-, three- or four-digit numbers (including sums of money) by a single-digit number and two- or three-digit numbers by two-digit numbers.
1Nc19 Double any single-digit number.	**2Nc21** Double two-digit numbers.	**3Nc8 (MS)** Work out quickly the doubles of multiples of 50 to 500.	**4Nc22** Multiply a two-digit number by a single-digit number.	**5Nc20** Multiply or divide three-digit numbers by single-digit numbers.	**6Nc10 (MS)** Divide two-digit numbers by single digit-numbers, including leaving a remainder.
1Nc20 Find halves of even numbers of objects up to 10.		**3Nc19** Understand the relationship between halving and doubling.	**4Nc23** Divide two-digit numbers by single digit-numbers (answers no greater than 20).	**5Nc21** Multiply two-digit numbers by two-digit numbers.	**6Nc19** Divide three-digit numbers by single-digit numbers, including those leaving a remainder and divide three-digit numbers by two-digit numbers (no remainder) including sums of money.
		3Nc22 Multiply teens numbers by 3 and 5.		**5Nc22** Multiply two-digit numbers with one decimal place by single-digit numbers, e.g. 3.6 × 7.	
		3Nc23 Begin to divide two-digit numbers just beyond 10x tables, e.g. 60 ÷ 5, 33 ÷ 3.		**5Nc23** Divide three-digit numbers by single-digit numbers, including those with a remainder (answers no greater than 30).	

Strand: Number | Sub-strand: Calculation – Multiplication and division, including Mental strategies (MS) Continued

Stage 1		Stage 2		Stage 3		Stage 4		Stage 5		Stage 6	
1Nc21	Try to share numbers to 10 to find which are even and which are odd	2Nc23	Understand that division can leave some left over.	3Nc24	Understand that division can leave a remainder (initially as 'some left over').	4Nc24	Decide whether to round up or down after division to give an answer to a problem.	5Nc24	Start expressing remainders as a fraction of the divisor when dividing two-digit numbers by single-digit numbers.	6Nc20	Give an answer to division as a mixed number, and a decimal (with divisors of 2, 4, 5, 10 or 100).
1Nc22	Share objects into two equal groups in a context.							5Nc25	Decide whether to group (using multiplication facts and multiples of the divisor) or to share (halving and quartering) to solve divisions.		
								5Nc26	Decide whether to round an answer up or down after division, depending on the context.		
				3Nc25	Understand and apply the idea that multiplication is commutative.	4Nc14 (MS)	Use knowledge of commutativity to find the easier way to multiply.	5Nc27	Begin to use brackets to order operations and understand the relationship between the four operations and how the laws of arithmetic apply to multiplication.	6Nc22	Know and apply the arithmetic laws as they apply to multiplication (without necessarily sing the terms commutative, associative and distributive).
				3Nc26	Understand the relationship between multiplication and division and write connected facts.	4Nc25	Understand that multiplication and division are the inverse function of each other.				

Tracking back and forward through the Cambridge Primary Mathematics Curriculum Framework

Strand: Geometry | Sub-strand: Shapes and geometric reasoning

Stage 1	Stage 2	Stage 3	Stage 4	Stage 5	Stage 6
1Gs1 Name and sort common 2D shapes (e.g. circles, squares, rectangles and triangles) using features such as number of sides, curved or straight. Use them to make patterns and models.	**2Gs1** Sort, name, describe, visualise and draw 2D shapes (e.g. squares, rectangles, circles, regular and irregular pentagons and hexagons) referring to their properties; recognize common 2D shapes in different positions and orientations.	**3Gs1** Identify, describe and draw regular and irregular 2D shapes including pentagons, hexagons, octagons and semi-circles. **3Gs2** Classify 2D shapes according to the number of sides, vertices and right angles. **3Gs6** Relate 2D shapes [and 3D solids] to drawings of them. **3Gs8** Identify right angles in 2D shapes.	**4Gs1** Identify, describe, visualise, draw [and make] a wider range of 2D [and 3D] shapes including a range of quadrilaterals, the heptagon [and tetrahedron]; use pinboards to create a range of polygons. Use spotty paper to record results. **4Gs2** Classify polygons (including a range of quadrilaterals) using criteria such as the number of right angles, whether or not they are regular and their symmetrical properties.	**5Gs1** Identify and describe properties of triangles and classify as isosceles, equilateral or scalene.	**6Gs1** Classify different polygons and understand whether a 2D shape is a polygon or not. **6Gs3** Identify and describe properties of quadrilaterals (including the parallelogram, rhombus and trapezium), and classify using parallel sides, equal sides, equal angles.
1Gs3 Recognise basic line symmetry.	**2Gs3** Identify reflective symmetry in patterns and 2D shapes; draw lines of symmetry.	**3Gs5** Draw and complete 2D shapes with reflective symmetry and draw reflections of shapes (mirror line along one side).	**4Gs3** Identify and sketch lines of symmetry in 2D shapes and patterns.	**5Gs2** Recognise reflective and rotational symmetry in regular polygons. **5Gs3** Create patterns with two lines of symmetry, e.g. on a pegboard or squared paper.	
1Gs2 Name and sort common 3D shapes (e.g. cube, cuboid, cylinder, cone and sphere) using features such as number of faces, flat or curved faces. Use them to make patterns and models.	**2Gs2** Sort, name, describe and make 3D shapes (e.g. cubes, cuboids, cones, cylinders, spheres and pyramids) referring to their properties; recognise 2D drawings of 3D shapes.	**3Gs3** Identify, describe and make 3D shapes including pyramids and prisms; investigate which nets will make a cube. **3Gs4** Classify 3D shapes according to the number and shape of faces, number of vertices and edges. **3Gs6** Relate [2D shapes and] 3D solids to drawings of them.	**4Gs4** Visualise 3D objects from 2D nets and drawings and make nets of common solids. **4Gs1** Identify, describe, visualise, draw and make a wider range of [2D and] 3D shapes including [a range of quadrilaterals], the [heptagon and] tetrahedron; [use pinboards to create a range of polygons. Use spotty paper to record results].	**5Gs4** Visualise 3D shapes from 2D drawings and nets, e.g. different nets of an open or closed cube.	**6Gs2** Visualise and describe the properties of 3D shapes, e.g. faces, edges and vertices. **6Gs4** Recognise and make 2D representations of 3D shapes including nets.

Tracking back and forward through the Cambridge Primary Mathematics Curriculum Framework

Strand: Geometry | Sub-strand: Shapes and geometric reasoning Continued

Stage 1	Stage 2	Stage 3	Stage 4	Stage 5	Stage 6
	2Gs4 Find examples of 2D and 3D shape and symmetry in the environment.	3Gs7 Identify 2D and 3D shapes, lines of symmetry and right angles in the environment.	4Gs5 Find examples of shapes and symmetry in the environment and in art.	5Gs5 Recognise perpendicular and parallel lines in 2D shapes, drawings and the environment.	
	2Gp2 Recognise whole, half and quarter turns, both clockwise and anti-clockwise.	3Gp3 Use a set square to draw right angles.	4Gp2 Know that angles are measured in degrees and that one whole turn is 360° or four right angles; compare and order angles less than 180°.	5Gs6 Understand and use angle measure in degrees; measure angles to the nearest 5°; identify, describe and estimate the size of angles and classify them as acute, right or obtuse.	6Gs5 Estimate, recognise and draw acute and obtuse angles and use a protractor to measure to the nearest degree.
	2Gp3 Recognise that a right angle is a quarter turn.	3Gp4 Compare angles with a right angle and recognise that a straight line is equivalent to two right angles.		5Gs7 Calculate angles in a straight line.	6Gs6 Check that the sum of the angles in a triangle is 180°, for example, by measuring or paper folding; calculate angles in a triangle or around a point.

Strand: Geometry | Sub-strand: Position and movement

Stage 1	Stage 2	Stage 3	Stage 4	Stage 5	Stage 6
		3Gp2 Find and describe the position of a square on a grid of squares where the rows and columns are labelled.	4Gp1 Describe and identify the position of a square on a grid of squares where rows and columns are numbered and/or lettered.	5Gp1 Read and plot co-ordinates in the first quadrant.	6Gp1 Read and plot co-ordinates in all four quadrants.
				5Gp2 Predict where a polygon will be after one reflection where the mirror line is parallel to one of the sides, including where the line is oblique.	6Gp2 Predict where a polygon will be after one reflection, where the sides of the shape are not parallel or perpendicular to the mirror line, after one translation or after a rotation through 90° about one of its vertices.
				5Gp3 Understand translation as movement along a straight line, identify where polygons will be after a translation and give instructions for translating shapes.	

Strand: Measure | Sub-strand: Length, mass and capacity

Stage 1	Stage 2	Stage 3	Stage 4	Stage 5	Stage 6
1MI1 Compare lengths and weights by direct comparison, then by using uniform non-standard units.	**2MI1** Estimate, measure and compare lengths, weights and capacities, choosing and using suitable uniform non-standard and standard units and appropriate measuring instruments.	**3MI1** Choose and use appropriate units and equipment to estimate, measure and record measurements.	**4MI1** Choose and use standard metric units and their abbreviations (km, m, cm, mm, kg, g, l and ml) when estimating, measuring and recording length, weight and capacity.	**5MI1** Read, choose, use and record standard units to estimate and measure length, mass and capacity to a suitable degree of accuracy.	**6MI1** Select and use standard units of measure. Read and write to two or three decimal places.
1MI2 Estimate and compare capacities by direct comparison, then by using uniform non-standard units.	**2MI2** Compare lengths, weights and capacities using the standard units: centimetre, metre, 100 g, kilogram, and litre.		**4MI3** Where appropriate, use decimal notation to record measurements, e.g. 1.3 m, 0.6 kg, 1.2 l.	**5MI3** Order measurements in mixed units.	
1MI3 Use comparative language, e.g. longer, shorter, heavier, lighter.					
		3MI2 Know the relationship between kilometres and metres, metres and centimetres, kilograms and grams, litres and millilitres.	**4MI2** Know and use the relationships between familiar units of length, mass and capacity; know the meaning of 'kilo', 'centi' and 'milli'.	**5MI2** Convert larger to smaller metric units (decimals to one place), e.g. change 2.6 kg to 2600 g.	**6MI2** Convert between units of measurement (kg and g, l and ml, km, m, cm and mm), using decimals to three places, e.g. recognising that 1.245 m is 1 m 24.5 cm.
				5MI4 Round measurements to the nearest whole unit.	
		3MI3 Read to the nearest division or half division, use scales that are numbered or partially numbered.	**4MI4** Interpret intervals/divisions on partially numbered scales and record readings accurately.	**5MI5** Interpret a reading that lies between two unnumbered divisions on a scale.	**6MI3** Interpret readings on different scales, using a range of measuring instruments.
				5MI6 Compare readings on different scales.	
		3MI4 Use a ruler to draw and measure lines to the nearest centimetre.		**5MI7** Draw and measure lines to the nearest centimetre and millimetre.	**6MI4** Draw and measure lines to the nearest centimetre and millimetre.

Tracking back and forward through the Cambridge Primary Mathematics Curriculum Framework

Strand: Measure | Sub-strand: Time

Stage 1	Stage 2	Stage 3	Stage 4	Stage 5	Stage 6
1Mt1 Begin to understand and use some units of time, e.g. minutes, hours, days, weeks, months and years.	**2Mt1** Know the units of time (seconds, minutes, hours, days, weeks, months and years).	**3Mt1** Suggest and use suitable units to measure time and know the relationships between them (second, minute, hour, day, week, month, year).		**5Mt1** Recognise and use the units for time (seconds, minutes, hours, days, months and years).	**6Mt1** Recognise and understand the units for measuring time (seconds, minutes, hours, days, weeks, months, years, decades and centuries); convert one unit of time into another.
1Mt3 Order the days of the week and other familiar events	**2Mt2** Know the relationships between consecutive units of time. **2Mt4** Measure activities using seconds and minutes. **2Mt5** Know and order the days of the week and the months of the year.				
1Mt2 Read the time to the hour (o'clock) and know key times of day to the nearest hour.	**2Mt3** Read the time to the half hour on digital and analogue clocks.	**3Mt2** Read the time on analogue and digital clocks, to the nearest 5 minutes on an analogue clock and to the nearest minute on a digital clock.	**4Mt1** Read and tell the time to nearest minute on 12-hour digital and analogue clocks. **4Mt2** Use am, pm and 12-hour digital clock notation.	**5Mt2** Tell and compare the time using digital and analogue clocks using the 24-hour clock.	**6Mt2** Tell the time using digital and analogue clocks using the 24-hour clock. **6Mt3** Compare times on digital and analogue clocks, e.g. realise quarter to four is later than 3:40.
		3Mt4 Read a calendar and calculate time intervals in weeks or days.	**4Mt3** Read simple timetables and use a calendar.	**5Mt3** Read timetables using the 24-hour clock. **5Mt5** Use a calendar to calculate time intervals in days and weeks (using knowledge of days in calendar months). **5Mt6** Calculate time intervals in months or years.	**6Mt4** Read and use timetables using the 24-hour clock. **6Mt6** Use a calendar to calculate time intervals in days, weeks or months. **6Mt7** Calculate time intervals in days, months or years.
		3Mt3 Begin to calculate simple time intervals in hours and minutes.	**4Mt4** Choose units of time to measure time intervals.	**5Mt4** Calculate time intervals in seconds, minutes and hours using digital or analogue formats.	**6Mt5** Calculate time intervals using digital and analogue times.

Strand: Measure | Sub-strand: Area and perimeter

Stage 1	Stage 2	Stage 3	Stage 4	Stage 5	Stage 6
			4Ma1 Draw rectangles, and measure and calculate their perimeters.	**5Ma1** Measure and calculate the perimeter of regular and irregular polygons.	**6Ma1** Measure and calculate the perimeter [and area] of rectilinear shapes. **6Ma3** Calculate perimeter [and area] of simple compound shapes that can be split into rectangles.
			4Ma2 Understand that area is measured in square units, e.g. cm².	**5Ma2** Understand area measured in square centimetres (cm²).	**6Ma1** Measure and calculate the [perimeter and] area of rectilinear shapes.
			4Ma3 Find the area of rectilinear shapes drawn on a square grid by counting squares.	**5Ma3** Use the formula for the area of a rectangle to calculate the rectangle's area.	**6Ma2** Estimate the area of an irregular shape by counting squares. **6Ma3** Calculate [perimeter and] area of simple compound shapes that can be split into rectangles.

Strand: Handling data | Sub-strand: Organising, categorising and representing data

Stage 1	Stage 2	Stage 3	Stage 4	Stage 5	Stage 6
1Dh1 Answer a question by sorting and organising data or objects in a variety of ways, e.g. – using block graphs and pictograms with practical resources; discussing the results – in lists and tables with practical resources; discussing the results – in Venn or Carroll diagrams giving different criteria for grouping the same objects	**2Dh1** Answer a question by collecting and recording data in lists and tables, and representing it as block graphs and pictograms to show results.	**3Dh1** Answer a real-life question by collecting, organising and interpreting data, e.g. investigating the population of mini-beasts in different environments. **3Dh2** Use tally charts, frequency tables, pictograms (symbol representing one or two units) and bar charts (intervals labelled in ones or twos).	**4Dh1** Answer a question by identifying what data to collect, organising, presenting and interpreting data in tables, diagrams, tally charts, frequency tables, pictograms (symbol representing 2, 5, 10 or 20 units) and bar charts (intervals labelled in twos, fives, tens or twenties). **4Dh2** Compare the impact of representations where scales have different intervals.	**5Dh1** Answer a set of related questions by collecting, selecting and organising relevant data; draw conclusions from their own and others' data and identify further questions to ask. **5Dh2** Draw and interpret frequency tables, pictograms and bar line charts, with the vertical axis labelled for example in twos, fives, tens, twenties or hundreds. Consider the effect of changing the scale on the vertical axis.	**6Dh1** Solve a problem by representing, extracting and interpreting data in tables, graphs, charts and diagrams, e.g. line graphs for distance and time; a price 'ready-reckoner' for currency conversion; frequency tables and bar charts with grouped discrete data.

Strand: Handling data | Sub-strand: Organising, categorising and representing data Continued

Stage 1	Stage 2	Stage 3	Stage 4	Stage 5	Stage 6
				5Dh3 Construct simple line graphs, e.g. to show changes in temperature over time.	
				5Dh4 Understand where intermediate points have and do not have meaning, e.g. comparing a line graph of temperature against time with a graph of class attendance for each day of the week.	6Dh2 Find the mode and range of a set of data from relevant situations, e.g. scientific experiments.
				5Dh5 Find and interpret the mode of a set of data.	6Dh3 Begin to find the median and mean of a set of data.
					6Dh4 Explore how statistics are used in everyday life.

Strand: Handling data | Sub-strand: Probability

Stage 1	Stage 2	Stage 3	Stage 4	Stage 5	Stage 6
				5Db1 Describe the occurrence of familiar events using the language of chance or likelihood.	6Db1 Use the language associated with probability to discuss events, to assess likelihood and risk, including those with equally likely outcomes.

Strand: Problem solving | Sub-strand: Using techniques and skills in solving mathematical problems

Stage 1	Stage 2	Stage 3	Stage 4	Stage 5	Stage 6
		3Pt2 Begin to understand everyday systems of measurement in length, weight, capacity and time and use these to make measurements as appropriate.	4Pt2 Understand everyday systems of measurement in length, weight, capacity and time and use these to solve simple problems as appropriate.	5Pt1 Understand everyday systems of measurement in length, weight, capacity, temperature and time and use these to perform simple calculations.	6Pt2 Understand everyday systems of measurement in length, weight, capacity, temperature and time and use these to perform simple calculations.
1Pt4 Decide to add or subtract to solve a simple word problem (oral), and represent it with objects.	2Pt4 Make sense of simple word problems (single and easy two step), decide what operations (addition or subtraction, simple multiplication or division) are needed to solve them and, with help, represent them, with objects or drawings or on a number line.	3Pt3 Make sense of and solve word problems, single (all four operations) and two-step (addition and subtraction), and begin to represent them, e.g. with drawings or on a number line.		5Pt2 Solve single and multi-step word problems (all four operations); represent them, e.g. with diagrams or a number line.	6Ps6 Make sense of and solve word problems, single and multi-step (all four operations), and represent them, e.g. with diagrams or on a number line; use brackets to show the series of calculations necessary. 6Ps7 Solve simple word problems involving ratio and direct proportion. 6Ps8 Solve simple word problems involving percentages, e.g. find discounted prices.
1Pt5 Check the answer to an addition by adding the numbers in a different order.	2Pt6 Check the answer to an addition by adding the numbers in a different order or by using a different strategy, e.g. 35 + 19 by adding 20 to 35 and subtracting 1, and by adding 30 + 10 and 5 + 9.	3Pt4 Check the results of adding two numbers using subtraction, and several numbers by adding in a different order.	4Pt3 Check the results of adding numbers by adding them in a different order or by subtracting one number from the total.	5Pt3 Check with a different order when adding several numbers or by using the inverse when adding or subtracting a pair of numbers.	6Pt3 Check addition with a different order when adding a long list of numbers; check when subtracting by using the inverse.
1Pt6 Check the answer to a subtraction by adding the answer to the smaller number in the question.	2Pt7 Check a subtraction by adding the answer to the smaller number in the original subtraction.	3Pt5 Check subtraction by adding the answer to the smaller number in the original calculation.	4Pt4 Check subtraction by adding the answer to the smaller number in the original calculation.		
		3Pt6 Check multiplication by reversing the order, e.g. checking that 6 × 4 = 24 by doing 4 × 6.	4Pt5 Check multiplication using a different technique, e.g. check 6 × 8 = 48 by doing 6 × 4 and doubling.	5Pt4 Use multiplication to check the result of a division, e.g. multiply 3.7 × 8 to check 29.6 ÷ 8.	

Tracking back and forward through the Cambridge Primary Mathematics Curriculum Framework

Strand: Problem solving | Sub-strand: Using techniques and skills in solving mathematical problems Continued

Stage 1	Stage 2	Stage 3	Stage 4	Stage 5	Stage 6
		3Pt8 Recognise the relationships between different 2D shapes. 3Pt9 Identify the differences and similarities between different 3D shapes.	4Pt7 Recognise the relationships between 2D shapes and identify the differences and similarities between 3D shapes.	5Pt5 Recognise the relationships between different 2D and 3D shapes, e.g. a face of a cube is a square.	6Pt4 Recognise 2D and 3D shapes and their relationships, e.g. a cuboid has a rectangular cross-section.
1Pt9 Make a sensible estimate of a calculation, and consider whether an answer is reasonable.	2Pt10 Make a sensible estimate for the answer to a calculation.	3Pt10 Estimate and approximate when calculating, and check working. 3Pt11 Make a sensible estimate for the answer to a calculation, e.g. using rounding.	4Pt8 Estimate and approximate when calculating, and check working.	5Pt6 Estimate and approximate when calculating, e.g. using rounding, and check working.	6Pt5 Estimate and approximate when calculating, e.g. use rounding, and check working.
	2Pt11 Consider whether an answer is reasonable.	3Pt12 Consider whether an answer is reasonable.		5Pt7 Consider whether an answer is reasonable in the context of a problem.	

Strand: Problem solving | Sub-strand: Using understanding and strategies in solving problems

Stage 1	Stage 2	Stage 3	Stage 4	Stage 5	Stage 6
		3Pt2 Begin to understand everyday systems of measurement in length, weight, capacity and time and use these to make measurements as appropriate.	4Pt2 Understand everyday systems of measurement in length, weight, capacity and time and use these to solve simple problems as appropriate.	5Ps1 Understand everyday systems of measurement in length, weight, capacity, temperature and time and use these to perform simple calculations.	6Ps2 Understand everyday systems of measurement in length, weight, capacity, temperature and time and use these to perform simple calculations.
1Pt1 Choose appropriate strategies to carry out calculations, explaining working out.	2Pt1 Choose appropriate mental strategies to carry out calculations and explain how they worked out the answer. 2Pt2 Explain methods and reasoning orally.	3Ps2 Explain a choice of calculation strategy and show how the answer was worked out.	4Ps2 Explain reasons for a choice of strategy when multiplying or dividing. 4Ps3 Choose strategies to find answers to addition or subtraction problems; explain and show working.	5Ps2 Choose an appropriate strategy for a calculation and explain how they worked out the answer.	6Ps1 Explain why they chose a particular method to perform a calculation and show working.

Strand: Problem solving | Sub-strand: Using understanding and strategies in solving problems Continued

Stage 1	Stage 2	Stage 3	Stage 4	Stage 5	Stage 6
1Pt2 Explore number problems and puzzles.	2Pt3 Explore number problems and puzzles.	3Ps3 Explore and solve number problems and puzzles, e.g. logic problems.	4Ps4 Explore and solve number problems and puzzles, e.g. logic problems.	5Ps3 Explore and solve number problems and puzzles, e.g. logic problems.	6Ps3 Use logical reasoning to explore and solve number problems and mathematical puzzles.
				5Ps4 Deduce new information from existing information to solve problems.	6Ps2 Deduce new information from existing information and realize the effect that one piece of information has on another.
		3Ps4 Use ordered lists and tables to help to solve problems systematically.	4Ps5 Use ordered lists and tables to solve problems systematically.	5Ps5 Use ordered lists and tables to help to solve problems systematically.	6Ps4 Use ordered lists or tables to help solve problems systematically.
1Pt7 Describe and continue patterns such as count on and back in tens, e.g. 90, 80, 70.	2Pt8 Describe and continue patterns which count on in twos, threes, fours or fives to 30 or more.	3Ps5 Describe and continue patterns which count on or back in steps of 2, 3, 4, 5, 10, or 100.	4Ps6 Describe and continue number sequences, e.g. 7, 4, 1, –2 ...; identifying the relationship between each number.	5Ps6 Describe and continue number sequences, e.g. –30, –27, ☐, ☐, –18...; identify the relationships between numbers.	6Ps5 Identify relationships between numbers and make generalized statements using words, then symbols and letters, e.g. the second number is twice the first number plus 5 (n, 2n + 5); all the numbers are multiples of 3 minus 1 (3n − 1); the sum of angles in a triangle is 180°.
1Pt8 Identify simple relationships between numbers and shapes, e.g. this number is ten bigger than that number.	2Pt9 Identify simple relationships between numbers and shapes, e.g. this number is double ...; these shapes all have ... sides.	3Ps6 Identify simple relationships between numbers, e.g. each number is three more than the number before it.			
		3Ps7 Identify simple relationships between shapes, e.g. these shapes all have the same number of lines of symmetry.	4Ps7 Identify simple relationships between shapes, e.g. these polygons are all regular because ...	5Ps7 Identify simple relationships between shapes, e.g. these triangles are all isosceles because ...	
		3Ps8 Investigate a simple general statement by finding examples which do or do not satisfy it, e.g. when adding 10 to a number, the first digit remains the same.	4Ps8 Investigate a simple general statement by finding examples which do or do not satisfy it.	5Ps8 Investigate a simple general statement by finding examples which do or do not satisfy it, e.g. the sum of three consecutive whole numbers is always a multiple of three.	

Strand: Problem solving | Sub-strand: Using understanding and strategies in solving problems Continued

Stage 1		Stage 2		Stage 3		Stage 4		Stage 5		Stage 6	
		2Pt2	Explain methods and reasoning orally.	3Ps9	Explain methods and reasoning orally, including initial thoughts about possible answers to a problem.	4Ps9	Explain methods and reasoning orally and in writing; make hypotheses and test them out.	5Ps9	Explain methods and justify reasoning orally and in writing; make hypotheses and test them out.	6Ps1	Explain why they chose a particular method to perform a calculation and show working.
										6Ps9	Make, test and refine hypotheses, explain and justify methods, reasoning, strategies, results or conclusions orally.
1Pt4	Decide to add or subtract to solve a simple word problem (oral), and represent it with objects.	2Pt4	Make sense of simple word problems (single and easy two step), decide what operations (addition or subtraction, simple multiplication or division) are needed to solve them and, with help, represent them, with objects or drawings or on a number line.	3Pt3	Make sense of and solve word problems, single (all four operations) and two-step (addition and subtraction), and begin to represent them, e.g. with drawings or on a number line.			5Ps10	Solve a larger problem by breaking it down into sub-problems or represent it using diagrams.	6Ps6	Make sense of and solve word problems, single and multi-step (all four operations), and represent them, e.g. with diagrams or on a number line; use brackets to show the series of calculations necessary.

Stage 5 Record-keeping charts

Stage 5   Class: _____   Year: _____

Strand: **Number** Sub-strand: **Numbers and the number system**	
5Nn1	Count on and back in steps of constant size, extending beyond zero.
5Nn2	Know what each digit represents in five- and six-digit numbers.
5Nn3	Partition any number up to one million into thousands, hundreds, tens and units.
5Nn4	Use decimal notation for tenths and hundredths and understand what each digit represents.
5Nn5	Multiply and divide any number from 1 to 10 000 by 10 or 100 and understand the effect.
5Nn6	Round four-digit numbers to the nearest 10, 100 or 1000.
5Nn7	Round a number with one or two decimal places to the nearest whole number.
5Nn8	Order and compare numbers up to a million using the > and < signs.
5Nn9	Order and compare negative and positive numbers on a number line and temperature scale.
5Nn10	Calculate a rise or fall in temperature.
5Nn11	Order numbers with one or two decimal places and compare using the > and < signs.
5Nn12	Recognise and extend number sequences.
5Nn13	Recognise odd and even numbers and multiples of 5, 10, 25, 50 and 100 up to 1000.
5Nn14	Make general statements about sums, differences and multiples of odd and even numbers.
5Nn15	Recognise equivalence between: $\frac{1}{2}$, $\frac{1}{4}$ and $\frac{1}{8}$; $\frac{1}{3}$ and $\frac{1}{6}$; $\frac{1}{5}$ and $\frac{1}{10}$.
5Nn16	Recognise equivalence between the decimal and fraction forms of halves, tenths and hundredths and use this to help order fractions, e.g. 0.6 is more than 50% and less than $\frac{7}{10}$.
5Nn17	Change an improper fraction to a mixed number, e.g. $\frac{7}{4}$ to $1\frac{3}{4}$; order mixed numbers and place between whole numbers on a number line.
5Nn18	Relate finding fractions to division and use to find simple fractions of quantities.
5Nn19	Understand percentage as the number of parts in every 100 and find simple percentages of quantities.
5Nn20	Express halves, tenths and hundredths as percentages.
5Nn21	Use fractions to describe and estimate a simple proportion, e.g. $\frac{1}{5}$ of the beads are yellow.
5Nn22	Use ratio to solve problems, e.g. to adapt a recipe for 6 people to one for 3 or 12 people.

A.	B.	C.

Stage 5   **Class:** _____   **Year:** _____

Strand: **Number** Sub-strand: **Calculation – *Mental Strategies***	
5Nc1	Know by heart pairs of one-place decimals with a total of 1, e.g. 0.8 + 0.2.
5Nc2	Derive quickly pairs of decimals with a total of 10, and with a total of 1.
5Nc3	Know multiplication and division facts for the 2× to 10× tables.
5Nc4	Know and apply tests of divisibility by 2, 5, 10 and 100.
5Nc5	Recognise multiples of 6, 7, 8 and 9 up to the 10th multiple.
5Nc6	Know squares of all numbers to 10×10.
5Nc7	Find factors of two-digit numbers.
5Nc8	Count on or back in thousands, hundreds, tens and ones to add or subtract.
5Nc9	Add or subtract near multiples of 10 or 100, e.g. 4387 – 299.
5Nc10	Use appropriate strategies to add or subtract pairs of two- and three-digit numbers and numbers with one decimal place, using jottings where necessary.
5Nc11	Calculate differences between near multiples of 1000, e.g. 5026 – 4998, or near multiples of 1, e.g. 3.2 – 2.6.
5Nc12	Multiply multiples of 10 to 90, and multiples of 100 to 900, by a single-digit number.
5Nc13	Multiply by 19 or 21 by multiplying by 20 and adjusting.
5Nc14	Multiply by 25 by multiplying by 100 and dividing by 4.
5Nc15	Use factors to multiply, e.g. multiply by 3, then double to multiply by 6.
5Nc16	Double any number up to 100 and halve even numbers to 200 and use this to double and halve numbers with one or two decimal places, e.g. double 3.4 and half of 8.6
5Nc17	Double multiples of 10 to 1000 and multiples of 100 to 10 000, e.g. double 360 or double 3600, and derive the corresponding halves.
A.	B.	C.

Stage 5   Class: _____   **Year:** _____

Strand: **Number**		
Sub-strand: **Calculation – *Addition and subtraction***		
5Nc18	Find the total of more than three two- or three-digit numbers using a written method.	
5Nc19	Add or subtract any pair of three- and/or four-digit numbers, with the same number of decimal places, including amounts of money.	
A.	B.	C.

Strand: **Number**		
Sub-strand: **Calculation – *Multiplication and division***		
5Nc20	Multiply or divide three-digit numbers by single-digit numbers.	
5Nc21	Multiply two-digit numbers by two-digit numbers.	
5Nc22	Multiply two-digit numbers with one decimal place by single-digit numbers, e.g. 3.6 × 7.	
5Nc23	Divide three-digit numbers by single-digit numbers, including those with a remainder (answers no greater than 30).	
5Nc24	Start expressing remainders as a fraction of the divisor when dividing two-digit numbers by single-digit numbers.	
5Nc25	Decide whether to group (using multiplication facts and multiples of the divisor) or to share (halving and quartering) to solve divisions.	
5Nc26	Decide whether to round an answer up or down after division, depending on the context.	
5Nc27	Begin to use brackets to order operations and understand the relationship between the four operations and how the laws of arithmetic apply to multiplication.	
A.	B.	C.

Stage 5   Class: _____   **Year:** _____

Strand: **Geometry** Sub-strand: **Shapes and geometric reasoning**		
5Gs1	Identify and describe properties of triangles and classify as isosceles, equilateral or scalene.	
5Gs2	Recognise reflective and rotational symmetry in regular polygons.	
5Gs3	Create patterns with two lines of symmetry, e.g. on a pegboard or squared paper.	
5Gs4	Visualise 3D shapes from 2D drawings and nets, e.g. different nets of an open or closed cube.	
5Gs5	Recognise perpendicular and parallel lines in 2D shapes, drawings and the environment.	
5Gs6	Understand and use angle measure in degrees; measure angles to the nearest 5°; identify, describe and estimate the size of angles and classify them as acute, right or obtuse.	
5Gs7	Calculate angles in a straight line.	
A.	B.	C.

Strand: **Geometry** Sub-strand: **Position and movement**		
5Gp1	Read and plot co-ordinates in the first quadrant.	
5Gp2	Predict where a polygon will be after reflection where the mirror line is parallel to one of the sides, including where the line is oblique.	
5Gp3	Understand translation as movement along a straight line, identify where polygons will be after a translation and give instructions for translating shapes.	
A.	B.	C.

Stage 5   Class: _____   Year: _____

Strand: **Measure** Sub-strand: **Length, mass and capacity**		
5Ml1	Read, choose, use and record standard units to estimate and measure length, mass and capacity to a suitable degree of accuracy.	
5Ml2	Convert larger to smaller metric units (decimals to one place), e.g. change 2.6 kg to 2600 g.	
5Ml3	Order measurements in mixed units.	
5Ml4	Round measurements to the nearest whole unit.	
5Ml5	Interpret a reading that lies between two unnumbered divisions on a scale.	
5Ml6	Compare readings on different scales.	
5Ml7	Draw and measure lines to the nearest centimetre and millimetre.	
A.	B.	C.

Strand: **Measure** Sub-strand: **Time**		
5Mt1	Recognise and use the units for time (seconds, minutes, hours, days, months and years).	
5Mt2	Tell and compare the time using digital and analogue clocks using the 24-hour clock.	
5Mt3	Read timetables using the 24-hour clock.	
5Mt4	Calculate time intervals in seconds, minutes and hours using digital or analogue formats.	
5Mt5	Use a calendar to calculate time intervals in days and weeks (using knowledge of days in calendar months).	
5Mt6	Calculate time intervals in months or years.	
A.	B.	C.

Stage 5   Class: _____   **Year:** _____

Strand: **Measure**		
Sub-strand: **Area and perimeter**		
5Ma1	Measure and calculate the perimeter of regular and irregular polygons.	
5Ma2	Understand area measured in square centimetres (cm²).	
5Ma3	Use the formula for the area of a rectangle to calculate the rectangle's area.	
A.	B.	C.

Strand: **Handling data**		
Sub-strand: **Organising, categorising and representing data**		
5Dh1	Answer a set of related questions by collecting, selecting and organising relevant data; draw conclusions from their own and others' data and identify further questions to ask.	
5Dh2	Draw and interpret frequency tables, pictograms and bar line charts, with the vertical axis labelled for example in twos, fives, tens, twenties or hundreds. Consider the effect of changing the scale on the vertical axis.	
5Dh3	Construct simple line graphs, e.g. to show changes in temperature over time.	
5Dh4	Understand where intermediate points have and do not have meaning, e.g. comparing a line graph of temperature against time with a graph of class attendance for each day of the week.	
5Dh5	Find and interpret the mode of a set of data.	
A.	B.	C.

Stage 5   Class: _____   **Year:** _____

Strand: **Handling data** Sub-strand: **Probability**		
5Db1	Describe the occurrence of familiar events using the language of chance or likelihood.	
A.	B.	C.

Strand: **Problem solving** Sub-strand: **Using techniques and skills in solving mathematical problems**		
5Pt1	Understand everyday systems of measurement in length, weight, capacity, temperature and time and use these to perform simple calculations.	
5Pt2	Solve single and multi-step word problems (all four operations); represent them, e.g. with diagrams or a number line.	
5Pt3	Check with a different order when adding several numbers or by using the inverse when adding or subtracting a pair of numbers.	
5Pt4	Use multiplication to check the result of a division, e.g. multiply 3.7×8 to check $29.6 \div 8$.	
5Pt5	Recognise the relationships between different 2D and 3D shapes, e.g. a face of a cube is a square.	
5Pt6	Estimate and approximate when calculating, e.g. using rounding, and check working.	
5Pt7	Consider whether an answer is reasonable in the context of a problem.	
A.	B.	C.

Stage 5   Class: _____   Year: _____

Strand: **Problem solving** Sub-strand: **Using understanding and strategies in solving problems**	
5Ps1	Understand everyday systems of measurement in length, weight, capacity, temperature and time and use these to perform simple calculations.
5Ps2	Choose an appropriate strategy for a calculation and explain how they worked out the answer.
5Ps3	Explore and solve number problems and puzzles, e.g. logic problems.
5Ps4	Deduce new information from existing information to solve problems.
5Ps5	Use ordered lists and tables to help to solve problems systematically.
5Ps6	Describe and continue number sequences, e.g. –30, –27, ☐, ☐, –18…; identify the relationships between numbers.
5Ps7	Identify simple relationships between shapes, e.g. these triangles are all isosceles because …
5Ps8	Investigate a simple general statement by finding examples which do or do not satisfy it, e.g. the sum of three consecutive whole numbers is always a multiple of three.
5Ps9	Explain methods and justify reasoning orally and in writing; make hypotheses and test them out.
5Ps10	Solve a larger problem by breaking it down into sub-problems or represent it using diagrams.

A.	B.	C.

Notes